NATURE WATCH

WOLVES

JEN GREEN

Consultant: Douglas Richardson

Zoological Director, Rome Zoo

CON

Published by Anness Publishing Ltd,
Blaby Road, Wigston, Leicestershire LE18 4SE
Email: info@anness.com
Web: www.annesspublishing.com

Anness Publishing has a new picture agency outlet for images for publishing, promotions or advertising. Please visit our website www.practicalpictures.com for more information.

Publisher: Joanna Lorenz
Managing Editor: Gilly Cameron Cooper
Senior Editor: Nicole Pearson
Designer: Ann Samuel
Picture Researcher: Cathy Stastny
Illustrators: Peter Bull, Vanessa Card, Sarah Smith
Production: Yolande Denny

ETHICAL TRADING POLICY
Because of our ongoing ecological investment programme, you, as our customer, can have the pleasure and reassurance of knowing that a tree is being cultivated on your behalf to naturally replace the materials used to make the book you are holding. For further information about this scheme, go to www.annesspublishing.com/trees

PICTURE CREDITS: b=bottom, t=top, c=centre, l=left, r=right
Bryan and Cherry Alexander Photography: back cover tr, 23tr, 26br, 27tl, 28t, 37br, 43c, 61cr; Ardea London: John Daniels: 10b & 11tl; Chris Martin Bahr: 11cl; Jean Paul Ferrero: 18b, 33t, 57c; Liz Bonford 26bl; Stefan Meyers: 44b; Francois Gohier: 46t; M. Krishnan: 47tl; The Art Archive: 55tl George Catlin; BBC Natural History Unit: Francoise Savigny: back cover cbl, 24tr, ; Simon King: back cover br, 36t, 37bl, 46br; C. Hamilton James: 2bl, 6b, 39br, Pete Oxford: 3t, 20bl, 47tr; Tom Vezo:4l, 19br, 63bl; Bernard Castelein: 2-3c; Louis Gagnon: 5cr; Jeff Foott: 6tr; Pete Oxford: 7cr; Tony Heald: 9c; R. Couper Johnston: 9b; Keith Scholes: 13br; Richard Tu Tout: 19tr; Lockwood and Dattatri: 25tl, 33br; Richard Du Toit: 25b; Andrew Harrington: 29tl; Christopher Becker: 29tr; Vadim Sidorovich: 30t; Lynn M. Stone: 43b; Bernard Walton 51t; Ron O'Connor: 57br; Lynn Stone: 61tr;The Bridgeman Art Library: 32t, 47br; Bruce Coleman Collection: Gunter Ziesler: back papers, 24b; Hans Reinhard: 3b & 5br; 15br; Rod Williams: 19tl, 41cl; Bruce Davidson: 34t & b, 35cl; Erwin & Peggy Bauer: 40t, 41b, 51bl; Staffan Widstrand: 42b; Rita Meyer: 46bl; Atlantide: 54b; Jeff Foot: 57t; Mary Evans Picture Library: 41cr, 54t, 55bl, 58bl, 58br;

INTRODUCING WOLVES

4 What is a Wolf?
6 The Wolf Family
8 Wild Dogs
10 Focus on Domestic Dogs

HOW WOLVES WORK

12 Body Shapes
14 Body Parts
16 On the Move
18 Fur Coats
20 Sight and Sound
22 Smell, Touch and Taste

Gettyone Stone Images: Roy Corral: cover cl; Tony Davis: cover cr; back cover cl; Art Wolf: back cover cr; 27tr, Rosemary Calvert: 21c; Kathy Bushue: 41tl; 42cr; The Kobal Collection: 55tr; NHPA: Gerard Lacz: cover tl, 19c; Andy Rouse: cover tr; back cover bl, 2tr; 26t, 39tl, Stephen Krasemann: cover br, 60t, 61bl, 63br; Martin Harvey: 2t, 7tr; Jany Sauvanet 17c, 45cl, 53tr & b; T. Kitchin & V. Hurst:17br, 33br, 49c & b; Rich Kirchner: 25c 36b; K. Ghani: 31t; Mirko Stelzner: 45t; Oxford Scientific Films: Daniel J. Cox: cover, back cover tc, back cover tr, : 2br , 13cr, 16tl, 16bl, 20tl, 23bl, 27b, 37cr, 38t, 49cl, 63t, 64t & b; Victoria McCormick: 1c, 20br, ; Nick Gordon 8t, 62bl;Charles Palek: 13tr; Stan Osolinski: 15tr; Steve Turner: 18t; Lon E. Lauber: 22t, 23cr, 33bl, 49tr, 62cr; Konrad Wothe: 22br, 37t; Alan and Sandy Carey: 25tl; Krupaker Senani: 29c; Peter Weiman: 29b; Richard Day: 30b; Michael Leach: 31cl; Anthony Bannister: 31cr; Michael Sewell: 31b; David Cayless; 35t,cr &b; M & C Tibbles: 38b; Matthews/Purdy: 39bl; Richard Packwood: 40b; Colin Willcock: 43tl; Tom Ulrich: 43tr; Mike Hill: 44t; Vivek Sinha: 45cr; Villarosa/Overseas: 45b; Joe McDonald: 47bl; Bob Bennett: 48b; Rafi Ben-Shahar: 56t, 61tl; Joel Bennett: 56b; Steve Turner: 57bl; Anna Walsh: 59bl; Claude Steelman: 59t; Owen Newman: 59cr; Matthias Breiter: 60b; Papilio Photographic: 48t; Peter Newark's Pictures: 13bl;Planet Earth Pictures: Ken Lucas: 15tl; Science Photo Library: George Bernard: 51c, 62t; Still Pictures: John Newby: back cover cbr, 52t; Klein/Hubert: 7cl; 52bl & br; Warren Photographic/Jane Burton: 21tl.

TENTS

BEHAVIOUR
24 Living Together
26 Focus on a Wolf Pack
28 Home Territory
30 A Meaty Diet
32 Going Hunting
34 Focus on African Hunting Dogs

LIFE CYCLES
36 Finding a Mate
38 Newborn Cubs
40 Growing Up

WHERE WOLVES AND WILD DOGS LIVE
42 Icy Wastes
44 In the Forest
46 Grassland and Desert
48 Focus on Coyotes

FACTS ABOUT WOLVES AND WILD DOGS
50 Relatives and Namesakes
52 Focus on Foxes
54 Fact or Fiction?
56 Natural Enemies
58 Wolves and People

60 Conservation
62 Glossary
64 Index

What is a Wolf?

Wolves are the wild members of the dog family, canids, with gleaming yellow eyes and lean, muscular bodies. The 37 different species of canids include wolves, jackals, coyotes, foxes and wild and domestic dogs. Canids are native to every continent except Australia and Antarctica. All of them share a keen sense of smell and hearing, and are carnivores (meat-eaters). Wolves and wild dogs hunt live prey, which they kill with their sharp teeth. However, many canids also eat vegetable matter and even insects. They are among the most intelligent of all animals. Some, such as wolves, are social in habit and live together in groups.

BODY FEATURES ► The wolf is the largest wild dog. It has a strong, well-muscled body covered with dense, shaggy fur, a long, bushy tail and strong legs made for running. Its muzzle (nose and jaws) is long and well developed and its ears are pricked up. Male and female wolves look very similar, although females are generally the smaller of the two.

▲ PRODUCING YOUNG
A female wolf suckles (feeds) her cubs. All canids are mammals and feed their young on milk. Females produce a litter of cubs, or pups, once a year. Most are born in an underground den.

◄ KEEN SENSES
The jackal, like all dogs, has very keen senses. Its nose can detect faint scents and its large ears pick up the slightest sound. Smell and hearing are mainly used for hunting. Many canids also have good vision.

The Big, Bad Wolf

Fairy tales often depict wolves as wicked, dangerous animals. In the tale of the Three Little Pigs, the big, bad wolf terrorizes three small porkers. Eventually he is outwitted by the smartest pig, who builds a brick house that the wolf cannot blow down, and all the pigs are safe.

▲ LIVING IN PACKS
Wolves and a few other wild dogs live in groups called packs of about 8 to 20. Each pack has a hierarchy (social order) and is led by the strongest male and female.

EXPERT HUNTERS ►
A wolf bares its teeth in a snarl to defend its kill. Wolves and other canids feed mainly on meat, but eat plants, too, particularly when they are hungry.

The Wolf Family

Wolves were once common throughout the northern hemisphere, right across North America, Europe and Asia. In the past, people hunted them mercilessly. In many areas they died out altogether. Across the north, wolves from various regions may look quite different, but they are all one species, *Canis lupus*. However, there are many subspecies (different types). Two of the main types are the grey wolf, also known as the timber wolf, and the Arctic, or tundra wolf. Many other subspecies are named after the area or habitat they come from, such as Mexican and steppe wolves. The same type of wolf may be known by different names in different areas. For example, the Arctic wolf is the same as the tundra wolf. The wolf's closest relatives are the coyote, the jackal, the dingo and the domestic dog.

▲ NEARLY EXTINCT
The red wolf (*Canis rufus*) is found only in the south-eastern USA. It was once widespread, but became extinct in the wild. Today's red wolves stem from captive-bred wolves. They have longer legs and larger ears than the grey wolf and are named after the reddish fur on their heads, ears and legs. The grey wolf is thought to be a hybrid (cross) between the coyote and the wolf.

▼ MOUNTAIN HOME
The Simien wolf is found only in the Simien Mountains of Ethiopia, East Africa. It is about the same size as a coyote (1m long) and has a mainly reddish-brown coat, with pale fur on its belly and throat. The Simien wolf was once thought to be a kind of jackal. Now scientists have found out that it is a small wolf.

Did you know? Wolves kill prey animals as large as bison and as small as mice.

Simien wolf
(Canis simensis)

dingo
(Canis familiaris)

◀ DOG WITH NO BARK

The dingo lives in Australia and South-east Asia. It is now wild, but its ancestors are descended from tame dogs that Aboriginal people brought to the region from Asia more than 8,000 years ago. Dingoes are found mainly in dry areas such as the dusty Australian interior. They live in small families, and sometimes join other dingo families to hunt.

JACKAL ▶

Jackals (*Canis lantrans*) are found on the grassy plains and in the woodlands of Africa, south-eastern Europe and southern Asia. They live in pairs and stay with the same mate for life. There are three species of jackal – side-striped and black-backed jackals (found only in Africa) and golden jackals.

black-backed jackal
(Canis mesomelas)

▲ COYOTE

The coyote (*Canis latrans*) is found in North and Central America. It mainly lives on prairies (grassy plains) and in open woodlands and is also known as the prairie or brush wolf. Full-grown coyotes are half the size of grey wolves. They live alone, or in pairs and small family groups.

miniature poodle
(Canis familiaris)

◀ DOMESTIC DOG

You may not have seen a wolf in the wild, but you will know one of its relatives, for all domestic dogs are descended from the wolf. Dogs were the first animals to be tamed by people over 12,000 years ago. Now there are over 400 breeds of dog, including the miniature poodle.

Wild Dogs

bush dog
(Speothos venaticus)

▲ **STURDY HUNTER**
The bush dog inhabits forests and marshes in Central and South America. It is stocky, with a brown coat and paler fur on its neck and head.

In addition to the wolf's immediate family, five species of wild dog are distantly related to wolves and to each other. Foxes are also canids but are not closely related to wild dogs or wolves. Of the wild dogs, the bush dog and maned wolf are both found in South America. The dhole and raccoon dog come from eastern Asia. The African hunting dog lives in central and southern Africa. Like other canids, the bodies of these animals are adapted to suit their way of life and their environment. Raccoon dogs and bush dogs are short-legged species that make their homes in underground burrows. Maned wolves and African hunting dogs have long legs to help them see over the long grass and scrub of savannas (grasslands) and woodlands. Human settlements are currently expanding in the areas where these wild dogs live and threatening their existence.

◄ **FOXY CORGI**
The dhole (*Cuon alpinus*) can be found in India and China, on the islands of Sumatra and Java, and also in parts Russia. Adult dholes reach about 1m long, but have short legs. With their reddish coats, they look a little like corgis with long, bushy tails. Dholes live in packs and do much of their hunting during the day.

Classification Chart

Kingdom	*Animalia*	all animals
Phylum	*Chordata*	animals with backbones
Class	*Mammalia*	hair-covered animals that feed their young on milk
Order	*Carnivora*	mammals that eat meat
Family	*Canidae*	all dogs
Genus	*Cuon*	
Species	*alpinus*	dhole

▲ CLASSIFICATION OF WILD DOGS

All wolves and wild dogs are canids (members of the dog family). Wolves and their close relatives belong to the genus (group) *Canis*, but the wild dogs shown here each belong to a different genus. The African hunting dog's scientific name (*Lycaon pictus*) is completely different from the dhole's (*Cuon alpinus*).

▲ MASKED HUNTER

The raccoon dog (*Nyctereutes procyonoides*) is found in eastern Europe, Siberia, China and Japan. Small in size, it reaches only 55cm long and has a chunky build. Its summer coat is grey-brown, with darker eye patches like a raccoon's mask. Its winter coat is white. These dogs live alone or in small groups.

African hunting dog *(Lycaon pictus)*

HAPPY IN A CROWD ►

The African hunting dog lives on the grassy plains south of the Sahara Desert. This wild dog has a mottled coat, long legs and rounded ears. It grows to about 1.1m long. African hunting dogs, like wolves, are social animals and live in packs. Males and females look very much alike and live to around the age of ten years old.

◄ SHY LONER

The maned wolf (*Chrysocyon brachyurus*) lives in grasslands and swampy regions in eastern South America. Despite its name, this animal is not a wolf. Although its long legs have earned it the nickname 'the fox on stilts', it is not a fox either. Maned wolves have mostly reddish fur with darker legs and a mane of long hair at the neck. They grow to about 1.2m long. Solitary and shy, the wolves usually hunt at night.

Focus on

ANCIENT BREED
The pharaoh hound is one of the world's oldest dog breeds. It was developed in Egypt around 4,000 years ago as a swift hunting dog. Other breeds, such as the fierce mastiff, were developed as guard dogs. By Roman times, around 500BC, many of the breeds we know today had already been developed.

The skeletons of dogs buried with their owners in ancient tombs show that dogs have been domesticated for at least 12,000 years. All domestic dogs are descended from the wolf. The first dogs were probably bred from wolves tamed by hunters. These may have been captured as cubs and brought back to camp to help with hunting. Later, ancient peoples began to develop different breeds of dogs by selecting animals with definite features that they valued. From these beginnings developed the 400 different breeds we know today. Domestic dogs come in an amazing variety of shapes and sizes, from the tiny chihuahua to the huge St Bernard. Modern breeders divide the breeds into six main families: hounds, gundogs, working dogs, terriers, toys and utility (useful) dogs.

WORKING DOG
Border collies are working dogs and are used to herd sheep. The Border collie came from the Borders, an area on either side of the boundaries of England and Scotland. Working dogs have been bred to perform many useful tasks, from guarding homes to pulling sledges. This group includes corgis, boxers, mastiffs and huskies.

Domestic Dogs

BRED TO FETCH AND CARRY
A golden retriever fetches a duck shot by its owner. Retrievers, setters, labradors and pointers are all gun-dogs, bred to help with hunting game birds. These naturally reliable and obedient dogs make good pets.

SPEEDY DOG
A greyhound's body is built for speed, with long legs, a flexible back and small, pointed head. It is bred in many countries for racing. People have bred sporting dogs for centuries. In the past, dogs were often bred for cruel sports such as bear-baiting. Today these sports are against the law in many countries.

HUNTING HOUNDS
A pack of foxhounds follow the scent of their quarry, a fox. Various breeds of hound have been developed to hunt different animals, including wolves, deer, rabbits, badgers and foxes. Foxhounds, beagles and bloodhounds track their prey by scent. Wolfhounds were bred to hunt wolves and deer. Afghan hounds were bred to chase antelope (similar to deer) and hunt by sight.

BURROWING EXPERT
The wire-haired fox terrier was very popular in England during the 1800s. The name terrier comes from the Latin *terra* (earth). All terriers were bred to hunt animals in burrows, such as foxes, rabbits, rats and badgers.

Body Shapes

Like all mammals, wolves and dogs are vertebrates (animals with backbones). The backbone protects the spinal cord, the main nerve of the body. The bones in the skeleton support the body and give it a distinct shape. The skeletons of wolves and dogs are different from other mammals. They have long skulls with large teeth, longish necks, and long, strong leg bones. Narrow collar bones help to make them slim and streamlined for speed over the ground, while their joints at the shoulders and hips pivot freely, giving them great agility. Wolves, foxes and wild dogs also generally have long tails. Many have a well-defined tail shape that can help to identify the animal at a distance.

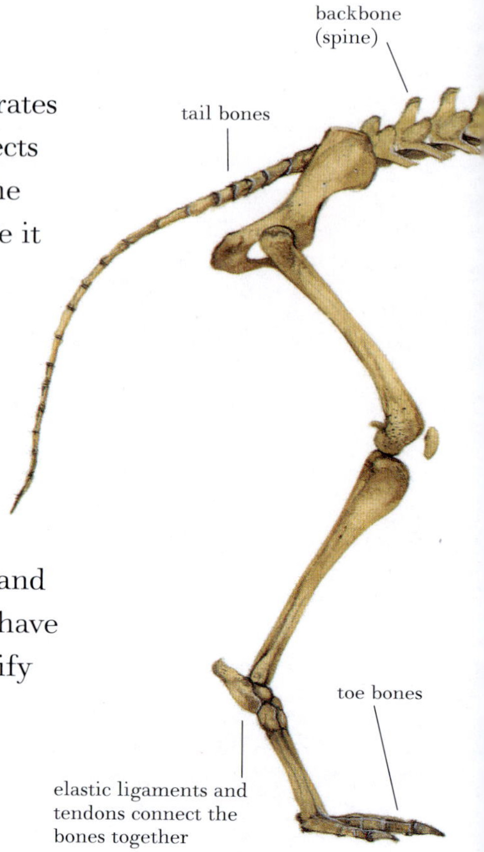

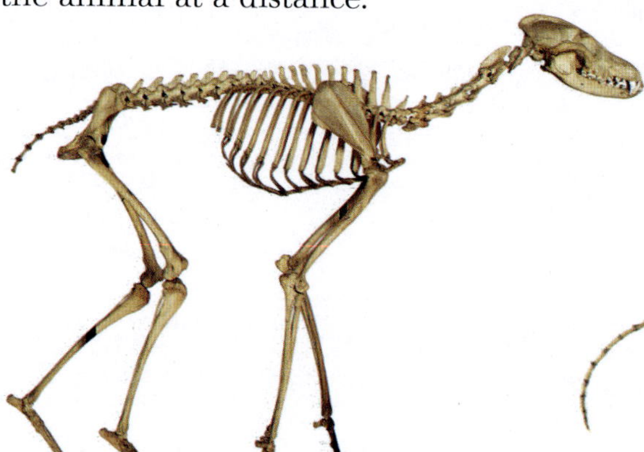

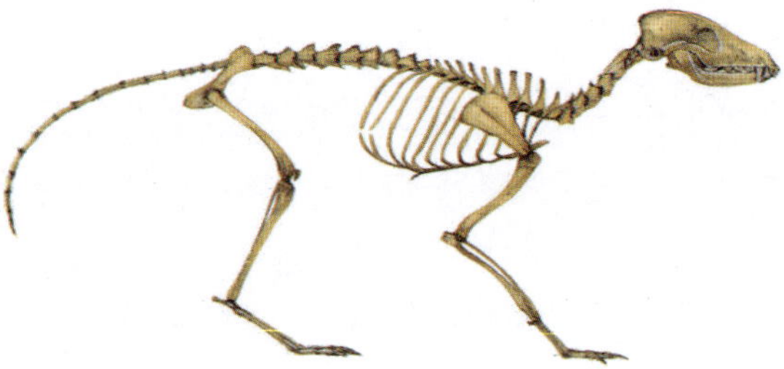

▲ TALL AND GRACEFUL
The maned wolf has longer leg bones than a true wolf. It uses them to hunt in the tall grass of its homeland. Its leg bones are weaker than a wolf's so it often lacks the power and strength to run down swift-moving prey. It has a short tail.

▲ BUILT FOR POUNCING
A fox's skeleton looks small and delicate compared to a wolf's. The leg bones are shorter in relation to its body size. Foxes spend much of their lives crouching and slinking through the undergrowth rather than running down prey.

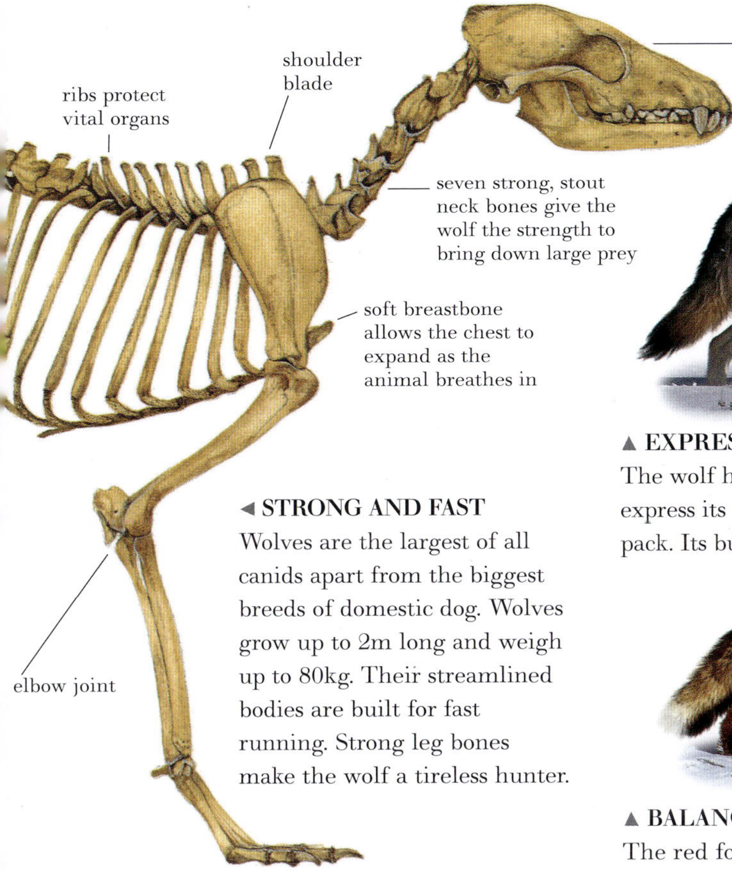

◄ **STRONG AND FAST**
Wolves are the largest of all canids apart from the biggest breeds of domestic dog. Wolves grow up to 2m long and weigh up to 80kg. Their streamlined bodies are built for fast running. Strong leg bones make the wolf a tireless hunter.

▲ **EXPRESSIVE TAIL**
The wolf holds its tail in different positions to express its feelings and show its position in the pack. Its bushy tail grows up to 48cm long.

▲ **BALANCING BRUSH**
The red fox's tail, known as a brush, is long and bushy. It grows up to 50cm long and helps to balance the animal when running and jumping.

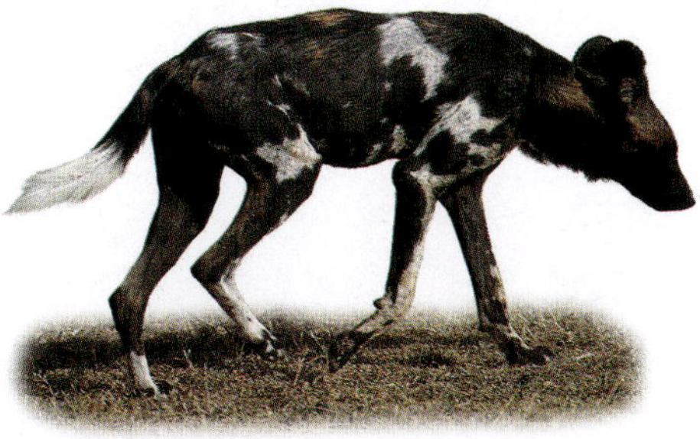

▲ **TAIL TALK**
The African hunting dog's tail is short compared to most canids, about 30-40cm long. Its white tip stands out clearly and is used to signal other members of the pack.

Strong Like the Wolf
Native Americans admired the strength, courage and intelligence of wolves and wild dogs. Plains tribes such as the Blackfoot and Mandan formed warrior-bands called Dog Societies to honour the loyalty shown by wild dogs to other dogs in their pack. This Hidatsa shaman (medicine man) is dressed for the Dog Dance. Dances were performed in celebration and for good luck.

Body Parts

Muscular, fast-running wolves and wild dogs are built for chasing prey in open country. Thick muscles and long, strong legs enable them to run fast over great distances. The long skull helps the wolf to seize prey on the run. The wolf has a large stomach that can digest meat quickly and hold a big meal after a successful hunt. Wolves, however, can also go without food for more than a week if prey is scarce. Teeth are a wolf's main weapon, used for biting enemies, catching prey and tearing food. Small incisors (front teeth) strip flesh off bones. Long fangs (canines) grab and hold prey. Towards the back, jagged carnassial teeth close together like shears to chew meat into small pieces, while large molars can crush bones.

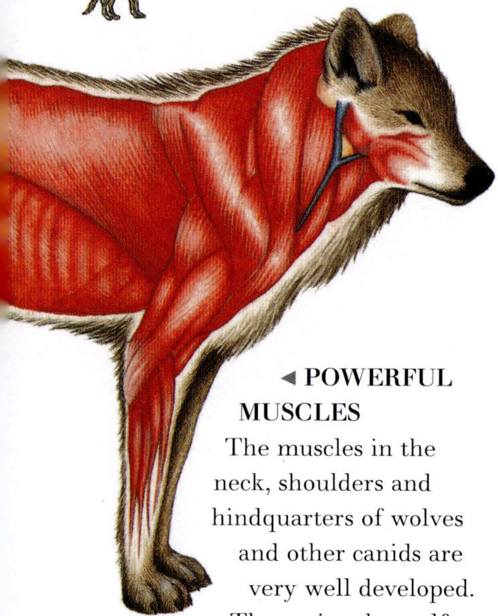

◄ **POWERFUL MUSCLES**
The muscles in the neck, shoulders and hindquarters of wolves and other canids are very well developed. These give the wolf strength and long-distance stamina as well as speed. Muscle-power alone is not enough, however, to catch prey. Wolves also need to use their cunning and stealth if the hunt is to succeed.

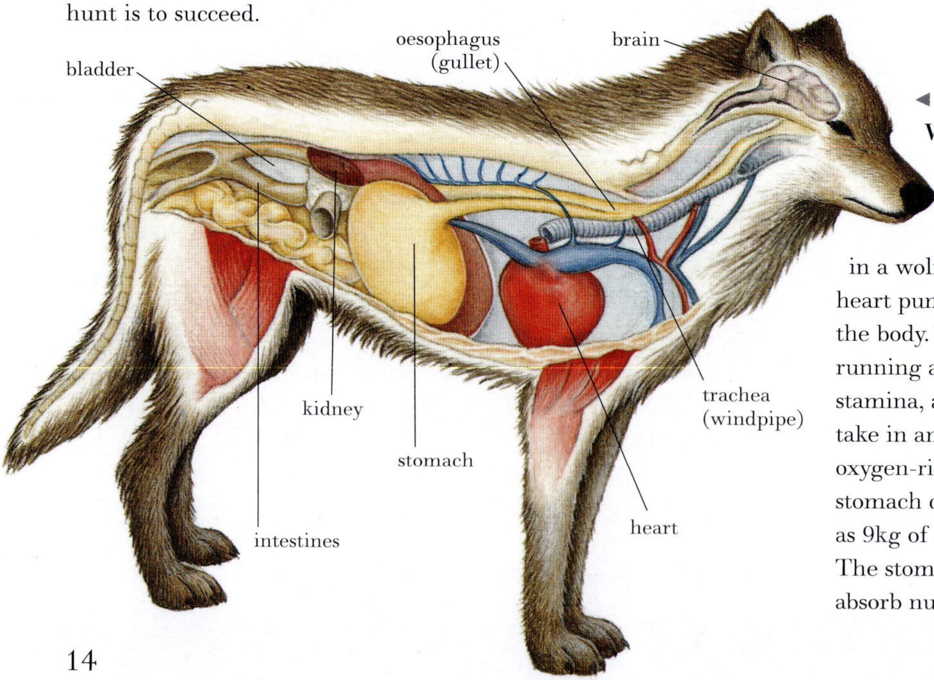

◄ **INSIDE A WOLF**
This diagram shows some of the main organs in a wolf's body. A strong heart pumps blood around the body. For long-distance running a wolf needs lots of stamina, and its large lungs take in ample supplies of oxygen-rich air. A wolf's stomach can hold as much as 9kg of food at a time. The stomach and intestine absorb nutrients in the food.

▼ WOLF'S SKULL

A wolf's head has a broad crown and a tapering muzzle. The bones of the skull are strong and heavy. They form a tough case that protects the animal's brain, eyes, ears and nose. The jaws have powerful muscles that can exert great pressure as the wolf sinks its teeth in its prey.

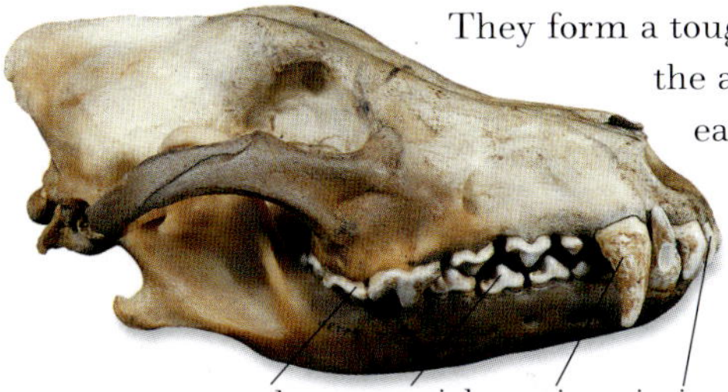

▼ BAT-EARED FOX SKULL

The bat-eared fox has a delicate, tapering muzzle. Its jaws are weaker than a wolf's and suited to deal with smaller prey, such as insects. This fox has 46–50 teeth, which is more than any other canid. Extra molars at the back of the animal's mouth enable it to crunch up insects, such as beetles, which have a tough outer casing on their bodies.

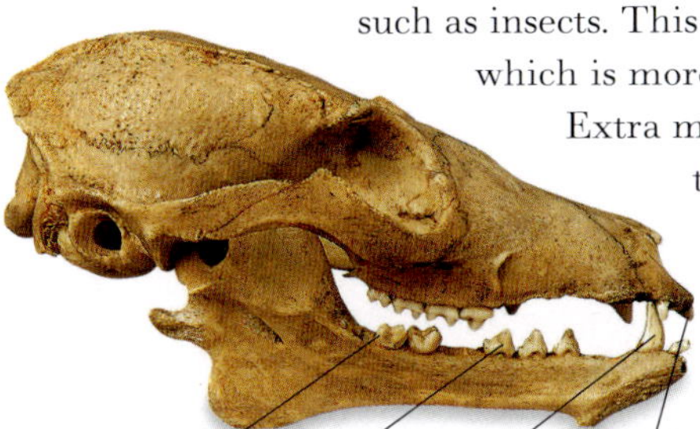

▲ TIME FOR BED

A wolf shows its full set of meat-eating teeth as it yawns. Wolves and most other canids have 42 teeth. In wolves, the four large, dagger-like canines at the front of the mouth can grow up to 5cm long.

COOLING DOWN ►

Like all mammals, the wolf is warm-blooded. This means its body temperature remains constant whatever the weather, so it is always ready to spring into action. Wolves do not have sweat glands all over their bodies as humans do, so in hot weather they cannot sweat to cool down. When the wolf gets too hot, it opens its mouth and pants with its large tongue lolling out. Moisture evaporates from the nose, mouth and tongue to cool the animal down.

On the Move

▲ **SPEEDING COYOTE**
Like wolves, coyotes are good long-distance runners. They run on their toes, like all canids. This helps them to take long strides and so cover more ground. If necessary, coyotes can trot along for hours in search of food.

Wolves are tireless runners. They can lope along for hours on end at a steady pace of 38km/h without resting. They have been known to cover an amazing 200km in a day searching for food. Compared to cheetahs, which can reach speeds of 96km/h over short distances, wolves are not fast runners. They can, however, put on a burst of speed to overtake fleeing prey.

Wolves and most other canids have four toes on their back feet and five toes on their front feet. The fifth toe on the front foot is called the dew claw, a small, functionless claw located a little way up on each front leg. They are more like pads than claws. They also have tough pads on the underside of their toes to help absorb the impact shock as the wolf's feet hit the ground.

Did you know? Studies of wolves in the USA show one pack travelled 1,125km in 40 days.

grey wolf
(Canis lupus)

◀ **IN MID-LEAP**
Strong leg muscles enable a wolf to leap long distances – up to 4.5m in a single bound. Wolves and other canids are very agile and can leap upwards, sideways and even backwards. As the wolf lands, its toes splay out to support its weight and prevent it from slipping.

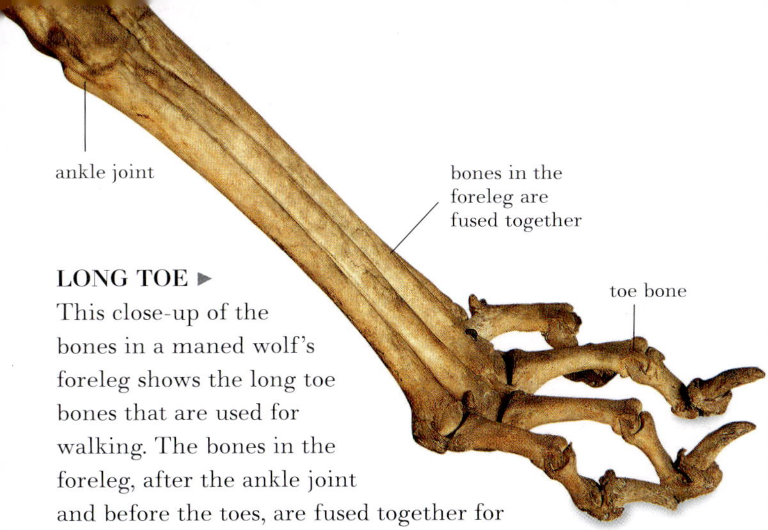

LONG TOE ►
This close-up of the bones in a maned wolf's foreleg shows the long toe bones that are used for walking. The bones in the foreleg, after the ankle joint and before the toes, are fused together for greater strength. Of all the canids, only the African hunting dog does not have dew claws.

▲ WOLF TRACK
Clawmarks show up clearly in a line of wolf prints in a snowy landscape. Unlike cats, wolves and wild dogs cannot retract (draw in) their claws. When walking, the wolf places its paws almost in a straight line, to form a single track. The pawprints of a running wolf are more widely spaced.

IN THE SWIM ►
Bush dogs make their homes near streams and rivers and spend much of their lives in water. They are strong swimmers, and water creatures such as capybaras (a type of rodent) form part of their diet. Wolves, dingoes and most other canids can also swim well.

A KEEN CLIMBER ►
Wolves and wild dogs are quick on the ground, but they cannot climb trees. Some foxes, however, climb well. The grey fox of North America is an expert climber. It scrambles up trees to steal birds' eggs and chicks. It also climbs to get a good view over surrounding countryside when searching for prey.

grey fox
(Urocyon cinereoargenteus)

black-backed jackal
(Canis mesomelas)

Fur Coats

Wolves and other members of the dog family have thick fur coats. This dense layer of hair helps to protect the animal's body from injury, and keeps it warm in cold weather. Wolves and other canids that live in cold places have extra-thick fur. Dingoes, jackals and wild dogs that live in warm countries close to the Equator have sparser fur. The fur is made up of two layers. Short dense underfur helps to keep the animal warm. Long guard hairs on top have natural oils that repel snow and rain to keep the underfur dry. A wild dog's fur coat is usually black, white or tan, or a mixture of these colours. Markings and patterns on the fur act as camouflage to disguise these animals, so they can creep up unseen on their prey.

▲ JACKAL COLOURS
The three species of jackal can be distinguished by their different markings. As its name suggests, the black-backed jackal has a dark patch on its back as well as brown flanks and a pale belly. The golden jackal is sandy-brown all over. The side-striped jackal is so named because of the light and dark stripes that run along its sides.

◄ WAITING FOR SPRING
Two raccoon dogs shelter under a bush waiting for the snow to melt. They already have their summer coats of grey with pale and dark patches. This will help to camouflage them among the summer grasses and vegetation. In autumn, they will moult (shed) these coats and grow a pure-white coat, ready for the winter snow.

◄ MANES AND RUFFS

The maned wolf gets its name from the ruff of long hairs on its neck. This may be dark or reddish-brown in colour. Wolves also have a ruff of longer hairs which they raise when threatened to make themselves look larger.

▲ HANDSOME CAMOUFLAGE

African hunting dogs have beautiful markings, with tan and dark-grey patches on their bodies, and paler, mottled fur on their heads and legs. The patterns work to break up the outline of their bodies as they hunt in the dappled light of the bush.

Arctic wolf
(Canis lupus tundarum)

▲ WARM COAT

The Arctic wolf has very thick fur to keep it warm in icy temperatures. Its winter coat is pure white so that it blends in with the snow. In spring, the thick fur drops out and the wolf grows a thinner coat for summer. This coat is usually darker to match the earth without its covering of snow.

VARYING COLOURS ►

Grey wolves vary greatly in colour, from pale silver to buff, sandy, red-brown or almost black. Even very dark wolves usually have some pale fur, often a white patch on the chest.

grey wolf
(Canis lupus)

Sight and Sound

Wolves have excellent hearing. They can hear the sound of a snapping twig up to 3km away and are alert to the smallest noise that might give away the presence of potential prey. Wolves hear a wider range of sounds than humans. They can hear ultrasounds (very high-pitched noise) that are too high for human ears to catch. This means they can track down mice and other rodents in the dark. Sight is less important than hearing for hunting. Wolves are good at spotting movement, even at a great distance, but find it harder to see objects that keep still. Canids that hunt at night rely on sound and smell rather than sight. African hunting dogs and dholes, however, hunt by day, often in open country, and have keener sight.

▲ **PRICKED EARS**
Wolves cock (turn) their ears in different directions to pinpoint distant sounds. Even a tiny noise betrays the hiding place of a victim.

▲ **LISTENING IN**
An African hunting dog's large, rounded ears work like satellite dishes to gather sound. Keen hearing is vital in the hunt and allows pack members to keep in touch amongst the undergrowth.

Coyote
(Canis latrans)

HOWLING HELLO ▶
Coyotes and other wild dogs keep in touch with distant members of their group by howling. The coyote call is actually a series of yelps that ends in a long wail.

Did you know? African hunting dogs chatter like monkeys as they play together.

◀ BARKING MAD

Some domestic dogs, such as this German shepherd, have been bred to bark loudly to warn their owners of approaching strangers. Wolves also bark if they meet an intruder near the den, but more quietly and less aggressively.

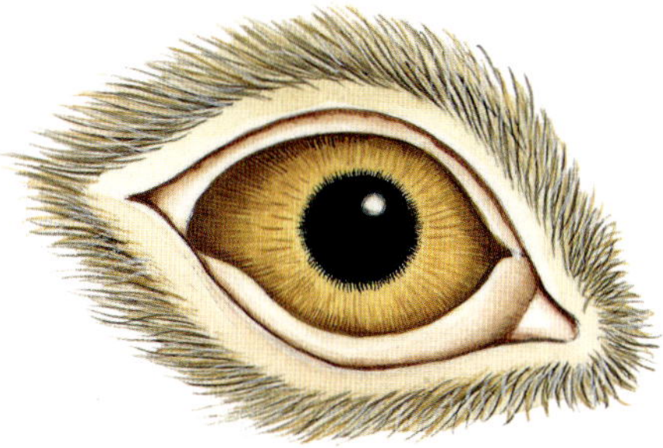

▲ A THIRD EYELID

Wolves' eyes have a third eyelid, called a nictating (blinking) membrane. This membrane is inside their upper and lower eyelids, and sweeps over the surface of the eye when the wolf blinks. It protects the eyes from dust and dirt that might damage it otherwise.

◀ GLOWING EYES

Wolves have round, yellow eyes. In dark conditions, a layer at the back of the eye, called the *tapetum lucidum*, intensifies what little light there is. This allows a wolf to see at night. The layer also reflects light, making the wolf's eyes glow in the dark if a strong light is shone into them.

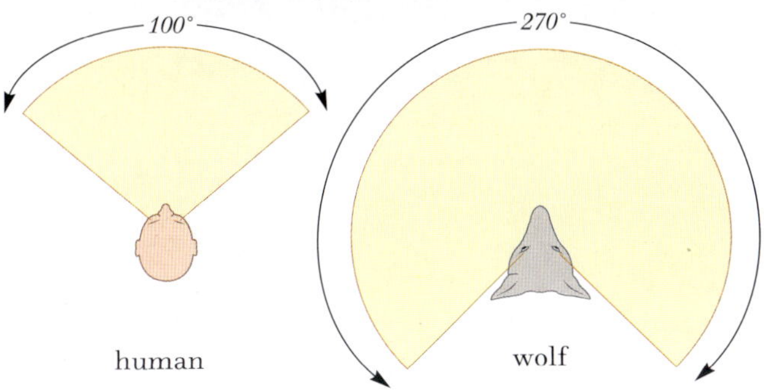

◀ WOLF VISION

Wide-set eyes in the front of its face give a wolf a very wide field of vision. As the view from each eye overlaps, binocular vision (using both eyes at the same time) allows a wolf to focus at close range and pounce on its prey.

Smell, Touch and Taste

Of all the senses, smell is the most important for wolves and wild dogs. These animals are constantly surrounded by different scents and their keen sense of smell can distinguish them all. They follow the scent trails left behind by other animals in their quest for food, and can pick up even faint whiffs of scent on the wind. This helps them to work out the direction of distant prey. Canids that hunt in packs use scent to identify and communicate with other pack members. They also communicate by sight and touch.

Like other mammals, wolves and wild dogs have taste buds on their tongues to taste their food. They eat foods they find the tastiest first. The tongue is also used to lap up water.

▲ **TRACKING PREY**
Nose to the ground, a wolf follows a scent trail in the snow. From the scent a wolf can tell what type of animal left it, whether it is well or ill, how far away it is and whether another wolf is following the trail.

► **ON THE SCENT**
Bloodhounds were specially bred as tracker dogs. They have a very acute sense of smell and can follow a scent that is several days old. They keep their noses very close to the ground. Their drooping ears help to channel scent into the nose.

▲ **PLEASED TO MEET YOU**
When two wolves meet they sniff the glands at the base of the tail. Pack members all have a familiar scent. Scent is also used to signal mood, such as contentment or fear, or if a female is ready to breed.

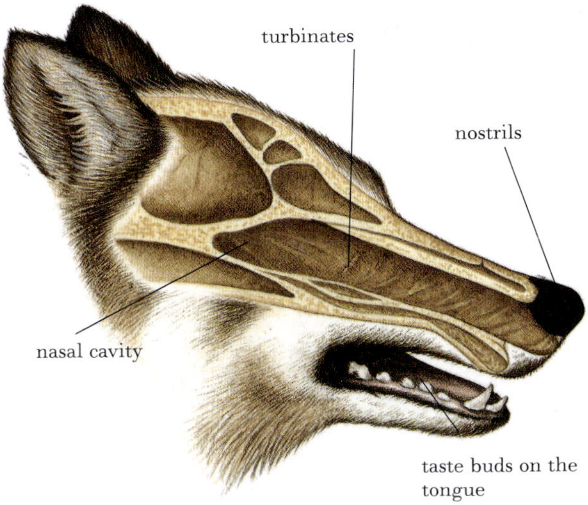

▲ INSIDE THE SNOUT
Inside a wolf's snout is a large nasal cavity used for smelling. Scent particles pass over tubes of very thin bone in the roof of the nasal cavity. These tubes, called turbinates, are connected to a nerve network that sends signals to the brain.

▲ TOUCHY-FEELY
Wolves use touch to bond with each other. They rub bodies, lick one another and thrust their noses into each other's fur when they meet. Pack members play-fight by wrestling with locked jaws, or chasing around in circles.

▼ SENSITIVE NOSE
The wolf's leathery outer nose is set right at the end of its snout. Two nostrils draw air laden with scents into the nasal cavity. The wolf may flare its nostrils to take in extra air. The animal may lick its nose before scenting, because a damp nose helps its sense of smell. Long, sensitive whiskers on either side of the snout are used for touching things at close range.

▲ WELL GROOMED
A wolf nibbles at the tufts of hair between its paw pads. It is removing ice that might cut and damage the paw. Wolves groom (clean) their fur to keep it in good condition. Licking and running fur through the teeth helps to remove dirt and dislodge fleas.

Living Together

Wolves are very social animals. A few may live alone, but most live in packs. A wolf pack may contain as few as 2 or as many as 36 animals. Most packs have between 8 and 24 members. The main purpose of the pack is to hunt. A team of wolves working together can hunt down and kill much larger and stronger prey than a wolf would be able to on its own. Only the strongest, healthiest pair in the wolf pack will actually mate. Every pack member then helps to feed and bring up the cubs. Bush dogs, dholes and African hunting dogs also live in packs, while jackals, and sometimes coyotes and raccoon dogs, live in smaller family groups. Maned wolves and foxes usually live alone.

▲ **TWO'S COMPANY**
A pair of jackals drinks from a waterhole in South Africa. Some jackals are solitary, but most pair up for life. The cubs are reared in small, close-knit family groups. Older brothers or sisters often help their parents to rear the small cubs.

Did you know? Foxes produce alarming screams when looking for a mate.

WOLF PACK ▶
A wolf pack is led by the strongest, most experienced animals. The rest of the pack often consists of their children – young cubs and older, half-grown wolves. The young wolves follow their leaders until they are old enough to leave the pack.

▲ A FAMILY AFFAIR
Dholes live in family packs of between 5 and 12 animals. Sometimes several families join together to form a very large dhole pack called a clan. Hunting in a big group helps these relatively small wild dogs to tackle large prey such as wild cattle and buffalo.

maned wolf
(Chrysocyon brachyurus)

▲ EACH FOR ITSELF
The maned wolf is mostly solitary, living and hunting on its own. Males and females pair up during the breeding season. The male helps to feed and rear his pups – as do all canids, except for the domestic dog.

◀ COYOTE COUPLE
Most coyotes live and hunt alone, in pairs or in small family units. Sometimes, several of these small groups band together to make a bigger hunting party to go after large prey.

DOG SOCIETY ▶
African hunting dogs are the next most social canids after wolves. They hunt co-operatively and all pack members help to raise the pups of the breeding pair. The males in the pack are all brothers. Females often join from a different pack.

Focus on

A wolf pack has a strict social order and each member knows its place. The senior male and female, known as the alpha male and female, are the only animals to breed. The alpha male takes the lead in hunting, defends the pack members from enemies, and keeps the other animals in their place. In most packs, a second pair of wolves, called the beta male and female, come next in the ranking order. The other pack members are usually the offspring of the alpha pair, aged up to three years old.

LEADER OF THE PACK
An alpha male wolf greets a junior pack member. Wolves use different body positions and facial expressions to show rank. The leader stands upright with tail held high. The junior has his ears laid back and his tail tucked between his legs.

IT'S A PUSHOVER
A junior wolf rolls over on its back in a gesture of submission to a more dominant pack member. A junior wolf can also pacify a stronger animal by imitating cub behaviour, such as begging for food.

SHOWING WHO IS BOSS
A wolf crouches down to an alpha male. The young wolf whines as it cowers, as if to say, 'You're the boss'. The pack leader's confident stance makes him look as large as possible.

a Wolf Pack

'I GIVE IN'
A male grey wolf lays its ears back and sticks its tongue out. Taken together, these two gestures signal submission. A wolf with its tongue out, but its ears pricked, is sending a different message, showing it feels hostile and rebellious.

SCARY SNARL
A grey wolf bares its canine teeth in a snarl of aggression. Studies have shown that wolves use up to 20 different facial expressions. Junior wolves use snarling expressions to challenge the authority of their leaders. The alpha male may respond with an even more ferocious snarl. If it does so, the junior wolf is faced with a choice. It must back down, or risk being punished with a nip.

REJECTED BY THE PACK
Old, wounded or sickly wolves are often turned out of the pack to become lone wolves. Although pack members may be affectionate with each other, there is no room for sentiment. Young wolves may also leave to start their own packs. Lone wolves without the protection of a pack are much more vulnerable to attack and must be more cautious.

▲ **KEEP TO THE TRAIL**
A pack of wolves runs along a snowy trail in single file. Each wolf treads in the tracks of the one in front. This saves a lot of energy that would be wasted if each animal broke a separate trail through the deep snow. Wolves use well-worn paths inside their territories. These connect meeting places with lookout points and good ambush sites.

Home Territory

Territories are areas that animals use for finding food or for breeding. An animal will defend its territory against others that might provide competition. Wolf packs use their territories as hunting grounds and also as safe places to raise their cubs. Wolf territories vary in size, depending on how much food is available to feed the pack. Small territories cover about 100 sq km. Large territories may be 10 or even 100 times this size. At the heart of the territory is the rendezvous, a meeting place where the wolves gather. This place also acts as a nursery where older cubs are left to play. The borders of the territory are patrolled by the pack on a regular basis, and marked with urine. Strong-smelling urine sprayed at marker sites lasts several days, while howling also serves as a long-distance warning. The pack will defend its territory fiercely if a rival pack tries to enter.

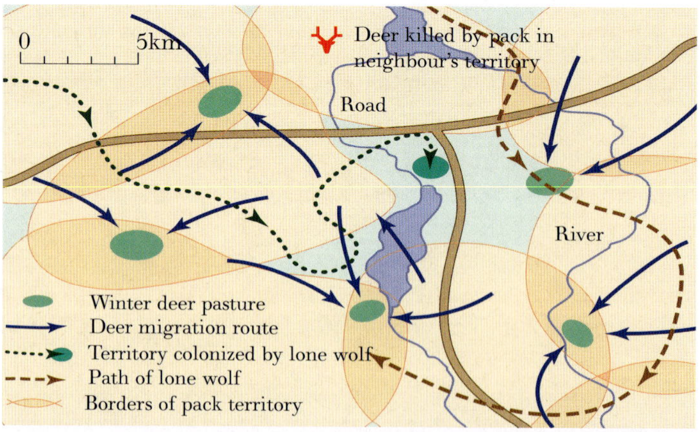

◀ **NEIGHBOURING TERRITORIES**
Several wolf territories may lie close to each other. Where food is plentiful, for example where deer breed, territories may overlap (see diagram). Rivers, towns and roads form natural boundaries. There may be a neutral 1km wide no-go area between territories. If food becomes scarce, however, wolves will enter this zone.

▲ **SCENT SIGNAL**
A wolf smells a rock that another pack member has marked with urine. The scent lingers for several days after the animal has moved on. A wolf smelling this signal can tell how many animals have passed this way and whether they belong to its pack. Wolves from a rival pack will mark the other side of the rock to reinforce territory boundaries.

▲ **BORDER DISPUTE**
A wolf snarls at a wolf from another pack on the boundary between their territories. When food is scarce, packs are more likely to trespass or raid each other's territory. In times of plenty, scent signals keep trespassers out, so border disputes are rare.

◀ **SCENT-MARKING**
A dhole marks the edge of its territory by spraying urine on a patch of grass. Some canids, such as African hunting dogs and foxes, leave piles of droppings too. In a wolf pack, it is usually the alpha male that marks the boundaries.

KEEP-OUT CALL ▶
Three wolves raise their heads in a group howl. Each animal howls a slightly different note, making an eerie harmony. The sound carries over a long distance – 10km or more. If it reaches a rival pack, the other wolves may howl back. Howling is also used to rally the pack, for example, before a hunt.

A Meaty Diet

Wolves and their relatives are carnivores (meat-eaters). They kill prey for fresh meat, but also eat carrion (dead animals). When no meat is available wolves eat plants, such as fruit and berries, and also grass to aid digestion. Large herd animals such as musk oxen, moose, deer and caribou are the favourite targets of the wolf pack. All canids target other kinds of prey if a particular creature becomes scarce. For example, they hunt a wide range of smaller animals, including birds, hares, mice and beavers.

Wolves swim well and chase fish, frogs and crabs, but still spend much of their lives with empty bellies. When food is scarce, they sometimes approach towns and villages, where they rifle through rubbish or kill domestic sheep, goats, cattle and horses.

▲ NOT-SO-FUSSY FEEDERS
Raccoon dogs of eastern Asia are omnivores – they eat all kinds of different foods, including rodents, fruit and acorns. They are strong swimmers and catch frogs, fish and water beetles in streams and rivers. They also scavenge carrion and scraps from people's rubbish tips.

▼ COYOTE PREY
Three coyotes tear at the carcass of a moose. Coyotes usually hunt small prey such as mice, but sometimes they band together to go after larger creatures such as moose. Teams of coyotes also gang up on other predators and steal their kills.

◄ **FAST FOOD**

A pack of dholes makes quick work of a deer carcass. Each one eats fast to get its share. A hungry dhole can eat up to 4kg of meat in an hour. Mammals form a large part of a dhole's diet, but if meat is scarce a dhole will also eat berries, lizards and insects.

▲ **MEAT AND FRUIT-EATER**

A maned wolf lopes off in search of prey. Without a pack to help it hunt, it looks for easy prey, including armadillos and small rodents such as agoutis and pacas. It also feeds on birds, reptiles, frogs and insects, fruit and sugar cane.

▲ **BEACH SQUABBLE**

Two black-backed jackals squabble over the carcass of a seal pup. Jackals eat almost anything – fruit, frogs, reptiles and a wide range of mammals, from gazelles to mice. Jackals also scavenge kills from other hunters.

CACHING FOOD ►

A wolf looks for a suitable spot in the snow to bury a freshly caught hare. After a pack has killed a large beast, or when a lone hunter has eaten its fill, it hides the remains of its food. Then, when food is scarce, the wolf can return to the hidden cache and retrieve its kill.

grey wolf
(Canis lupus)

Did you know? All canids are quick feeders, but dholes in particular bolt their food down at a great rate.

Going Hunting

Wolves and wild dogs do not all hunt at the same time of day. Maned wolves, bush dogs and raccoon dogs are mainly nocturnal (active at night). They rely on smell and hearing to find their prey. Dholes and African hunting dogs are daytime hunters, and track their prey by sight as well as smell and sound. Wolves hunt at any time of day or night. Members of a wolf pack work together like players in a sports team. Each animal has particular strengths that help the group. Some wolves are good trackers, others are particularly cunning, fast, or powerful, and so help to bring down large animals such as moose. Wolves spend a lot of time searching for food. A hunt may last for several hours, but nine out of ten hunts are unsuccessful and the wolves go hungry. If they strike lucky, they might kill a beast large enough to provide meat for all the pack.

Little Red Riding Hood

In the story of Little Red Riding Hood, a cunning wolf eats Red Riding Hood's grandmother. The wolf then steals the old woman's clothes to prey on the little girl. Fortunately a wood cutter rescues Red Riding Hood in the nick of time. As he kills the wolf, the grandmother emerges alive from inside the wolf.

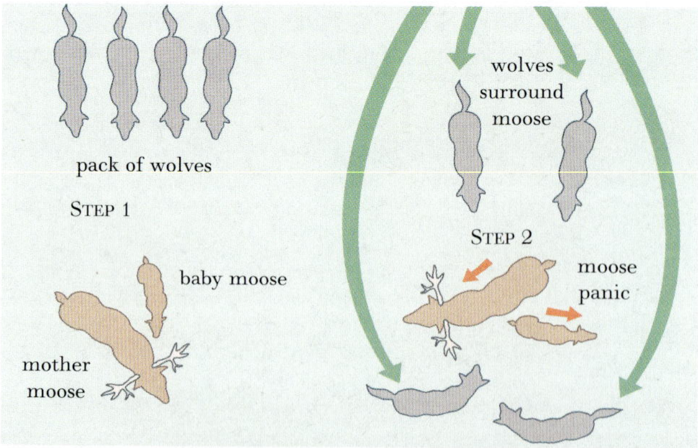

◄ WOLF PACK IN ACTION

Wolves use skill as well as strength to hunt large creatures such as moose. A moose calf is an easier target than a full-grown animal. The wolves stalk their prey, fanning out and running ahead to surround the victim. Pack members dash forward to panic the animals and separate the mother from her baby. Once the young calf is separated, the wolves run it down and kill it with a bite to the neck.

DINGO KILL ►
Two dingoes have just caught a kangaroo. In the Australian outback, dingoes hunt a wide range of creatures, from tiny grasshoppers and lizards to large prey such as wild pigs and kangaroos. Sheep, introduced by settlers in the 1800s, are a favourite target.

◄ GROUP KILL
A large pack of wolves has killed a white-tailed deer. This amount of meat will not satisfy the group for long. Where food is scarce, territories are often much larger than in places where the hunting is easy. The pack will always hunt the largest game it can find.

Coyote
(Canis latrans)

CLEVER TACTICS ►
A coyote plays with a mouse it has surprised in the snow. Coyotes often hunt mice, leaping high in the air to pounce on their victims. Coyotes have a more varied diet than wolves, feeding on fruit, grass, berries and insects, as well as mammals such as rabbits, deer and rodents. They take to water to catch fish and frogs, and also steal sheep and chickens, which makes them unpopular with farmers.

Did you know For peak condition, a wolf needs to eat 4kg of meat a day.

▲ TEAMWORK
Working as a team, dholes hunt large prey such as sambar (a type of deer). Dholes whistle to keep in touch with one another as they surround their prey. Teamwork also helps the pack to defend the kill from scavengers such as vultures.

Focus on African

African hunting dogs eat more meat than any other canid. One in every three of their hunts ends in a kill, a very high success rate. They live on the savanna (grassy plains) of central and southern Africa, which is also home to vast herds of grazing animals such as zebra, wildebeest and gazelle. The pack wanders freely over a huge area of savanna, looking for herd animals to prey on. They rely on sight to find their quarry, so they hunt during daylight hours or on bright moonlit nights. They mainly hunt at dusk or dawn when the air is coolest and rest in the shade during the hottest time of day.

1 A pack of wild dogs begins to run down their quarry, a powerful wildebeest. On the open plains of the Serengeti in East Africa, there is little cover that would allow the dogs to sneak up on their prey. The hunt is often therefore a straightforward chase. A junior dog may lead the hunt at first. The pack also targets many kinds of antelope, such as kudu and gazelle.

2 The dogs run along at an easy lope at first. They have tested out the wildebeest herd to find an easy target. They look for weak, injured, or young and inexperienced animals that will make suitable victims. This wildebeest is an older animal whose strength may be failing.

Hunting Dogs

3 A hunting dog tries to seize the wildebeest's tail. Hunting dogs with different strengths and skills take on different roles during the hunt. The lead dogs are super-fit and strong. They dodge out of the way if the wildebeest turns to defend itself with its sharp hooves and horns. Fast runners spread out to surround the victim and cut off its escape.

4 As the wildebeest tires, two dogs grip its snout and tail, pinning it down. Hunting dogs can run at 50km/h for quite a distance, but their prey is much swifter. While the lead dogs follow the fleeing animal's twists and turns, backmarkers take a more direct line, saving their strength. The rear dogs take over the chase as the leaders tire.

5 More dogs arrive and the strongest move in for the kill. While some dogs hold their victim by the snout and flanks, others jump up to knock it off balance. The dogs attack their victim's sides and rump and soon the animal is bleeding freely. It begins to weaken through shock and loss of blood.

6 The wildebeest crashes to the ground and the dogs rip at its underparts to kill it. There is little snapping and snarling as they eat, but the kill is fiercely defended if a scavenger such as a jackal comes close. Half-grown cubs feed first, then the carcass is ripped apart and bones, skin and all are eaten. Back at the den young cubs are fed with pre-chewed meat.

Finding a Mate

Wolves and wild dogs breed once a year, towards the end of winter. The cubs are born roughly two months later, in the spring. The size of the litter varies from species to species. Maned wolves give birth to the fewest young, usually only two cubs. Each pregnant wolf produces between three and eight cubs per litter. African hunting dogs have the largest litters: up to 16 pups at a time.

About six weeks after mating, a female wolf prepares a den in a cave, a hollow tree trunk or an underground burrow. Here her cubs will be born. The pack gathers outside the den as she begins to give birth. They howl as if to encourage the mother as the young are born.

▲ PURE PEDIGREE
In some parts of Australia, dingoes live in packs, in which only the dominant female breeds. In other areas, they live in smaller family groups. Because there are no other wild dogs to breed with, dingoes are in fact the most pure bred dogs in the world. They are directly descended from prehistoric domestic dogs.

Did you know? Wolf cubs take about 63 days to develop in their mother's womb before they are born.

COURTING COUPLE ▶
A male wolf sniffs his partner to find out if she is ready to mate. Tensions run high in the wolf pack during the breeding season, as all the adult wolves are ready to breed. The alpha male and female must dominate the rest to make sure they are the only ones to mate. The alpha female temporarily drives the other females from the pack. Once the alpha pair have mated, all the wolves can relax.

◄ **MATING WOLVES**
When they mate, the male wolf mounts the female and grasps her sides with his front paws. The female becomes ready to mate as the periods of daylight grow longer. This means that her pups will be born in spring as the weather warms up and food becomes plentiful. In North America, in the south of their range, if mating is in February the cubs will be born in late April or May. Farther north, wolves give birth in May or June.

LOCKED TOGETHER ►
Directly after mating, canids often stay tied (locked together) for many minutes. Tying helps to ensure that the alpha male is the father of the cubs, rather than an alpha male from a rival pack that may mate with the alpha female afterwards. It also helps strengthen the bond between parent wolves.

◄ **DINGO DEN**
A rocky cave overlooking a stony desert makes a good den site for this mother dingo. The cave will be a safe, cool and shady place for the cubs to be born.

▲ **GIVING BIRTH**
A mother wolf licks her newborn cub to clean it and stimulate it to begin breathing. Wolf cubs are born at intervals of about 15 to 30 minutes, and it may take up to six hours for a large litter to be born.

Newborn Cubs

At birth the young of wolves and dogs are tiny and helpless. Newborn wolves are only about 21cm long from the tip of their short noses to the end of their thin, little tails. With eyes tightly closed, they cannot see or hear, or even stand on their weak legs. They squirm around and huddle close to their mother for warmth. Like all mammals, their first food is their mother's rich milk, which she encourages them to suck from the moment they are born.

After one or two weeks, the cubs' eyes open and they take notice of their surroundings. They take their first wobbly steps and scramble over each other in the den. At about five weeks, the cubs begin to take solid food as well as milk. Half-chewed meat, stored in the stomach of an adult wolf, is brought to the den and coughed up when the cubs beg for food.

▲ AT THE DEN
Wolf cubs take a first look at the big world outside their den. For nearly eight weeks their only experience has been the burrow – a 4m long tunnel dug in soft earth with room for an adult wolf to creep along. The cubs sleep in a cosy chamber at the end, while their mother sleeps in a hollow nearer the entrance.

Did you know? Young wolf cubs need to feed every few hours and drink 300ml of milk a day.

NURSING MOTHER ►
The mother wolf hardly leaves her cubs for the first weeks of their lives. Her mate, and the other members of the pack, bring food so she does not need to leave the den to go hunting. As a nursing mother, she is always hungry and needs large quantities of food to produce enough milk to feed her hungry cubs.

▼ CUBS IN DANGER

These wolf cubs are about six or seven weeks old. Not all cubs are born in a den. Some are born in a hollow sheltered from the wind, or in a nest flattened in long grass. There are many dangers for cubs in the open. They may be snatched by predators such as bears or eagles. Many do not survive to adulthood.

grey wolf cubs
(Canis lupus)

Romulus and Remus

According to ancient Roman legend, Romulus and Remus were twin brothers who were abandoned as babies on a remote hillside. A she-wolf found them and brought them up, feeding them on her milk. Both brothers survived and Romulus went on to found the city of Rome.

▲ HUNGRY PUPPIES

An African hunting dog suckles her pups. They suck milk from two sets of nipples on her underside. Female hunting dogs often have more nipples than other canids, because they have the biggest litters and therefore the most mouths to feed. As the pups' sharp teeth begin to hurt she will wean them on to meat.

▲ RARE CUBS

In the mountains of Ethiopia, a Simien wolf guards her litter of five cubs. Simien wolves have similar breeding habits to other wolves, but are much rarer. These cubs look healthy, so have a good chance of surviving long enough to breed as adults.

Growing Up

At eight weeks old, wolf cubs are very lively. Their snouts have grown longer, their ears stand up and they look much more like adult wolves. They bound about on long, strong legs. Now weaned off milk, they live on a diet of meat brought by the adults. As they leave the safety of the den, the other pack members gather round and take great interest in the cubs. The cubs' new playground is the rendezvous, the safe place at the heart of wolf-pack territory where the adults gather. This is usually a sheltered, grassy spot near a stream where the cubs can drink. Here they develop their hunting skills by pouncing on mice and insects. In play-fights they establish a ranking order that mirrors the social order in the pack.

▲ **CARRIED AWAY**

A wolf carries a cub to safety by seizing the loose skin at the scruff of its neck in its teeth. This adult is most likely the cub's mother or father, but it may be another member of the pack. All the adult wolves are very tolerant with the youngsters to begin with. Later, as the cubs grow up, they may be punished with a well-placed nip if they are naughty.

Did you know? Father wolves make squeaking noises to call their cubs.

SHARING A MEAL ▶

A young African hunting dog begs for food by whining, wagging its tail and licking the adult's mouth. The adult responds by arching its back and regurgitating (bringing up) a meal of half-digested meat from its stomach. The pups grow quickly on this diet. At the age of four months, they are strong enough to keep up with the pack when they go hunting.

▼ PRACTICE MAKES PERFECT

Two Arctic fox cubs practise their hunting skills by pouncing on one another. Wolf cubs play-fight to establish a ranking order. By the age of 12 weeks, one cub has managed to dominate the others. He or she may go on to become leader of a new pack.

Wolfchild

Rudyard Kipling's Jungle Book, *which was published in 1894, is set in India. The book tells the story of Mowgli, a young boy who is abandoned and brought up by wolves in the jungle. When Mowgli becomes a man he fights his arch-enemy, the tiger Shere Khan. Kipling's tale was inspired by many true-life accounts of wolf-children who grew up in the wild in India during the 1800s.*

▲ YOUTHFUL CURIOSITY

Young maned wolves investigate their surroundings. Females usually bear three cubs at most. Newborn young have grey-brown fur, short legs and snouts. Later they develop long legs and handsome red fur.

ALMOST GROWN ▶

These two young wolves are almost full-grown. Cubs can feed themselves at about ten months, but remain with the pack to learn hunting skills. At about two or three years old, many are turned out. They wander alone or with brothers or sisters until they find a mate and start a new pack.

Icy Wastes

Wolves were once widespread throughout the northern hemisphere. As human settlements have expanded, so wolves have been confined to more remote areas such as the far north. The Arctic is a frozen wilderness where very few people live. Wolves and Arctic foxes are found here. On the barren, treeless plains known as the tundra, harsh, freezing winter weather lasts for nine months of the year. Both land and sea are buried beneath a thick layer of snow and ice. Few animals are active in winter, so prey is scarce. During the brief summer, the ice and snow melt, flowers bloom and birds, insects and animals flourish, so prey is abundant. Arctic wolves and foxes rear their cubs in this time of plenty. Another harsh, remote habitat, the windswept grass steppes of Asia, is home to the small steppe wolf.

Arctic Legend

Native Americans named natural phenomena after the animals that lived around them. The Blackfoot called the Milky Way the Wolf Trail. In northern Canada, the Cree believed the Northern Lights, shown below, shone when heavenly wolves visited the Earth. In fact these spectacular light shows in the Arctic are caused by particles from the Sun striking the Earth's atmosphere.

Arctic wolf
(Canis lupus tundarum)

◄ **WHITE WOLF**
Arctic wolves are larger than most other wolf species. They scrape under the snow to nibble plant buds and lichen if they are desperate for food.

Did you know? The largest Arctic wolf territories cover 13,000^{2} km – an area about the same size as Northern Ireland.

▲ WARM FUR
This Inuit girl is wearing a hood trimmed with wolf fur. The fur is warm and sheds the ice that forms on the hood's edge as the wearer breathes. The Inuit and other peoples of the far north traditionally dressed in the skins of Arctic animals. Animal skins make the warmest clothing and help to camouflage the wearer when hunting.

▲ ARCTIC HUNTER
A grey wolf feeds hungrily on a caribou carcass. In the icy north, wolves need very large territories to find enough prey. They will follow deer for hundreds of kilometres as the herds move south for the winter.

◄ ARCTIC HELPER
One crack of a whip brings a team of huskies under control. Tough and hardy huskies, with their thick fur coats, are the working dogs of the far north. They are used by the Inuit and other Arctic peoples to pull sledges and help to hunt.

SNOWY BED ►
A grey wolf shelters in a snowy hollow to escape a howling blizzard. With its thick fur, it can sleep out in the open in temperatures as low as -46°C. Snow drifting over its body forms a protective blanket.

▲ **WOODLAND JACKAL**
A side-striped jackal keeps a wary look-out for danger. In Africa, the three different kinds of jackal are found in different types of terrain. Side-striped jackals keep mostly to woods and swampy areas. Golden and black-backed jackals live in more open countryside.

In the Forest

South of the treeless Arctic tundra, a belt of dense evergreen forests rings the northern hemisphere. It covers large parts of Canada, northern Europe and Russia. South of this belt lie the broad-leaved woodlands of temperate (warm) regions. Yet farther south, tropical rainforests grow in the hotter regions around the Equator.

Wolves and other canids flourish in forests and woods where there is a plentiful supply of prey and dense undergrowth in which to hide and stalk. Wolves are perhaps most at home in temperate northern regions, where large game such as deer abound. In tropical rainforests, most creatures live high in the tree-tops, where wild dogs cannot reach them. However, canids such as bush dogs and raccoon dogs are found near streams and rivers, which also teem with life.

◀ **HIDDEN HUNTERS**
In dark pine forests and dappled broad-leaved woodlands, the grey or blackish coats of wolves blend in with the shadows. This helps them to creep up on deer, moose, and other forest prey. Dense foliage also protects the wolves from the worst of the north's drenching rain.

Raccoon dog
(Nyctereutes procyonoides)

◄ SOUND SLEEPERS

Raccoon dogs live in thickly wooded river valleys in eastern Asia. They are the only canids that hibernate in winter. In autumn raccoon dogs gorge themselves on fruit and meat to put on a thick layer of fat. Then they retreat to their burrows. They sleep right through the harsh winter and wake in spring.

JUNGLE PACK ►

A dhole moves through the thick undergrowth of a forest in India. Packs of dholes hunt deer such as chital and sambar. They call to one another to surround their prey as it moves through the dense jungle. The pack will guard its kills against bears, tigers and scavengers.

◄ RODENTS BEWARE

Bush dogs live in the dense forests and marshlands of South America. In the wetlands, their main prey are aquatic rodents such as pacas and agoutis. These fierce dogs will even plunge into the water to hunt capybaras – the world's largest rodents, 1.3m long.

Did you know? Raccoon dogs are the only canids that cannot bark.

A SCARCE BREED ►

A wolf surveys the snowy landscape in the Abruzzo region of central Italy. Wolves are common in remote forests in Canada and Russia, but in western Europe they are scarce. They survive in small pockets of wilderness, hiding in the hills by day and creeping down to villages to steal scraps at night.

Grassland and Desert

Grasslands are found on every continent except Antarctica. Savanna (grassland) is home to several canids, including maned wolves, black-backed jackals and African hunting dogs. Other canids, including coyotes, dingoes and several kinds of fox, live in deserts. In these harsh, barren places, the sun beats down mercilessly by day, but at night the temperature plummets. Scorching daytime temperatures may cause animals to overheat. Desert foxes keep cool during the hot days by hiding under shady rocks or in dark burrows, emerging to hunt only at night. Another big problem for desert animals is lack of water. Wild dogs and foxes can survive for long periods with little water, or derive most of the liquid they need from their food.

▲ DESERT WOLVES
Wolves are found in deserts and dry areas in Mexico, Iran and Arabia. With little vegetation to provide cover, they stalk prey by hiding behind boulders or rocky outcrops. Desert wolves often have pale or sandy fur, to blend in with their surroundings.

▲ GIVING OFF HEAT
A black-backed jackal's large ears contain a network of fine veins. Blood flowing through these veins gives off heat, keeping the animal cool.

◄ HUNTERS OF THE OUTBACK
A pair of dingoes waits at a rabbit warren in central Australia. Dingoes have lived wild in the outback for more than 8,000 years. Their reddish-brown coats, with paler fur on their legs and bellies, are perfect camouflage in the desert landscape.

▲ OUTCAST DOGS

Feral dogs are the descendants of domestic dogs that have become wild. In Asia they are known as pariah (outcast) dogs. Feral dogs are very adaptable and change their behaviour to suit any situation. In India, pariah dogs hang around villages and sneak in to scavenge scraps.

▲ SMALLER PACKS

A pack of African hunting dogs tears a carcass apart. In the past, hunting dogs were numerous and widespread throughout central and southern Africa. Packs were large, containing 100 animals or more. Now these wild dogs are much more scarce and their packs are also smaller, usually containing between 6 and 30 animals.

▲ SLY HUNTER

A maned wolf hunts in the long grass in Argentinian marshland. Despite its long legs, this canid is not a fast runner. It also lacks the stamina needed to chase prey over great distances. Instead, it stalks animals such as rodents by slowly sneaking up on them before making a sudden pounce.

The Jackal-headed God

In ancient Egypt, Anubis, the god of the dead, was shown with a human body and the head of a jackal. This god was believed to be responsible for the process of embalming, which preserved the bodies of the dead. Anubis often appears in wall paintings and sculptures found in burial places. Here he is shown embalming the body of an Egyptian king.

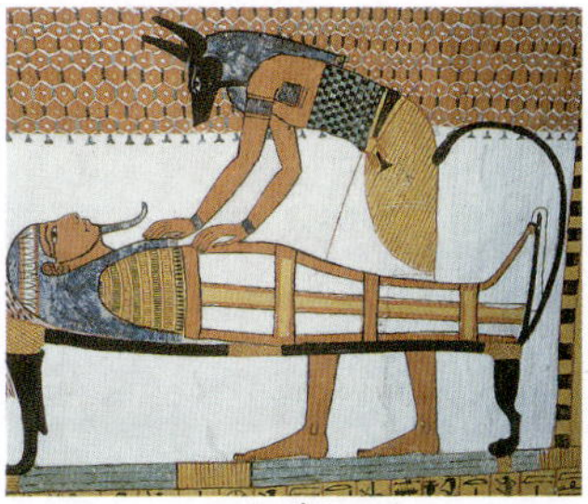

Focus on

The name coyote is from the ancient Aztec word *coyotl*, which means barking dog. These canids are found in most parts of North and Central America, from Costa Rica right up to Alaska and northern Canada. Coyotes eat the same sorts of food as wolves, and the two species are rarely found together. Coyotes are only half the size of wolves, with bodies measuring about 95cm long. They can be recognized by their narrow muzzles, large, pointed ears and long legs. Their tails have a black tip. Coyotes were once thought to be solitary animals, but research has shown that they often live in pairs or small family groups.

COYOTE COUNTRY
A coyote prowls the rocky desert landscape of Death Valley in California, USA. Coyotes are also known as prairie wolves because they are usually associated with grassy habitat. In fact they inhabit all kinds of different terrain, including mountains, forests, woodlands and desert.

TERRITORIAL MARKING
A coyote marks the boundary of its territory by spraying urine. Coyotes also howl to keep other coyotes off their patch. Like wolves, coyotes use their territory both for hunting and for breeding. Coyote territories, however, are generally much smaller than wolf territories – usually between 14 and 65 sq km in size.

Coyotes

CUNNING HUNTERS
Birds, such as this pheasant, make a tasty meal for coyotes. Coyotes often work in pairs to surround ground-nesting birds or grazing rabbits. They also team up to flush burrowing animals, such as ground squirrels, from their underground homes.

CLOSE TO HOME
A coyote cub surveys the world from the safety of its den entrance. Dens may be abandoned skunk or badger burrows. Coyotes mate in late winter. Around nine weeks later, the female gives birth to about five cubs. The young feed only on their mother's milk until the age of three weeks, when they start to eat meat as well.

GROWING UP
Two cubs sniff at a freshly killed pheasant brought by their mother. When the cubs are very young, the father brings all the food while the mother stays in the den. Later, she goes out to hunt as well. Most young coyotes leave their parents at the age of one year. A few youngsters stay on to help their parents rear a new litter of cubs.

CHICKEN RAID
A coyote peers into a chicken coop, hoping to steal a hen. Coyotes mainly feed on rabbits and rodents. In a way coyotes are actually useful to farmers. They kill large numbers of the unwanted small animals that graze on farmland pasture.

Relatives and Namesakes

Wolves, foxes and other members of the dog family have taken millions of years to evolve (develop). Scientists can trace their history through bones and other fossils preserved in the Earth from ancient times. Somewhere around 65 million years ago, a family of tree-dwelling mammals called miacids developed. These carnivorous mammals tore the flesh from their prey using jagged carnassial teeth similar to those of modern dogs and wolves. Later, around 30 million years ago, one branch of the miacid family evolved into fast-running, mongoose-like creatures called cynodictis. In turn, cynodictis and relatives such as hesperocyon developed into the first distinctively wolf-like creatures. From such ancestors, modern canids evolved around 300,000 years ago.

▲ ANCIENT ANCESTOR
Hesperocyon was the prehistoric ancestor of wolves and other carnivores. It had a long, dog-like snout and sharp carnassial teeth. Over millions of years, hesperocyons had evolved from earlier tree-dwelling creatures, called miacids. Miacids gradually evolved strong legs and feet to make them fast runners.

◄ DIRE WOLVES
The long-extinct dire wolf was an ancient relative of the modern wolf. Remains of these prehistoric creatures have been found in fossil pits in California, USA. Dire wolves lived on Earth about two million years ago. They were fierce predators, much bigger than today's wolves, that preyed on large camelids (camels) and American rhinos (both now extinct) and (the now rare) bison.

◄ WOLF-DOG HYBRID

A wolf-dog skirts farmland in Italy. This animal is the offspring of a grey wolf and a German shepherd dog. Coy-wolves, red wolves and coy-dogs are other hybrids that resulted from interbreeding between coyotes, wolves and domestic dogs. These animals may resemble domestic dogs, but they cannot usually be tamed.

TASMANIAN WOLVES ►

This engraving of the 1800s shows two thylacines, or Tasmanian wolves. As the name suggests, these creatures lived on the island of Tasmania, off Australia. With their long snouts and strong legs, thylacines looked a lot like wild dogs and barked like dogs, too. However, they were marsupials, a group of mammals that carries its young in pouches. Their natural prey was kangaroos and wallabies, but when they started to take sheep farmers hunted them to extinction.

Tasmanian wolf *(Thylacinus cynocephalus)*

Did you know? Dogs have been domesticated for at least 12,000 years.

▲ SCAVENGERS AT WORK

Hyenas gather at a Kenyan waterhole to finish off a buffalo killed by lions. Hyenas look like African hunting dogs, but are from the *Hyaenidae* family. They can crunch through the toughest bones with their powerful jaws and strong, sharp teeth.

▲ DESCENDED FROM ONE ANCESTOR

All domestic dogs are descended from the wolf. Human and dog came together thousands of years ago for mutual benefit. By Roman times, around 500BC, many of the breeds we know today had already developed.

Focus on

BIG EARS
The fennec is the world's smallest fox, but it has the biggest ears. Fennec foxes live in the sandy deserts of Africa and Arabia. Their very large ears help to give off heat and keep the animals cool in the scorching desert. The fox rests in its cool, sandy burrow by day, and comes out to hunt rodents, lizards and insects at night.

With their large, pointed ears and long snouts, foxes look a lot like coyotes, but have shorter legs. There are more than 20 species of fox. They are found in most places, but are not native to Australia, South-east Asia or Antarctica. Like other canids, foxes are very adaptable and live in a wide variety of habitats, from the frozen Arctic and bare mountainsides to scorching deserts and crowded cities. Foxes eat almost anything, including small mammals, birds, reptiles, insects, worms and fruit. Most are solitary hunters that keep other foxes away from their territories. In the breeding season, they yelp harshly to find a mate. Litters of up to six cubs are born in a burrow or rocky crevice.

Arctic fox in winter coat *(Alopex lagopus)*

Arctic fox in summer coat *(Alopex lagopus)*

SNOWY WANDERER
Arctic foxes roam over the freezing, wind-swept Arctic tundra. They feed on lemmings (rodents) and sometimes follow polar bears to steal the leftovers from their kills.

BRAND NEW COAT
In spring, the Arctic fox's thick white fur falls out. A new dark brown coat grows in its place. This blends in well with the rocks and plants of the tundra after the snow has melted.

Foxes

Crab-eating fox
(Cerdocyon thous)

NIGHT RAIDER

Red foxes live in woods and open country in North America, Europe and Asia, and were introduced into Australia after 1780. They kill their prey by springing upwards and then pouncing, to trap their victims in their front paws. Red foxes have moved into towns, where they find food by raiding rubbish bins at night.

CRAB-EATING FOX

The crab-eating fox lives in the grasslands, woods and swamps of South America. As well as crabs, these foxes hunt a wide range of other animals, including lizards, frogs, rodents and chickens. They also steal reptiles' eggs. The species earned its rather misleading name because the first animal to be spotted by a scientist had a crab in its mouth.

COLPEO

A pair of colpeo foxes rest by a rocky outcrop in the Andes Mountains of Bolivia. This South American fox lives mainly on grasslands and in mountains. Its tawny-black fur provides camouflage among the boulders, where it hunts for mice, birds, eggs and snakes.

Colpeo fox
(Dusicyon culpaeus)

Fact or Fiction?

Myths, fairytales and even modern films depict wolves as bold, wicked hunters that prey on humans, especially young children. In fact, there have been very few confirmed accounts of wolves attacking people, and none in North America, one of the animal's main strongholds. Wolves were more likely to have been a menace in Europe before guns were invented. Then, they probably had little fear of humans and may have prowled close to villages in the hope of stealing sheep and other domestic beasts. Wolves may have acquired their reputation as man-eaters by scavenging meat from dead bodies on battlefields. In modern times, the rare attacks on people have probably been made by wolves suffering from rabies, a disease that drives animals mad.

▲ **TALL TALE**
In 1901, a pack of wolves was reported to have attacked and eaten a group of five Romanian soldiers. Suspiciously, no trace of the men was left except their blood-stained weapons. This event is hard to believe, since even starving wolves would be unlikely to attack a band of well-armed men.

◀ **ALWAYS ALERT**
A modern-day shepherd tends his flock in the remote Spanish mountains. He has only his dogs and a stout stick to scare away any wolves that threaten his flock. An experienced shepherd, however, knows that wolves are easily frightened.

▲ **WOLF MAGIC**
Native Americans thought of wolves as magical creatures. This shaman (medicine man) is dressed in a wolf skin to take on the creature's power. He calls upon the wolf spirit to give his patient the strength to recover from illness.

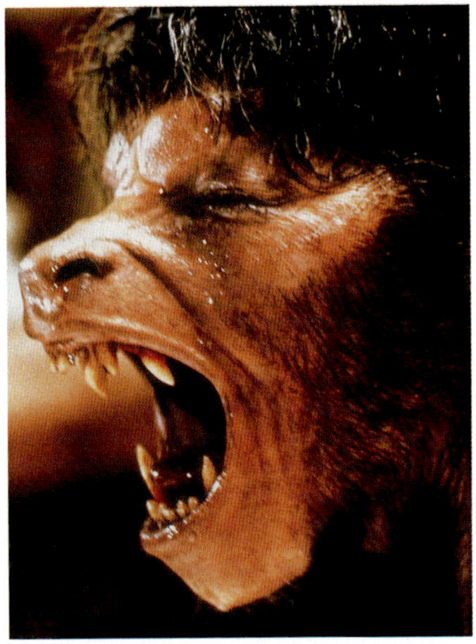

▲ **HOWLING AT THE MOON**
Werewolves are a popular subject of horror films. In folk tales, these are humans who turn into wolves when there is a full moon. They can only be killed by a silver bullet.

▲ **WOLVES BLAMED FOR ATTACK**
In 1907, wolves were said to have attacked a group of Hungarian gypsies. There was no real proof, however, and the attack may have been carried out by a gang of robbers.

◄ **SLY OLD FOX**
In this scene from one of *Aesop's Fables*, a clever vixen (female fox) tries to outwit a cockerel crowing at dawn. She says to the cockerel that she would like to embrace the owner of such a fine voice. Wolves and foxes are portrayed as villains in many folk tales and legends. They are frequently described as sly and cunning, terms that show them in a negative light.

Natural Enemies

▲ **HUNTING THE HUNTERS**
A lioness has killed an African hunting dog. Groups of lions are known to stalk and ambush hunting dogs while they feed on a kill, or drink at waterholes. Hyenas and jackals also prey on hunting dogs. They sneak up to the den and steal young pups if the adult dogs are not keeping a careful watch.

Wolves and wild dogs may be powerful predators, but they face many threats in the wild. Their natural enemies include the largest creatures that they hunt, such as moose, bison and musk oxen. Their sharp hooves and horns can fatally wound a predator. One careless slip, and a wolf may be gored or trampled to death. Wolves and wild dogs are also threatened when their habitats are disturbed or destroyed. In many areas, the territories where wild dogs can roam free are getting smaller and smaller. Land is needed for crops or to graze herds of sheep and cattle. Forests are cut down for timber or to make way for new roads and towns. The survival of wolves and wild dogs is also threatened by deadly diseases such as distemper, anthrax and rabies.

▼ **DEFENCE TACTICS**
On the Arctic tundra, large, shaggy musk oxen form a defensive ring around their young. Their long, fierce horns face outwards, keeping young and weak members of the herd safe from wolves.

◄ **DANGEROUS TARGET**
A moose browses among the tall lakeside shrubs in Wyoming, USA. Moose are powerful beasts. An adult male stands 2m tall at the shoulder and weighs as much as ten wolves. Its antlers and hooves can inflict great damage. A moose can crack a wolf's skull with one mighty kick, or gore it with its antlers and toss it high in the air. Wolves must be very wary when hunting such dangerous prey.

DINGOES KEEP OUT ►
A dingo attacks a sheep, snapping at its hindquarters. Farmers have built a great fence across 5,300km of south-eastern Australia to keep dingoes out of sheep country. Any dog caught inside the fence is shot.

▲ **DEADLY RABIES**
A black-backed jackal lies dying of rabies. Rabies is a fatal disease that attacks the brain and nervous system. It is passed on by saliva from an infected bite. Wild dogs have been wiped out in many areas to prevent rabies spreading to humans, even though a vaccine is available.

▲ **KEEP OFF THE ROAD**
African hunting dogs roam along a newly built road that cuts across their territory in the bush. They are scavenging for road-kills, but may well become victims themselves. The rapid human development of the dogs' territory threatens their way of life and very survival.

Wolves and People

Wherever wolves and wild dogs come into contact with people, the animals are regarded as dangerous pests who will – given the chance – kill livestock. They are poisoned, trapped and shot, not only for their skins, but for sport. Wolves were once the most widespread carnivores in the northern hemisphere. Now they survive in a much reduced area, often in small, scattered groups. Several species are endangered, including the Simien wolf, the red wolf and the African hunting dog. Much of the land where these animals once lived is now being farmed. Dholes and bush dogs have also become very rare as their forest habitat is destroyed. Some species, such as the Falkland Island wolf, a kind of fox, are already extinct.

▲ **WOLF HUNT**
In the Middle Ages domestic dogs were often used to kill wolves, as this Dutch engraving of 1880 shows. The last wolves were wiped out in England by 1500 and in Ireland by 1800.

Tame Wolf
This book is a first edition of the popular novel White Fang *by American writer Jack London. Set in northern Canada, it describes how a wolf, named White Fang, is tamed and becomes a pet. In general it is not against the law to keep a wolf as a pet, but countries with restrictions require owners to have a special permit.* The Call of the Wild *by the same author describes how a pet dog joins a wolf pack and becomes wild.*

▲ **WOLF TERRITORY**

Grey wolves once had the greatest range of any wild land mammal. In the past, wolves were once common right across North America, throughout Europe, the Middle East and most of Asia. Their present range shows they have been exterminated in most of Mexico and the USA, in almost all of western Europe and over much of Asia.

◀ **NOWHERE TO RUN**

A hunter in Colorado, USA, shoulders a coyote he has shot. In country areas, farmers shoot or poison coyotes because they steal sheep and other livestock and spread disease. Elsewhere, when coyotes and other wild dogs enter towns to scrounge scraps, they risk being shot as pests.

▼ **AN UNKIND LUXURY**

Fox fur was very fashionable in the early 1900s, mainly for coats and for trimming garments. The fox fur stole (scarf) shown here uses the pelt (fur and skin) of an entire animal. In the past, furs were worn mainly to keep warm in winter. Today, however, man-made fabrics are as warm as fur, making it unnecessary and cruel to kill these animals for their pelts.

▲ **UNDER THREAT**

A Simien wolf howls high in the Ethiopian mountains. As the human population grows, more land is farmed and the animal's range is restricted. Simien wolves are shot for fur and killed by farmers as pests. There may be only 500 Simien wolves left in the wild.

Conservation

▲ **SUCCESS STORY**
Conservationists release a red wolf that was bred in captivity into a reserve in North Carolina, USA. Red wolves were once found throughout the south-eastern states. They almost became extinct, but breeding programmes have saved the species.

In many parts of the world, efforts are being made to save threatened wolves and wild dogs. Grey wolves have recently been reintroduced into areas where they had died out. Conservationists working to protect wolves face opposition from local farmers who fear that wolves will kill their livestock. In some reserves, wolves and wild dogs have begun to be promoted as tourist attractions. This helps people to learn about these animals and the entrance fees help to finance conservation work. Today, wolves and their relatives are gradually losing their bad image. More and more people are appreciating their admirable qualities – intelligence, loyalty and strong family ties. In the wild, these predators actually help to improve stocks of prey animals. By hunting mostly weak or sickly individuals, they help to ensure the survival of the fittest.

◄ **RADIO TRACKING**
A red wolf has been fitted with a radio collar. The collar allows scientists to track the animal as it roams the wilds. Radio tracking helps to provide scientists with valuable information about the wolf's habits and range. Increasing such knowledge also helps conservationists with their work.

▲ **STAR OF THE SHOW**

Tourists on safari photograph African hunting dogs in a reserve. In recent years, such tourist attractions have earned much-needed cash for remote villages. The money helps to persuade local people not to hunt the dogs, but to see them as a valuable asset instead.

▲ **HELPING TO KEEP THE BALANCE**

A pack of wolves feeds on a deer carcass. By targeting old and sick animals, the wolves actually help the rest of the herd to survive. They may be removing a deer whose sickness could infect others in the herd, or an old animal whose share of food could be better used to rear healthy young.

▼ **SOUND OF THE WILD**

For many people, the wolf is a symbol of the wilderness. Now in some countries, wolves are becoming a tourist attraction. At some centres, members of the public can even walk alongside tame wolves, petting them if they wish, accompanied, of course, by expert handlers.

▲ **WOLF RESEARCH**

Scientists check the teeth of a drugged Arctic wolf. Researchers sometimes capture the same wolves several times over the course of a number of years to study their life histories. This work helps to provide evidence of the strong family ties and keen intelligence of the wolf.

GLOSSARY

adapt
When an animal or group of animals changes – physically or in behaviour – in order to survive in new conditions.

alpha pair
The top male and female in a wolf pack. Only this strong, healthy pair of animals breeds.

binocular vision
The ability to see things with both eyes at the same time, which helps animals to judge distances well and pounce on prey.

camouflage
The colours and patterns on an animal's body that help it to blend in with its surroundings so it can hide from enemies or sneak up on prey.

canid
A member of the dog family. Wolves, jackals, coyotes and foxes are all canids.

canines
The sharp, pointed teeth near the front of a wolf's jaw, which are used for seizing and killing prey.

carnassials
The jagged teeth towards the back of a wolf's mouth, that are used for cutting up meat.

carnivore
An animal that feeds mainly on the flesh of other animals.

carrion
The body of a dead animal.

conservationist
A person who works to protect the Earth's natural resources, including animals and plants.

cub
A young wolf or dog.

den
The home of an animal such as a wolf or fox. A den may be an underground burrow or a cave, or simply a nest in the long grass.

dew claw
The toe found high on a wolf's foreleg, which does not touch the ground. It has no obvious use, so it is probably used for grooming.

diet
The range of food an animal eats.

domesticated
Describes animals that have been tamed by people. Cows, sheep and horses are all domesticated.

Equator
An imaginary line running around the centre of the Earth, separating north from south.

evolve
When a species of animals or plants changes gradually over time, to become better suited to the conditions in which it lives.

extinct
When a whole species of animals has died out completely, so that none is left.

fossil
The remains of plants or animals that have turned to stone over thousands or millions of years.

gland
An organ of the body that makes chemicals for a particular use.

habitat
A type of place where certain animals and plants live, such as a tropical rainforest or a desert.

hierarchy
A strict social order within a group of animals.

hybrid
The offspring of two animals from different species or subspecies.

litter
A group of young animals born to a mother at one time.

mammal
A warm-blooded animal with a bony skeleton and hair on its body. Female mammals produce milk from mammary glands (equivalent to human breasts) to feed their young.

moult
To shed fur or feathers.

muzzle
The jaws and nose of an animal such as a wolf.

nocturnal
Describes an animal that rests by day and is active during the hours of darkness.

pack
The name given to a group of wolves or wild dogs that lives and hunts together.

predator
An animal that hunts other creatures for food.

prey
An animal that is hunted by another animal for food.

range
The area in which an animal species lives.

regurgitate
When an animal brings up half-digested food to feed its young.

rendezvous
A meeting place that lies in the heart of a wolf pack's territory. This is a safe place where wolves gather and cubs play.

rodent
An animal with chisel-shaped front teeth used for gnawing.

savanna
Open grassland found in warm, wet tropical regions around the Equator.

scavenger
An animal that feeds on rubbish, dead animals and the remains of carcasses left behind by other predators.

social
Describes animals that live together in large family groups.

stalk
When a hunting animal follows its prey cautiously to avoid being seen until close enough to pounce.

submissive
When a junior animal gives way to a more powerful animal.

territory
An area that an animal uses for feeding or breeding, and defends against others of its species.

tundra
The cold, treeless lowlands of the far north. Summers are very short here. During the winter the sun hardly rises in the sky and it is dark for most of the day.

wean
When a young animal moves from drinking only its mother's milk to a diet of solid food.

INDEX

A
Aboriginal Australians 7
African hunting dog 8–9, 13, 19, 20–1, 25, 29, 32, 34–5, 36, 39, 40, 47, 57, 61
alpha pair 26–7, 29, 36–7
Arctic fox 41, 42, 52
Arctic wolf 6, 19, 42, 61

B
bat-eared fox 15
beta pair 26
black-backed jackal 7, 18, 31, 46, 57
blood 14, 46
body language 26–7
bones 12–13, 50
brain 14
breeding 24–5, 26, 36–9, 42, 48–9
bush dog 8, 17, 32, 58

C
camouflage 18–19, 43, 46, 52–3
canids 4–5, 9, 50
carnivores 4–5, 30–1, 34, 39, 40
colpeo fox 53
communication 20–3, 26–7, 28–9, 40
conservation 60–1
coy-dog 51
coy-wolf 51
coyote 4, 6, 7, 16, 20, 24–5, 30, 33, 46–9, 59
crab-eating fox 53
cubs and pups 4, 24–5, 28, 36–41, 42, 49
colour 12–13, 44–9
conservation 60–1
Coppers 30, 50, 58
courtship 12, 16, 30–1
crochets 20

D
dens and burrows 4, 36–37, 38–9, 46
desert wolf 46
dew claws 16
dhole 8, 20, 25, 29, 31, 32–3, 58
diet 4–5, 30–1, 34, 38–9, 40
dingo 6–7, 17, 18, 33, 37, 46, 57
dire wolf 50
diseases 54, 56–7
dog 4, 6–7, 8–11, 18, 21, 43, 46–7, 51
drinking 22, 24, 46

E
ears 4, 5, 20, 46
endangered species 58–61
evolution 50
extinct species 6, 50–1, 58, 60–1
eyes 5, 21, 38

F
facial expressions 27
Falkland Island dog 58
fangs 14
fennec fox 52
feral dog 47
fossils 50
fox 4, 12–13, 17, 24, 29, 46, 52–3, 55, 59
fur 4–5, 18–19, 23, 42–3, 52, 59

G
golden jackal 7, 18
grey fox 17
grey wolf 6, 16, 19, 26–7, 31, 39, 43, 59, 60
greyhound 11
grooming 23
guard hairs 18

H
hearing 4, 5, 20
heart 14
hesperocyon 50
hound 11
howling 20, 29, 36, 48, 55
hunting 4–5, 6, 12–13, 14, 17, 20–3, 30–5, 56–7
husky 43
hybrids 6, 51

I
intelligence 4, 14, 60
intestines 14

J
jackal 4–5, 6–7, 18, 24, 47, 56
jaws 4, 15, 22

L
legs 4, 12–13, 16–17
litter 4
lone wolf 27

M
mammal 4, 38–9
maned wolf 8–9, 12, 17, 19, 24–5, 31, 32, 36, 41, 46–7
mating 36–7
Mexican wolf 6
miacid 50
milk 4, 38–9, 40
muscles 14–15, 16
muzzle 4, 23

N
Native Americans 13, 42, 55
nictating membrane 21
nocturnal species 32

P
pack 5, 22, 24–7, 28–9, 32–5, 36, 40–1, 47
paws 5, 16–17, 23
play 21, 23, 40–1
predators 56–9

R
rabies 54, 56–7
raccoon dog 8–9, 18, 24, 30, 32
red fox 53
red wolf 6, 51, 58, 60
rendezvous 28, 40
running 4, 12–13, 14, 16–17

S
senses 4–5, 20–3, 28–9, 32
side-striped jackal 7, 18
Simien wolf 6, 39, 58, 59
skeleton 12–13
social groups 4–5, 22–7, 40–1, 48–9, 60
steppe wolf 6, 41

T
tail 4–5, 12–13, 22
teeth 4, 12, 14–15
territory 28–9, 42, 48
toes 5, 16–17
tongue 22–3
tracking 22–3, 32
temperature 14, 36–7

W
werewolf 55

NATURE WATCH

MONKEYS

TOM JACKSON

Consultant: Rachel Hevesi

The Monkey Sanctuary Trust

C O N

Published by Anness Publishing Ltd,
Blaby Road, Wigston,
Leicestershire LE18 4SE

Email: info@anness.com

Web: www.annesspublishing.com

Publisher: Joanna Lorenz
Managing Editor: Linda Fraser
Consultant Editor: Gilly Cameron Cooper
Project Editors: Rebecca Clunes, Leon Gray
Designer: Caroline Reeves, Aztec Design
Picture Researcher: Veneta Bullen
Illustrator: John Francis
Production Controller: Claire Rae
Editorial Reader: Jonathan Marshall

INTRODUCING SMALL PRIMATES
4 Prosimians – The First Primates
6 Monkeys Take Over
8 The Last of the Lemurs

HOW PRIMATE BODIES WORK
10 Outward Appearances
12 Inner Power
14 Sensing the Surroundings
16 On the Move
18 The Fast, Acrobatic Langurs
20 Brainpower

ENVIRONMENT
22 Changing Habitats
24 Survivors in the Snow
26 Diets for Living
28 Sharing Space
30 Territorial Rights
32 Trumpeters of the Swamps
34 Predator and Victim

T E N T S

BEHAVIOUR
36 Living Together
38 Communication
40 Bright Colours of the Mandrills
42 Forming Bonds
44 A Hamadryas Baboon Troop
46 Mating Time
48 Growing Up
50 A Young Macaque Matures

THE HUMAN CONNECTION
52 Monkey Relations
54 Habitat Destruction
56 Life in a Concrete Jungle
58 Use and Abuse
60 Saving Species

62 Glossary
64 Index

Prosimians – The First Primates

The earliest primates roamed the world about 25 million years before monkeys evolved. They were the prosimians – which means 'pre-monkeys'. Amazingly, there are still many species of prosimians alive today, including aye-ayes, bush babies, lemurs, lorises and tarsiers.

The prosimians' hands mark them out as primates. They have flexible fingers, and thumbs that work in the opposite direction, making it possible for them to hold and pick things up.

Prosimians rely heavily on their sense of smell to communicate and to find food. The advanced primates – the monkeys and apes – use their sight more. In areas where monkeys and prosimians lived together and competed for the same food, the monkeys were more successful and so prosimians had to develop specialized tactics to survive.

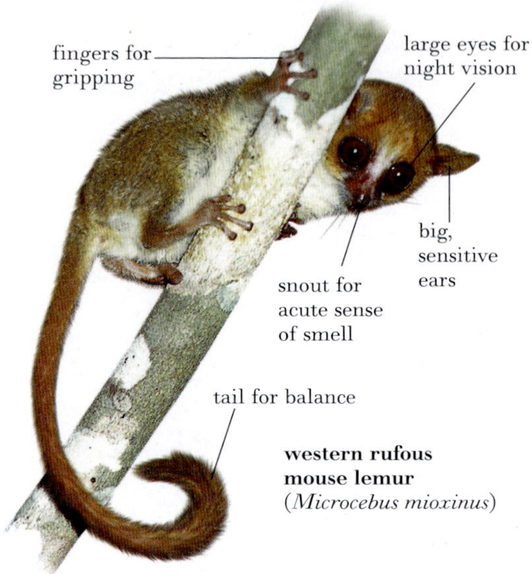

western rufous mouse lemur
(*Microcebus mioxinus*)

▲ TYPICAL PROSIMIAN
This lemur is the smallest primate of all. It weighs only 30g, about the same as a tablespoonful of sugar. Lemurs have characteristics common to all prosimians, but are a distinct group of their own. They live only on the island of Madagascar. Monkeys never crossed the sea to the island once it was cut off from Africa 100 million years ago, so prosimian lemurs survived and evolved into many different species.

EXCLUSIVE OLD WORLDERS ▶
Prosimians live only in parts of Asia and Africa. No species are found in the Americas or in Australia. Bush babies and lorises are found in central and southern Africa. Lemurs live off the east coast of Africa, on the island of Madagascar. The islands and peninsulas of South-east Asia are home to lorises and tarsiers.

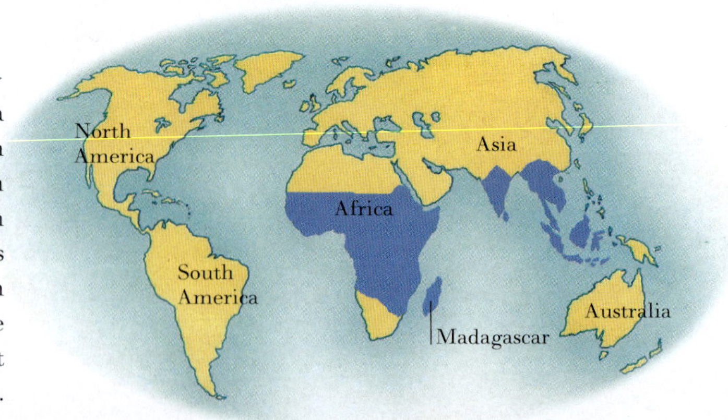

◄ USING YOUR EYES
Despite its goggle-eyed appearance, the tarsier from South-east Asia, is more closely related to monkeys than to other prosimians. Most prosimians rely heavily on their sense of smell to communicate and find food. With the tarsier, however, eyesight has become more important. It can swivel its head 180 degrees in each direction to look over its back.

▲ NIGHTWATCH
Bush babies, or galagos, are nocturnal. In daytime, they snuggle up with their family. Some prosimians survived after monkeys evolved by becoming nocturnal and feeding when the monkeys slept. Bush babies live in African forests, feeding on fruit and insects. This South African galago is only about 16cm (a relaxed adult handspan) long, but it can leap up to 5m.

SCENTED PATH ►
Like other prosimians, the slow loris from South-east Asia stakes out its territory with scent. It rubs urine on its hands and feet so that it can imprint its smell exactly where it wants. It also uses the smell to mark its path. The loris moves slowly to avoid attracting attention from predators in the forest. If something frightens it, the loris can easily return to the safety of its nest by following the scented path.

Did you know? Bush babies can grip tightly to a branch for hours without getting tired.

◄ SOCIAL LEMURS
Ring-tailed lemurs are almost as versatile as monkeys. They are agile climbers, and scamper like cats on the ground. They live in large groups, in which females do all the leading and fighting for dominance. Male ring-tailed lemurs fight only over females.

Monkeys Take Over

▲ **DOWN TO EARTH**
Unlike many monkey species, baboons do not live in trees. This olive baboon, which lives in grassland, is typical. Baboons are Old World monkeys, mostly from Africa. They live in large social groups that are among the most complicated and interesting of all monkey societies. All four baboon species are bigger than most monkeys.

In the fierce, overcrowded world of the ancient forests, monkeys had many advantages over prosimians. They had bigger brains and keener eyesight, which enabled them to adapt to different habitats and conditions.

Apes and monkeys evolved around 35 million years ago. At first, apes dominated the forests, but the versatile monkeys soon reigned supreme because they could access foods all over the forest, from the top to the bottom. Because of this, monkeys are still generally built for life in the trees. Most species continue to inhabit warm, wet forests around the Equator, although some have adapted to life on the ground, in desert scrub, on mountains, or even in bustling cities.

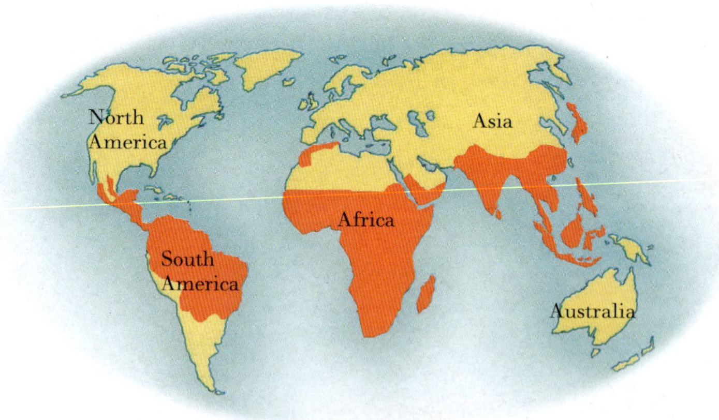

◄ **OLD WORLDS AND NEW**
Monkeys evolved in Africa, and spread to the Americas and Asia. The monkey populations on each side of the ocean developed in different ways. Those that remained in Africa and Asia became known as the Old World monkeys. They evolved into many different species but all have narrow noses and downward-pointing nostrils. The American, or New World, monkeys have wider noses with outward-facing nostrils.

◀ NEW WORLD NOSE STYLE

Like all other monkeys that evolved in the New World, this cotton top tamarin has a flat nose with outward-facing nostrils. New World monkeys live in the forests of Central and South America. Most of them have tails that, unlike those of Old World monkeys, can wrap around and grip on to things, and are used as an extra limb.

WALKING FRAME ▶

This Old World monkey from India has many characteristics that are common to all monkeys. Long, slender limbs of about equal length make for easy walking on all fours. Tails are long and flexible – especially on New World monkeys. Hands and feet are designed for climbing, walking on the ground and holding leaves and fruits.

hanuman langur
(*Semnopithecus entellus*)

Did you know? Monkeys usually have ridged palms to help them grip on to trees.

OLD WORLD PORTRAIT ▶

Look at the downward-pointing, closely spaced nostrils of this pig-tailed macaque. They are typical features of an Old World monkey. These monkeys are more closely related to the great apes such as gorillas and chimpanzees that also live in the Old World. The wide space between lip and nose is typical of many monkeys and apes but unlikely to be seen on prosimians.

The Last

One particular group of prosimians flourished in just one place in the world. They were the lemurs of Madagascar. Millions of years ago the island split away from the east coast of Africa, leaving its population of prosimians to evolve separately from those on the mainland. With no competition from monkeys and apes, the lemurs of Madagascar evolved into many different species. Today there are around 40, from the tiny mouse lemur to the athletic sifakas and indris, which are about the weight of a medium-sized monkey. Once people arrived on Madagascar, many species of lemur became extinct, including a huge gorilla-sized lemur.

GHOST DANCE
A sifaka hops daintily over open ground. It spends most of its time leaping athletically through the springy leaves of cactus-like trees. When humans first came to Madagascar, they heard the eerie calls of sifaka troops, glimpsed the white-furred figures and thought that the island was haunted. The animals became known as *lemures* – Latin for 'spirits of the dead'.

ENJOYING THE NIGHTLIFE
The weasel lemur feeds only at night and has a solitary nature, like most of the smaller lemur species. A couple of times a night, a male and female might meet up and groom each other. Most small lemurs eat insects, seeds, gums and fruit. Weasel lemurs, however, are specialist leaf-eaters. Their large intestines break down tough, indigestible plant matter. As an extra aid to digestion and to extract every bit of goodness, they sometimes eat their droppings.

of the Lemurs

COOLING OFF

Ring-tailed lemurs bare their chests and splay their legs to lose as much heat as possible in the cool breeze during the day. They are active at night, too, stopping for short breaks every few hours. This helps to keep up their energy levels. These larger lemurs are quite monkey-like in their behaviour. They live in permanent social groups of up to 30, with a female leader. The smaller, nocturnal species tend to lead more solitary lives, like the mainland prosimians.

SOLE SURVIVOR

The aye-aye is the only surviving lemur of its type, and is itself threatened with extinction. It has survived as long as this by being highly specialized – a long middle finger with a claw-like nail hooks out grubs from wood, or the pulp from fruit, as in this picture. Aye-ayes are very secretive and move about only at night. During the day, they curl up in a nest of leaves.

HIGH-LEVEL LIVING

Ruffed lemurs rarely touch the ground. Their short legs are more suited to running and jumping through the trees than to climbing like the longer-limbed sifakas and indris. As they eat more fruit than other lemurs, they do a useful job of spreading seeds via their droppings. All lemurs, and the ruffed lemur in particular, have a snout like that of a wild dog. This provides more surface area for picking up smells.

Outward Appearances

Monkeys and prosimians became experts in forest survival. As different species evolved, each adapting to a particular way of life, a great variety of shapes and sizes emerged. All, however, had the primate speciality of flexible hands and fingers. Thumbs work in the opposite direction to the fingers so that the digits can close together and grasp. Long arms and legs are geared for climbing. On many species the arms and legs are of similar length – which means the animals can walk comfortably on all fours. On others, strong hind legs provide power for leaping. These primates can stand or sit on their haunches, leaving both hands free.

▲ **UNDER THE THUMB**
The pygmy marmoset from South America is small enough to fit on the palm of your hand. It is only about 30cm from its head to the tip of its tail, and weighs about 100g – 500 times less than a mandrill.

▲ **THE FIFTH LIMB**
A black-handed spider monkey has a muscular tail that can wrap around a branch. A hairless stretch of ridged skin on the underside gives extra grip, like the ridges on the palm of a primate hand. Only New World monkeys have tails with this prehensile (grasping) power. The tails of prosimians and Old World monkeys simply help to balance them when leaping through the trees.

▼ **BIG AND BULKY**
Male mandrills are the world's biggest monkeys. They weigh 50kg, about the same as a small adult human. Mandrills walk on all fours and have limbs of roughly the same length. They do not live in trees so they don't need a tail.

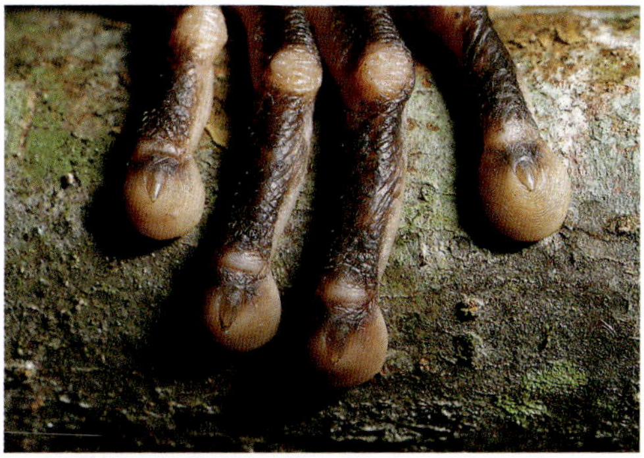

▲ **HANDS ON**

The long, flexible fingers of the tarsier are padded for extra grip when climbing. The tarsier is a prosimian from South-east Asia. As with almost all primates, it has fingernails. For animals that forage for plant food, fingernails are more useful than claws. They can be used for picking, peeling and gouging as well as for scratching and grooming.

▲ **GETTING A GRIP**

A young vervet makes the most of its hands and feet to climb. Its opposing thumbs curve in to help it grip. Balanced securely on two feet and one hand, the monkey has one hand free for eating. Monkey hands need to hold small items, so tend to be smaller and more flexible than their feet.

◄ **PROBLEM HAIR AND SKIN**

A rhesus macaque pulls ticks from her friend. The fur of monkeys is home to many tiny insects that suck the host's blood or make the skin flaky. Monkeys deal with the problem by grooming each other. Grooming is also an important way of making friends. Monkey skin darkens with age because of the effects of sunlight and so many monkeys avoid spending too much time in the sun.

A HANDY FUR COAT ►

For a new-born chacma baboon, mother's fur is handy to cling on to. In the sun-baked savannas, baboons do not need to keep warm, so the baby's fur will become coarse and sparse like its mother's. Fur also gives some necessary protection against the burning sun, and its grey-brown colour provides good camouflage on the African plains.

Inner Power

In terms of brainpower, monkeys and other primates are often considered more 'advanced' than other animals. Their bodies, however, are basically the same as most other mammals. Primates did not develop special physical features like animals that grew hooves or horns.

Small primate bodies are built for flexibility and agility, with hinged joints supported by long, elastic muscles to allow maximum range of movement. Different species have body shapes adapted to whether they climb and leap through trees, or move along the ground. Head shape and size depend on whether brainpower or the sense of smell or sight is top priority. Digestive systems vary according to diet.

▲ DENTAL PRACTICE
Monkeys and prosimians have four types of teeth. There are incisors for cutting, canines for stabbing and ripping, and molars and premolars for grinding tough leaves and fruit into a mushy paste.

▼ INTERNAL VIEW
A monkey's bones and strong muscles protect the vital organs inside its body. Its facial muscles allow it to make expressions. Monkey legs are generally shorter in relation to their bodies than prosimian legs. This gives them more precise climbing and reaching skills, especially in the treetops.

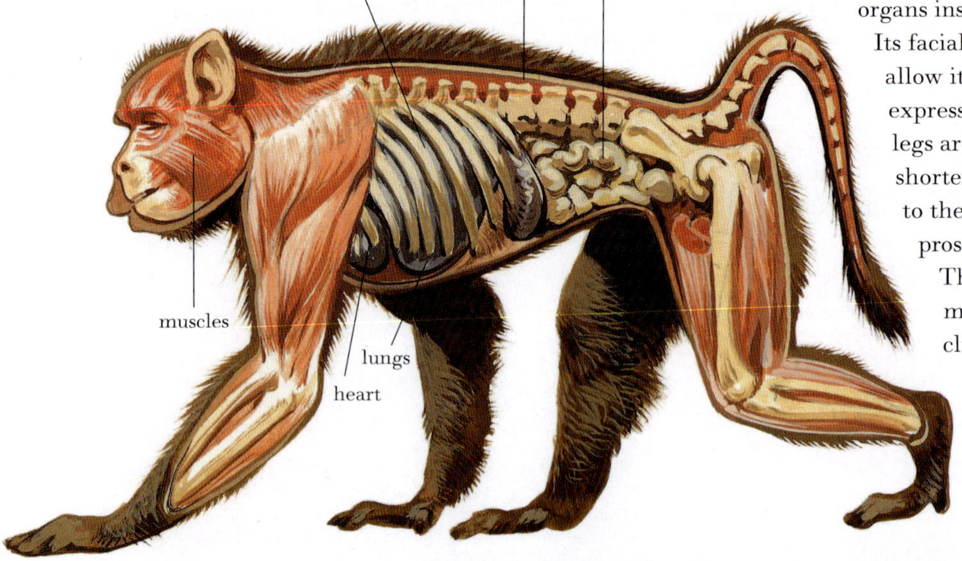

▼ BIG EYES

The tarsier has the biggest eyes of any animal in relation to its body size. There is not much room left in its skull for a brain and each of the eyes is heavier than the brain. Prosimians have simpler lives than monkeys and do not need big brains. Instead, they have an array of sharp senses – more sensitive than a monkey's – that help them to survive. As well as its huge eyes, this tarsier has large ears that pick up the slightest sounds in the quiet of the night.

▲ A STRONG STOMACH

A baboon's digestive system can cope with raw meat as well as plant food. Although meat is nutritious, it can be harmful if it stays in the body too long. Meat-eaters have shorter guts than vegetarians, so that food passes through the body quickly. Most monkeys and prosimians eat mainly plants and have longer guts because leaves and other plant foods are hard to digest.

INSIDE LOOK AT A LEMUR ►

The skeleton of this prosimian, a ruffed lemur, is quite different from that of a monkey. It has a long, narrow head, which has less space for the brain, and its legs are very long compared to the length of its body. Their long legs help lemurs to leap great distances from tree to tree. Because their legs and feet are designed for climbing, they can also cling in a relaxed fashion to sheer, upright tree trunks.

Sensing the Surroundings

▲ **COLOUR SELECTION**
Thanks to its colour vision, an African vervet monkey can select its favourite flowers when they are at the peak of perfection. Many leaf-eating monkeys have eyes that are particularly sensitive to different shades of green. This means that they can easily identify the fresh green of tender young leaves which are good to eat.

Monkeys are usually active in the daytime, and make the most of their excellent colour vision. Most prosimians are nocturnal, and have eyes to help them see at night. The eyes of all primates are lined with a combination of cone cells and rod cells. Monkeys have more colour-sensitive cone cells. Prosimians have plenty of rod cells, designed to see in dim light. Their eyes glow in the dark like cats' eyes because they are backed by a mirror-like layer that reflects every bit of available light.

Sight is very important to monkeys, but prosimians rely much more on the senses of smell and hearing. The ears of prosimians are constantly on the alert. The slightest rustle in the dark could identify an insect snack or an approaching predator.

▲ **EYES LIKE SAUCERS**
Unlike the eyes of many prosimians, tarsier's eyes have no reflective layer, but their size means they catch as much light as possible. Like monkeys, their eyes have a sensitive area called the fovea, which picks out very sharp detail.

MIDNIGHT MONKEY ▶
Douracouli monkeys are the only nocturnal monkeys. They live in South America where there are no prosimians to compete with. Like prosimians, their eyes are big to catch maximum light. They are able to pick up detail but not colour.

▼ **A KEEN SENSE OF SMELL**

An emperor tamarin monkey marks its territory with scent. New World monkeys and prosimians have a smelling organ in the roof of their mouths that Old World monkeys do not have. They use their sense of smell to communicate with each other and to identify food that is good and ready to eat.

▲ **WET NOSED SMELLING AIDS**

Look at the shiny nose of this ruffed lemur. It is more like that of a cat or dog than a monkey. Most prosimians have this moist nostril and lip area, called the rhinarium. It gives them a better sense of smell. The nose has a layer of sensitive cells to detect chemicals in the air. The cells work better when they are wet.

▼ **MUFFLED SENSE OF HEARING**

The furry ears of squirrel monkeys probably muffle sound. But although these and other monkeys use sound to communicate with each other, hearing is not as important for them as keen eyesight. Nocturnal primates, however, have highly sensitive, delicate-skinned ears.

On the Move

Primates are the most versatile movers of the animal world. Many can walk and run, climb and swim. Small monkeys and the larger lemurs can leap between branches. This method is risky for larger animals, in case a branch breaks beneath their weight. Heavier primates play safe by walking along branches on all fours and avoid jumping if possible.

When branches are bendy, climbers move with caution. They may use their weight to swing from one handhold to the next, but they do not let go of one handhold until the other is in place.

Did you know? A spider monkey's tail is so strong it can support the monkey's entire weight.

◄ CLIMBING POWER New World monkeys rarely move at ground level. They swing from tree to tree, stop and whip their tail around a branch, let go with both hands and grab something to eat. Old World monkeys do not have a prehensile (grasping) tail. Many have lost their tails altogether or have very short ones, which they use to help balance in the trees.

◄ PADDED MITTS The palm of a sifaka lemur is just one of the features that makes it a star jumper. The wrinkles and flesh pads are like a baseball glove, giving extra holding power. Sifakas can push off from their long hind legs to leap up to 5m high. Their arms are short, making it impossible to walk on all fours. Instead, sifakas hop on both feet.

◄ BABOONS ON THE MOVE
A troop of baboons makes its way down a track in the African savanna. They are strong and tough because they have to walk long distances to find enough food to feed the troop. Baboons and other monkeys that walk on the ground, such as mandrills, geladas and macaques, put their weight on the fingertips and palms of their hands and feet. This is different from apes, such as gorillas and chimpanzees, who walk on their knuckles.

▲ SLENDER LORIS
Lorises from Sri Lanka and southern India are known for their slow, deliberate movements. They have long, thin arms and legs with very flexible ankles and wrists. A loris can wriggle its way through and get a grip on dense twigs and small branches. Due to its strange appearance, some people describe the animal as a banana on stilts.

LIGHT AS AIR ►
A pygmy marmoset can perch on the flimsiest of twigs. It has to be small, light and quick so that it can catch insects to eat. Marmosets scurry along branches rather like squirrels, and for this reason are the only primates to have claws instead of nails. They are also able to sit up on their hind legs to free their hands for collecting food as well as for feeding.

The Fast,

COMING TO GROUND
A golden langur sits and waits for other members of the troop. Gatherings usually occur in the mornings. They take place on or close to the ground, as there is more space than in the trees.

Long limbs, strong back muscles and a super-long tail make langurs the supreme all-round athletes of the monkey world. They are masters of balance, suspension and speed; as agile on the ground as at the very top of a forest canopy. They can leap farther than most other monkeys, and spend up to 70 per cent of their time moving around the trees.

Langurs live in big troops in south Asian countries, such as Bangladesh, India and Bhutan. The males in a troop try bossing the females around, but the females are really in charge of everything apart from mating. The females also help each other out by sharing childcare.

CLING ON
Long limbs make langurs great at rope-climbing – or in this case climbing a liana (trailing plant). Where trees are more spaced out, langurs are happy to move at ground level. If danger arises, they can speed overland at around 40km/hour.

BALANCING ACT
A langur reaches out to grab some flowers. The branch sways precariously, but the langur is perfectly balanced. Its heavy tail hangs like a plumb weight beneath to secure its position. Langurs have a vegetarian diet of fruits, flowers, berries, shoots and leaves.

Acrobatic Langurs

WELL SPOTTED
Juicy fruit on a neighbouring tree has caught this golden langur's eye. A super-sensitive patch on each retina enables it to focus sharply. The langur's vision also helps it to measure distances for leaping. Their eyes are almost identical in strength and ability to human eyes.

LEAPING FROM TREE TO TREE
As they fly from one tree to another, langurs whoop to other members of the troop to follow. They can leap or drop a dizzying 20m – the height of a two-storey house. Brainpower combined with learning when young, by watching older members of the troop, mean that the monkeys judge distances very accurately.

TIGHTROPE WALKING
A langur walks along a long branch on its hands and feet in the same way it would if it was on solid ground. The monkeys find it easy to do this because they spend so much of their time in the trees.

Brainpower

To survive in the complicated world of forest branches, monkeys developed intelligence. They have excellent hand–eye co-ordination to make the most of their grasping hands and forward-facing eyes. In the forest, monkeys move fast and make split-second decisions such as: "Will that branch hold my weight?" They can work out what the likely effect of their actions might be.

As they live in groups, monkeys have developed their brainpower so that they can negotiate with each other. Another skill for group living is a good memory, and monkeys are certainly good at remembering things. They rely on their memory to identify areas where they found last year's choicest food supplies.

◀ **PROSIMIAN TEAM**
Prosimians that are active in the day, such as these ring-tailed lemurs, are usually cleverer than nocturnal prosimians. Like monkeys, they live in groups and have to work together to control territory and find food.

▼ **THE EVIDENCE FOR INTELLIGENCE**
The skull of a sifaka (*below*) has little brain space, even though it is probably the most intelligent prosimian. The brain makes up less than half the length of the sifaka's skull. The front section – the nose – is long, which reflects the importance of the sifaka's sense of smell. The vervet monkey skull (*right*) is a different shape, with a much larger area behind its eye sockets for the brain.

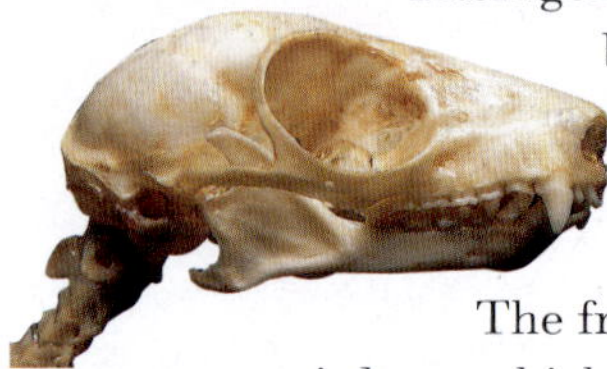

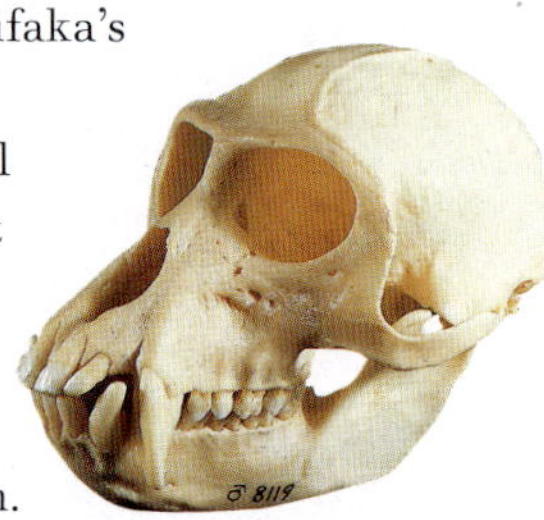

Monkey Dance
In Mexico, people dress up as monkeys for some traditional dances. The Mayans believed that when the world began, monkeys were types of humans. The monkeys fled to live high in trees because they were tired of being laughed at by other humans.

▲ **FINE JUDGEMENT**
Thanks to its large brain, this red colobus monkey can make death-defying leaps with absolute precision. Even on jumps of over 20m – the length of a tennis court – it can work out that the jump is possible and where to land safely. The monkey's sense of sight is also vital for gauging the distance between take-off and landing.

▲ **LOOK AND LEARN**
Toque macaques probably learned to raid rubbish bins for food by watching humans. Their big monkey brains give them the capacity to learn. This ability to learn means that monkeys can adapt to new environments. Macaques have worked out how to survive in a greater variety of habitats than any other monkey.

▲ **SOCIAL CONSCIENCE**
The members of this troop of chacma baboons all know each other very well. Because they are intelligent, each baboon can remember how other baboons have behaved and reacted in the past. As a result some are good friends, while others are not so close. Some baboons are even enemies. A monkey's intelligence means that it can live in complicated societies like this.

▲ **LIMITED NEEDS, SMALL BRAIN**
Most prosimians do not need to move and think quickly. They do not have the stimulation of group living, so they do not need as big a brain as a monkey.

Changing Habitats

Nearly all monkeys and prosimians live in steamy rainforests, like their early ancestors. Many monkeys may never touch the ground. Some live right at the top of the tallest trees, leaping from branch to branch. Others live lower down, where branches are so thick that they form an almost solid platform. Species that live in the forest understorey are agile both on the ground and in the trees.

Ground-based baboons and monkeys evolved later than the forest-dwellers. Sometimes, they live far from forests. There are fewer species and smaller social groups in open areas than in the forest, as food is more scarce.

▲ **GROUND-BASED**
Gelada baboons come from the open grasslands in the Ethiopian highlands in Africa. They are the only monkeys that eat mostly grass, and they live in groups for safety from predators. Like other ground-based monkeys, they have stout bodies for walking long distances in search of food.

Did you know? Some lemurs don't eat in the dry season, living off the fat they store in their tails.

EXTREME VERSATILITY ►
A Tibetan macaque has a thick winter coat and must rest more than other monkeys to save energy for keeping warm. It gets very cold high in the mountains and food is scarce. The macaques grow thick winter coats to trap heat. Macaques are among the most adaptable of monkeys. Different species have learned to survive farther north than any other monkey, in mountain forests and in cities.

Monkey Gods

Black-faced langurs are named after the Hindu monkey god, Hanuman. According to legend, Hanuman stole a mango plant and faced punishment by fire. When the fire was lit, the god tried to put it out. His face and paws were burnt in the process. According to the story, this is why hanuman langurs have black faces and paws. Although the monkeys are often pests in modern India, they are considered sacred. On the feast day of the god Hanuman, the monkeys are fed offerings at Hindu temples.

▲ LIFE IN THE FOREST
A baby squirrel monkey hitches a ride on an adult. These monkeys are part of a big troop on a fruit-hunting expedition through the lower levels of the trees. Because there is such a rich variety of food all year round, large numbers of monkeys live in the tropical forests.

◄ POISONOUS SNACK
The favourite food of the rare golden bamboo lemur is a variety of giant bamboo that contains enough cyanide to kill almost any other animal. These lemurs eat only bamboo and live only where they can find it, in the forests of south-eastern Madagascar.

ADAPT AND SURVIVE ►
De Brazza's guenons are excellent swimmers and have carved out a niche for themselves in dense, flooded rainforests in Africa, often on islands. Guenons are one of the most varied monkey species. Most are tree-living, but some, such as the vervet, have adapted to life on the African plains.

Survivors

1 In autumn, the snow monkey feeds on fruit, such as apples. In spring, there are plenty of shoots and young leaves. Then as summer comes, the monkeys can feast on flowers and insects.

In the forested mountains in the north of Japan, winter temperatures can drop to -15°C. Japanese macaques have learned to survive farther north and in colder weather than any other primate – except people, of course. Local people call them 'the snow monkeys'. Macaques are generally well known for their ability to adapt and pick up new survival skills. Snow monkeys have learned to make the most of local resources, such as hot springs, and of any food that is available in each season. They live in groups and rarely fight among themselves – for that would waste precious energy. Some migrate south from the very coldest parts of the country.

2 As winter draws near, the monkeys' fur thickens. Long, fine hairs trap body heat close to the skin like a cosy blanket. Farther south, where winters are milder, the monkeys have lighter winter coats.

3 Some snow monkey troops move south when the winter snows come, but this snow monkey braves it out. They mate three months later than their southern relations, so that the young are not born until late spring. Then the mothers have plenty to eat and can provide milk for the babies.

in the Snow

4 A snow monkey eats bark stripped from a tree. It will also feed on tree buds and seeds if it can find any. In the very coldest areas, snow monkeys only find about half the food they need to stay alive in the winter months. The rest of the nutrition their bodies need comes from fat stores built up during the previous summer.

5 One of the ways in which snow monkeys are able to survive the extreme cold is by having a hot bath. Some years ago, scientists observed a few monkeys taking a trial bath in a hot volcanic spring. Others in the troop followed suit, and now it is a regular snow monkey habit. Because they are using less energy to stay warm, they are conserving vital stores of fat.

6 Japanese macaques cannot sleep in the water, so they have to get out when they are tired. To stay warm the monkeys huddle into large groups. Together, their bodies lose heat more slowly during the icy nights. Long hairs on the outside of the monkeys' coats prevent snow from reaching their skin and keeps them dry.

Diets for Living

Very few small primates eat just one type of food. Most eat whatever's available through the seasons. Some species, such as mandrills and baboons, range over a wide territory to find enough food to feed the troop. Some become specialists, such as the leaf-eating colobus monkeys of Africa and Asia. They have sharp molar teeth to chop the leaves finely, and extra-big stomachs to extract as much goodness as possible from the tough fibres. Most monkeys eat a varied diet of fruit, nuts and insects, and a few eat eggs and small animals. In general, small monkeys eat more high-energy foods than larger monkeys because they burn up energy faster and need more efficient fuel.

▲ CLEAN FOOD
A Japanese macaque washes meat before eating it. It does not understand that unclean meat may make it ill, but knows that dirt spoils the taste. Once one monkey started to wash food, others in the group copied it. The habit will be passed on to their children. Other troops have not learned to do this.

▲ LAID-BACK EATERS
Red colobus monkeys usually eat leaves, although they also like flowers and fruits. Colobus monkeys do not need to look far for food as leaves are plentiful. They spend a long time sitting still, digesting the leaves to make the most of the limited goodness in them.

▲ SAP EXTRACTION
A bare-faced tamarin licks the thick, sweet juice from a fruit. Tamarins and their marmoset cousins eat a lot of resin from trees, including liquid rubber. Marmosets have pointed bottom teeth with which they pierce the bark so that the liquid flows out.

Greedy Monkey

A fable from Pakistan tells of a monkey that found some wheat inside a small hollow in a rock. The monkey thrust in its hand and filled it with the grain. The opening was too narrow for it to pull out its hand without letting go of some of the food. The monkey was too greedy to let go of any of the food, with the result that it had no food at all.

▲ **BABOON ATTACK**

A male olive baboon normally eats seeds, insects and grasses, but will not refuse the carcass of a baby animal. This African antelope fell victim to a baboon troop. The monkeys edge towards a herd until a young, weak animal is isolated. Then they rip it to shreds.

▼ **SAVING FOR LATER**

A drill has filled its cheek pouches with chewed food. Some other Old World monkeys do this too. They do this when they are eating leaves so that chemicals in the monkeys' spit begin to break the leaves down. The monkey's stomach will be able to absorb the nutrients more easily from the resulting mush.

▲ **EAT UP YOUR PETALS**

Flowers are a good supply of food, as this ring-tailed lemur is discovering. The pollen has a higher concentration of protein than any other part of the plant. Many insects are also attracted to flowers, and they are another good source of protein for monkeys and prosimians.

Sharing Space

When monkeys evolved, many prosimians became extinct, because the two groups of animals were competing for the same food. One way to survive in such circumstances is to become a specialist in a particular environment. Gradually, different species of monkeys and prosimians evolved that were able to share space in the same patch of forest. They had different diets, ate at different times, or operated at different levels in the trees.

Monkeys also share the forest with many other animals and plants. Together they make up an ecosystem. If an imbalance occurs in any part of the ecosystem, everything is affected. Too many leaf-eating monkeys living in one area, for example, might strip a tree of its leaves. The tree, and all the insects that rely on it then die, and the leaf-eaters, as well as any insect-eating monkeys, lose their food supply.

◄ **MY PATCH!** All species of the South American dusky titi monkey eat seeds and fruits. Each one, though, feeds at a different level of the forest. They do not compete with each other, but may be chased from food by larger monkeys. When this happens, dusky titis switch to eating leaves.

◄ **OPPORTUNISTS** Baboons are the only monkeys to live on the dry, open grasslands of Africa. They have adapted to eating anything they can find – even grass. The monkeys share the savanna with grazing animals such as wildebeest. There are no other monkeys here as there is not enough food. Even the baboons have to walk a long way each day to find enough to eat.

▲ **SECURITY SERVICE**

This pig-like animal is a peccary from South America. Peccaries are often found living close to monkeys, to the benefit of both. The hogs root around the bases of trees, feeding on insects, roots and whatever monkeys may drop from the branches. In return, peccaries will attack snakes heading up the trunk, which may be after small monkeys.

▲ **MIDNIGHT FEAST**

Night monkeys live in the same parts of the South American forest as other types of monkey. They eat many of the same foods, but feed at night, so do not compete. In the same way, nocturnal prosimians can hold their own against monkeys in the same forest.

SPECIALIST LEAF-EATER ▶

A black howler monkey feeds on its main diet of leaves in the South American rainforest. As the only leaf-eaters, troops of howlers can share space with fruit- or insect-eating monkeys. However, other monkeys tend to give the large, noisy howlers a wide berth.

▼ **MILITARY TACTICS**

Sharing space means developing self-protection. Baboon troops move in a defensive column with the largest males at the front and back. Females walk in the centre with their young. When a predator, such as this leopard, attacks, the males rush forward to fight off the big cat. The females gather with the young baboons at a safe distance.

▲ **CALLING CARD**
The female leader of a troop of ring-tailed lemurs from Madagascar marks her territory with scent from a gland on her chest. In ring-tailed lemur society, it is the females who fight to defend territory. Males only fight over females.

Territorial Rights

Troops of monkeys spend their lives in one clearly defined area. They create mental maps of their territories, knowing exactly where seasonal food supplies are, for example. Monkeys and social prosimians, such as sifakas and ring-tailed lemurs, defend their territory from neighbouring bands. These group-living primates mark their territory with scent and fiercely defend it against rivals of the same species who may be looking for more space or better food.

Small nocturnal prosimians live solitary lives and do not worry when a neighbour crosses their patch. They are more likely to mate with it or even ignore it altogether.

A primate's territory varies in size according to how many other animals there are and on food supply. A rainforest rich in food can support many animals in small territories. On the plains, animals need to range over a wide area to find enough food.

◄ **FOREST RANGERS**
Squirrel monkeys live in large societies of up to 50 members. Groups move around a home range of 20km², but territorial boundaries are blurred. One troop may merge with another without a hint of a fight. These bands of monkeys have no permanent troops, leaders or clearly defined territories. They roam through the South American forest looking for fruit, and keeping track of each other by high-pitched calling.

◀ **ON THE MOVE**
Baboons live on the African savanna where food supplies are very spread out. They live in big troops of up to 150 animals. This gives them the power to defend massive territories of $40km^2$. The monkeys walk about 5km each day to collect enough food for the troop. Most other monkeys move through only a few trees each day.

UNITED WE STAND ▶
Langurs live in large troops that are controlled by a single male. They patrol big home ranges throughout India, but do not live in rainforests. The troop's leader is always wary of attack from rival males wishing to steal his females and land. The whole langur troop may become involved in a battle with neighbouring troops for control of territory.

◀ **KEEP OFF MY LAND**
A howler monkey gives a loud, roaring howl warning other troops to keep out of their territory. Both males and females make this noise, and fight off invaders. In the thick forest, sound is a more effective way to assert territorial rights than scent. Fiercely defensive species, such as howler and proboscis monkeys, are the noisiest monkeys of all. The males are the most aggressive, as they also fight for the right to mate with females.

Trumpeters of

If you are a male proboscis monkey, the larger and more bulbous your nose, the better. It makes you look more fearsome to a rival male and it impresses the females. Most importantly, it swells the volume of your territorial honking sounds.

A single dominant male proboscis monkey controls a group of several females and their young. Bachelor males live in separate bands. They all make a lot of noise, screaming, growling and shrieking to each other.

The monkeys spend most of their time in the upper branches, feeding on various leaves, fruits and seeds.

LAZY DAYS
This family group of proboscis monkeys spends most of its day eating, high in the trees. They do take time off, though, to groom and to play with each other.

THE PROTECTOR
A male takes up a prominent position to watch out for rival males who may try to steal his harem. Male proboscis monkeys are nearly twice the size of females so that they can look as tough as possible to other males.

SAFE AND SOUND
Females do not defend territory or fight. They have smaller probosces (noses). Newborn babies have bright blue faces to alert other members of the troop to look after them. As they grow up and fend for themselves, their faces take on adult colours.

the Swamps

LOUD AND CLEAR
An adult male honks loudly to announce his territorial rights. Inside the enormous nose are hollow spaces that act like loudspeakers. As he honks, his proboscis (nose) straightens and stiffens. The sound echoes around the spaces and is amplified (made louder) so that it travels farther through the forest. The males also honk to frighten away predators, and snort to keep the family group in order.

DANGEROUS CROSSING
A proboscis monkey makes a solo voyage through water. Proboscis monkeys have to be very good swimmers because their territory is laced with rivers and inlets. They cross at the narrowest possible point, one at a time, then if a gharial (a type of small crocodile) strikes, it will get only one monkey. It is safer still to find a suitable tree with bouncy branches that the monkeys can use as a springboard to jump across the water.

NOWHERE TO GO
The proboscis monkey's home in the mangrove swamps of Borneo in South-east Asia is being taken over by roads and towns. Many troops are trapped in small islands of mangrove forest with territories too small for their needs. In addition, the coastal plants contain minerals that the monkeys need to stay alive. They cannot survive on similar food that grows inland. Many starve to death.

Predator and Victim

Danger is a way of life for small primates. One false move for a tree-living monkey could mean a long fall to its death on the forest floor. There are hungry predators, too, waiting for a tasty prosimian or monkey.

Small primates that move in the upper branches are in a good position to see any predators approaching from the ground and have plenty of time to move away. However, they are at greater risk from attack by birds of prey, such as eagles. Living in the middle branches offers protection from attack by air and from many ground predators, but tree snakes are a threat. They may spend days waiting motionless for prey to pass close by before striking. Prosimians that shuffle along at ground level by night are in danger from wild cats, silent snakes and swooping owls.

▲ **EAGLE-EYES**
Harpy eagles can swoop between tree branches to grab prey. The South American eagles make nests high in dead, leafless trees. From this exposed position, they watch for food such as monkeys, sloths and birds.

◀ **DANGEROUS CROSSINGS**
In the rivers and lakes of the African grasslands, crocodiles often lie in wait for prey to come near the water. They may snap up a few baboons as troops cross rivers in their quest for food. Swamp monkeys, crab-eating macaques and capuchins spend a lot of time near water and they, too, provide welcome snacks for various local species of crocodiles.

▲ **GANG WARFARE**

In the African jungle, chimpanzees form hunting parties to attack monkeys such as red colobuses. The monkeys are light and agile, but the chimps are clever and work as a team. One chimp separates a couple of monkeys from the colobus troop and drives them through the forest. Other chimps are stationed nearby to block any escape routes. Chasers force the victims into an ambush, where an old, experienced chimp lies in wait, ready to pounce. Victims are ripped apart in a matter of minutes, and the meat is shared out among the adult chimpanzees in order of importance.

▲ **SNAKE IN THE GRASS**

A python squeezes the life out of a grey langur monkey. Snakes are a danger to monkeys both on the ground and in the trees. Some snakes squeeze their prey to death, others inject fast-acting venom to kill. Tree-living snakes hold on to their prey as it dies to prevent it from falling out of the branches.

LONE HUNTER ►

A leopard holds a young baboon in its massive jaws. These predators of the African plains hunt alone. They use stealth to catch their prey off guard, preferring to attack at dusk or at night. The cat often quietly positions itself in a tree or high on some rocks and leaps down on to its prey from above. It will then drag its meal back to the vantage point, where it can eat in peace.

Living Together

All primates communicate in some way with other members of their species, and most of them live in social groups. Community living has advantages. There are more eyes to spot predators, and several animals can work as a team to fight off attack or forage for food. As a community, they stand a better chance of survival if there is a problem with the food supply, during a drought, for instance. The leaders will feed themselves and their young first to make sure that they survive. Solitary animals may have a hard time finding a mate, but within a group, there's plenty of choice. And, when there are young to be cared for, a community can provide many willing helpers.

Nocturnal prosimians rely on being solitary and silent to avoid being noticed by predators. Among bush babies and lorises, even couples live independently of each other, but they occupy the same patch, and their paths often cross. Babies stay with their mother until they are old enough to live alone.

▲ **TREE-SHARING**
Sifakas work together in groups of about seven adults to defend their territory. They are generally led by the females, and males may swap between groups. These prosimians gather in the higher branches of the trees in western Madagascar. If danger threatens, they all start a hiccuping groan.

HAPPY COUPLE ▶
In the jungles of South America, male and female sakis mate and usually live as a couple for a year. The female cares for the young. The father may not spend the day with his family, but does return to them at night. If there is plenty of food, families mingle with each other, forming large, loosely-knit groups.

◄ **FEMALE RULE**

A female is in charge of this troop of black tufted-ear marmosets. As with most other New World marmosets and tamarins, there may be several other females, but only the head female breeds. She mates with all the males to make sure her top-level genes are passed on. As none of the males knows who is the father, they all help rear and protect the young.

EQUAL SOCIETY ►

The relationship between a woolly monkey mother and her children can last for life. Woolly monkeys live in troops – there may be 20 to 50 of them in each group, with roughly equal numbers of males and females. Adult males often cooperate, and all the males and females can mate with each other. Individual females care for their own young. Woolly monkey groups are bound by an intricate web of relationships between all members that is hard for outsiders to understand.

◄ **MALE POWER**

These female hamadryas baboons are just two in a harem of several females. One top male mates with all the females to make sure he fathers all the children. Males with no harem live in separate bachelor groups of two or three. They try to mate with a member of the harem when the leader is not looking. When the leader gets old, a few young males will team up to depose him. Once he has been chased away, the victors fight for control of the harem.

▲ **DON'T HURT ME**
This toque macaque is showing by its posture and expression that it is no threat. If an adult monkey wants to make friends, it may make a sound like a human baby gurgling. The other monkey will usually respond gently.

Communication

Attracting attention in a noisy forest is a challenge. Groups of tree-living monkeys and prosimians lose sight of each other and keep in touch by calling. Nocturnal prosimians, however, need to keep a low profile, and so they leave scent messages that are easier to place accurately.

All prosimians and, to a lesser extent, monkeys, send messages of ownership, aggression or sexual readiness with strong-smelling urine, or scent from special glands. Monkeys can also express their feelings with facial expressions and gestures, and some use their ability to see in colour. African guenons, for example, have brightly coloured patches on their bodies that can be seen by their companions when they are hurtling through the trees. Even fur and tails can be useful. Ring-tailed lemurs swish their tails menacingly at rivals, and at the same time, fan evil smells over them.

Did you know? Monkeys have more face muscles than prosimians and pull more expressions.

PERSONAL PERFUME ▶
A black spider monkey smears a strong-smelling liquid on a branch. The liquid is produced by a gland on the monkey's chest. Its smell is unique to this monkey. When other monkeys smell it, they know that another of their kind has been there. If they meet the particular monkey that left the scent, they will recognize it.

◀ **TELL-TAIL SIGNS**

In complicated langur societies, high-ranking males hold their tails higher than lesser members of the group. Primates that live in complex social groups have a wider range of communication skills than solitary species. More information has to be passed around among a greater number of individuals.

▲ **YOU SCRATCH MY BACK...**

Grooming a fellow monkey not only gets rid of irritating fleas and ticks, but also forges a relationship. A lot of monkey communication is about preventing conflicts among group members. Forming strong personal bonds hold the troop together.

▲ **BE CAREFUL**

A mandrill has mobile face muscles to make different expressions. Here, he pulls back his gums and snarls. This makes him look very menacing to other males.

▲ **I'M ANGRY**

When a mandrill becomes angry, he opens his mouth in a wide yawn to show the size of his teeth and roars. Another monkey will hesitate before confronting this male.

LOOKING FIERCE ▶

This marmoset is literally bristling for a fight. Its fur stands on end like an angry cat's to make it look much bigger. It may scare its rival into withdrawing. Marmosets look cute, but they squabble a lot among themselves.

Bright Colours

LIGHT SNACK
A mandrill spends much of its day rooting for food such as fruit, seeds, or sometimes bird's eggs and small animals. The troop breaks up into small foraging parties to feed.

The startling colours of the male mandrill's face and rump make practical sense. They act like beacons in the gloom of the dense tropical forests of Central Africa. Troops of mandrills roam over large territories of around 50km², so they need to stay in touch with each other.

Groups of about 20 monkeys, made up of a ruling male, several females and more lowly males, may join forces to make armies of up to 250 or more. The monkeys are strongly built with limbs of equal length for moving around at ground level. At night, they climb into the lower branches of trees to sleep.

LOOKING COOL
You can tell that this male is in a relaxed mood, because his colours are quite muted. If he becomes agitated, though, the colours on his broad and boney muzzle flare up. Mandrills are the largest monkeys. The male weighs about 50kg – about the same as a small woman.

MODESTY MOST OF THE TIME
Female mandrills are about half the size of the males, and duller in colour. When ready to mate, a female sends out a colour signal. Her bottom swells and turns bright red. Females have one baby at a time. Other members of the troop may help find food for a nursing mother.

of the Mandrills

FOLLOW THE LEADER

All male mandrills have bottoms covered in blue, purple and red skin, with no long tail to reduce the impact of the strong colours. This makes sense in a dark forest, where it is easy to lose the way. Mandrill troops simply spot and follow the colourful male bottoms in front of them. The leading mandrill has the most vividly coloured rump. He also decides when to eat, and when to move on. He grunts to announce to the other monkeys when time is up.

BIG TEETH

The male mandrill's vampire-like canine teeth, which are around 6cm long, are nothing to do with its eating habits. Instead they are used as fearsome weapons during fights. The males of a troop of mandrills are constantly fighting to improve their rank because the top male is the only one who gets to mate with all the females.

CHOOSING A MATE

Female mandrills want their children to be as healthy as possible, so when they look for a mate, they choose a strong male leader. A male's large size and quality of colour are signs of strength and are very attractive to females. This is why males have evolved to look so different from the females. When a male is sexually active, he puts on weight and his colours become more vibrant.

Forming Bonds

From the moment it is born, a monkey must learn to make friends and influence other members of its troop. Without friends, a monkey will get less food and have a harder time rearing its young. Grooming is the social glue that bonds monkey groups together. Monkeys groom each other while they are lazing around, waiting for their food to digest. Grooming is relaxing, and relaxed monkeys are less likely to get into fights. Instead, they work together to find food and raise their young.

Many prosimians use their body scent to avoid confrontations with others. The smells are a signal to other monkeys coming into its territory. Once everyone knows who is in charge in each area, the animals do not get into fights very often.

▲ **MARKING A MESSAGE**
A dominant male lemur marks a branch with a scent gland on his bottom. The other males in the band will smell wherever he's been and make sure they are friendly when they meet him.

◄ **TUG OF LOVE**
The mothering instinct has got the better of the young, inexperienced baboon on the right. She is snatching a baby from its mother. The baby will probably come to no harm, but baby-snatchers sometimes just abandon infants, who then have to be saved by their mums.

PATERNAL INSTINCTS ►
Everyone likes babies, and that's true for monkeys, too. Male monkeys are always jostling with each other for dominance. When an argument gets too much for one male, he may hold up a baby to his rival. Both males forget their disagreements and look after the infant.

▲ SPIDER GROOMING SESSION
A spider monkey checks its friend's fur for lice and ticks. Spider monkeys are very social animals and live in big groups. When resting, these monkeys groom as many other monkeys as they can. If there are too many animals in the group for everyone to groom everyone else, the troop is too large, and the members split up into smaller groups.

▲ GIVING MUM A BREAK
The females in most monkey societies baby-sit for each other. A baby monkey may be suckled by several females as it grows. This tiny grey langur baby is being cared for by an aunty. Young female monkeys learn how to look after babies by caring for their relatives' young.

Did you know? Many monkeys spend almost half their days grooming each other.

HANDS UP FOR GROOMING ►
A male chacma baboon grooms a female during a break from feeding. Males do a lot of grooming to make sure everyone in the troop likes them. When the breeding season arrives, the females that have enjoyed being groomed by a male will probably be happy to mate with him as well.

◄ LAID-BACK LANGUR
When a monkey is being groomed, its brain releases chemicals, called endorphins, that make the animal very relaxed. Relaxed monkeys are less stressed and will catch fewer diseases, give birth to more babies and be better at avoiding predators. Monkeys will groom even when they are clean and fresh.

A Hamadryas

TOGETHERNESS
A huddle of females keeps warm in the chilly desert night. The male is at the back on the right. At night, a troop of hundreds of hamadryas baboons may spend the night together for warmth and safety.

Male hamadryas baboons are constantly posing threateningly to each other in attempts to win the right to mate with females. The baboons live in small groups known as harems, made up of a single male and several females. All the other males are constantly arguing and fighting to try to take control of another male's harem. The males are more fearsome looking than the females. They have long manes and dark red faces. Dominant males jealously guard their females, always herding them into a single group.

Hamadryas baboons have adapted to survive in near-desert conditions in Ethiopia and Somalia in Africa, and in southern Arabia on the other side of the Red Sea. They are smaller than other baboons, with brown rather than greyish coats.

CLAN SOCIETY
A female hamadryas baboon rests on a rock. During the day, these monkeys spread out to look for food. The troop divides into smaller groups called clans. A clan is made up of the harems of several related males. Large troops contain two or three clans. Smaller troops have just one. The hamadryas community is the most orderly of all monkey societies.

Baboon Troop

SHOW OF POWER
This is not a yawn of tiredness, but a serious warning to other males not to get too close to this top baboon's harem. Male hamadryas are armed with powerful jaws and long teeth. The layout of their teeth means that the dagger-like canines rub against teeth in the lower jaw, and are kept good and sharp. Like its red face and silver mane, the baboon's teeth are a sign of power.

BANISHMENT
A young male baboon flees from its family group, chased away by its father. As soon as males are old and big enough to be a threat, the top male banishes them from his harem. The young males band together in twos and threes, and join forces to attack a harem in a bid to steal the females. Then they will fight each other for control. The losers of this battle are banished back to the bachelor life.

THE GENTLE SEX
Female hamadryas baboons do not get involved in fights. They spend their time being friendly. They look after the children and groom each other. Most harems have four or five females and their children, but there may sometimes just be a single female. Because they neither compete nor fight, the female baboons are much smaller than the males.

Mating Time

For all animals, mating is a fiercely competitive business. Males want to be sure that it is their genes, rather than those of any other male, that carry on to the next generation. Male monkeys and prosimians fight for the right to mate with females, and bring into play many different strategies to win their favour.

In rainforests, where there is a year-round food supply and the weather is always warm, the young can be cared for at any time of the year. The female primates that live there are able to breed throughout the year. For monkeys and prosimians that live in seasonal climates, however, there may be only one short period in the year when the females are ready to mate. The breeding season has to be timed so that when the young are born there will be plenty of food available and the weather is not too extreme.

▲ **FEMALE CHOICE**
Two red-bellied lemurs share a flower. The male on the left can be identified by the white patches under his eyes. Among tamarins, marmosets, capuchins and leaf-eating monkeys, it is the females who choose their mates. Tamarins and marmosets lick and cuddle their chosen mate. Capuchins flirt by raising their eyebrows, and leaf-eating monkeys pucker their lips invitingly.

◄ **FIGHT FOR FEMALES**
The colobus monkey on the left is the dominant male of his society. Only he has the right to mate with his group of females. The dominant monkey is attacking another male who has challenged him. Female colobus monkeys have a short breeding time, so the competition to claim a group of females is intense.

▲ ONCE A YEAR ONLY

Vervets live a comparatively tough life on the savanna. They have longer pregnancies than forest monkeys in the same family, so that the baby develops more in the safety of the womb. Once they are born, the babies have to grow up quickly so that they can learn to look after themselves and not be a burden on their parents.

▲ READY AND WAITING

A female black macaque is ready to breed. Her genitals have swollen into large, red, balloon-like pads. These swellings indicate to a male macaque that she may allow him to mate with her. Many female monkeys have a visual signal like this to show they are ready to mate. Prosimians advertise by scent and calling.

▲ TILL DEATH DO US PART

A woolly lemur family huddles together affectionately in the Madagascan night. These lemurs form strong pair bonds. A male and a female mate only with each other throughout their lives. To keep in touch, they call to each other with long, high-pitched whistles.

Nazca Monkey

The Nazca people lived in southern Peru about 2,000 years ago. They are famous because they produced a set of huge etchings 120 metres long in the deserts in that area. These etchings include pictures of animals, including one of a monkey. Nobody knows why the Nazca people made these etchings, especially since the pictures cannot be seen from the ground. They were only discovered when an airplane flew over them.

Growing Up

▲ **CARING FATHER**
Young saddleback tamarins hitch a lift from their father. Because tamarins and marmosets have two or three babies at a time, the fathers get involved in childcare. The children go back to their mother for feeding. Most monkeys only have one baby, which is carried by its mother.

Monkeys and lemurs spend a lot of time looking after their young – it is a way of giving them a better chance of survival. In many social groups, adults like to gather around babies. They may find food for the mother, and help look after the child. Langur mothers move away from their troop to give birth. They feed and bond with the baby before introducing it to its relations and friends. All this interaction means that the youngsters soon pick up communication skills. Baby monkeys call out to their mothers in high-pitched voices.

Most primates have only one baby at a time. Nocturnal prosimians tend to have shorter, more frequent pregnancies than monkeys and lemurs. The babies develop less in the womb but grow up quickly after birth. Bush babies start life in a nest, but by 9 months are mature enough to have babies of their own.

◀ **PLAY AND LEARN**
Young chacma baboons play-fight together. They are learning how to use their growing bodies and practise pulling faces and making noises. They will use these tactics when they are older to threaten enemies or make friends. When young baboons are two or three months old, they start to explore away from their mothers.

◀ **HEALTHY DRINK**
Baby monkeys start suckling as soon as they are born. The infants of some leaf-eating species are not strong enough to eat solid food until they are 15 months old. Smaller species with more varied diets, such as marmosets, suckle their young for about two months. Lorises, which are nocturnal prosimians of South-east Asia, feed their babies infrequently but with very rich milk, to give the mothers more time to forage for food.

▲ **BACHELOR BAND**
Young hanuman langur males form gangs and wait for a chance to take over the troop. They have been chased away from the group by their father, the ruling male, because they are a threat to his leadership. In some primate societies that are controlled by a single male, a new leader may kill all the infants because he does not want to look after someone else's children. The females, deprived of their children, are soon ready to breed again and bear the new leader's children.

▲ **LOOK AFTER ME!**
Adult silver leaf monkeys of the Old World are looking after a couple of youngsters. Other members of a monkey troop may help a mother with childcare. The orange fur of the baby is a signal to the rest of the troop that they must treat the baby very gently. As the monkey grows up, its fur becomes grey and black, like its older friend on the right.

Did you know? Langur mothers hide their young high in trees to protect them from murdering males.

A Young Macaque

1 The newborn baby, a few hours old, is suckled by his mother. She gave birth alone in the early hours of the morning and spent a few hours bonding with her child, before bringing the new baby back to meet her friends in the troop.

Female macaques are pregnant for about five months before giving birth. All the adult females in a troop become pregnant at the same time. The babies are born at the time of year when there is most food.

Although the troop is led by a single dominant male, female macaques each mate with several males. This way, when the babies are born, all the males think they might be their fathers, and everyone works together to look after them.

A young macaque spends most of its time playing with friends. It learns how to express its feelings, and can soon spot when a monkey is angry, happy or hurt. These social skills are essential for adult macaques.

2 The growing macaque clings to his mother. She must feed a lot while she is suckling her baby, so she can make enough milk. Females lose up to one third of their body weight while they are producing milk.

3 Mother and child must follow the troop long distances each day. Now he is bigger, the baby rides all day on his mother's back. That's thirsty work for mum, and while she stops for a drink of water, her son takes a look around. He is just learning to walk, and stumbles and trips close to his mother while she takes a rest or feeds.

Matures

4 For several months the young macaque has taken nothing but his mother's milk. Now he is independent enough to begin to seek out his own food. By his first birthday, the young macaque knows where to find and how to recognize ripe and tasty food. He feeds himself, munching on nuts, fruits and plant shoots.

5 The young macaque is getting increasingly adventurous. Here he sets off to explore a watering hole. However, danger is all around, and he rarely strays far from his mother. Although she probably has a new baby by now, the young macaque will continue to live with his mother until he is four years old.

6 While still with his mother, the macaque begins to make friends with other monkeys. By grooming and being groomed, the young monkey takes his place in the troop. He learns which macaques are in charge, and, while under the watchful eye of his mother, he begins to get involved in the politics of the troop.

7 The macaque is now a fully grown male and he is less and less welcome in the troop. Soon he will have to leave home and set out to find another troop to join. He will need all his social and political skills to be accepted by them. Once in the new troop, he will have to spend a few years being kind to the females, and eventually they will let him father their children.

Monkey Relations

Prosimians, monkeys and apes are all members of the order of animals called primates. Each evolved in turn, adapting first to a life in the trees, and gradually relying on their intelligence to survive. Prosimians and less intelligent monkeys are called lesser primates. The advanced primates, or anthropoids (man-like), are the bigger-brained monkeys and apes. There are 19 ape species (compared with 242 species of monkey). Gibbons, known as the lesser apes, are the largest group and the most monkey-like. The great apes are gorillas, chimpanzees, orangutans, bonobos and humans. Of all primates, the great apes are the best at holding and manipulating small objects in their hands – and people are the most skilled of all.

KING OF SWING ►
A lar gibbon swings through the forests of Asia. Gibbons are expert swingers, with arms as long as the rest of their body and legs together. Tails were not necessary for balance and so disappeared. Gibbons look for grubs on the ground, and have to raise their arms in the air to stop them dragging. Male and female gibbons live as a couple for life.

◄ KINDRED SPIRITS
The white back of this gorilla shows he is a full-grown male. Like people, older gorillas, especially males, get white hair. It's a sign of their strength and experience. Gorillas are the largest primates. They are around twice the weight of a well-built adult man, and over 2m tall. The animals live in permanent family groups in the forests of central Africa. Their senses of sight and hearing, and many of their gestures, are similar to ours.

TOO CLEVER BY HALF ►

Like other apes, humans are descended from the primates that developed in the Old World. People are the most intelligent animals on Earth. As a result, humans can live in any environment. They can shape – and often destroy – the natural world to suit their needs. This young girl is only a few years older than the woolly monkey she is holding, but she has always been cleverer than the monkey.

▲ OLD MAN OF THE FOREST

The people of South-east Asia thought that orangutans were old men who had left their villages to live quietly in the trees. Orangutans are the world's largest tree-living animal, and have to climb slowly, choosing the strongest branches. They spend much of their time wandering alone through the forests looking for fruit.

◄ CLEVER PRIMATE

Bonobos are the closest ape relative to humans. They are the only other primate that regularly walks upright. Standing on two legs enables them to see over long distances, reach food on branches and have both arms free so that they can catch fish or hold their young.

▲ LESSONS WITH MOTHER

A chimpanzee baby is dependent on its mother for about five years. It has plenty of time to learn about how to behave in the complicated chimpanzee society. Chimps live in small patches of rainforest in central Africa. They form close relationships with each other and can make and use simple tools. They spend a lot of time on the ground, but climb trees to find fruit and to sleep.

Habitat Destruction

▲ **TRIBAL PREDATORS**
The traditional weapon of hunters in South American forests is a blow pipe. The hunters fire poisoned darts from the long pipe with a quick out-breath. The poison is a paste made from crushed plants or frog skins. In Africa, traditional hunters catch monkeys in nets, or fire arrows from a small bow.

Tribal peoples who live in the tropical rainforests and other wild habitats exist in harmony with monkeys and prosimians. They form part of the natural balance of their environment, along with the plants and other animals. The people take only what they need from their surroundings for building materials, food and medicines, and do not threaten the ecosystem.

On a global scale, though, humans are the world's dominant species and they exploit the Earth's resources for their own ends. They destroy habitats to make room for cities and roads, and for farmland to feed their huge populations. As a result, the habitats of many prosimian and monkey species are either chopped up into fragments too small to support them, or lost altogether. The animals then face extinction.

MONKEY MEAL ►
A man cooks a monkey in a central African forest. Traditional hunters do not kill more than they need for their families, so do not threaten the monkey populations. In some parts of the world, however, more and more people want to eat monkey meat. Hunters, armed with guns, kill monkeys for the bush meat trade. If the demand for monkey meat continues to increase, some areas may soon have no wild monkeys left at all.

◀ **DEFORESTATION**
This barren, South American landscape was once thickly forested, and home to howler and black squirrel monkeys, and capuchins. The trees were cut and burned to make space for grazing cattle. The ash fertilizes the soil, briefly. Soon, though, without a rich variety of vegetation growing on it, the soil becomes poor and dusty. More forest is then destroyed to make new farmland.

LAST REFUGE ▶
These Verreaux's sifaka lemurs are one of many species that are unique to Madagascar. All lemur species are in danger as people cut down vast tracts of forest, mainly for fuel. More than 14 lemur species are known to have become extinct since the humans began settling the island around 2,000 years ago.

▲ **TALKING DESTRUCTION**
Mobile phones contain a rare metal called tantalum. It is found in only a few places around the world, including central Africa. Here, where mandrills and mangebeys struggle for survival, people are fighting over control of the tantalum. The war has made it very hard to protect the monkeys.

◀ **HIGHWAY ROBBERY**
A lorry transports logs that have been cleared to make way for a road. Roads and villages in a forest can make it difficult for monkey populations to meet each other. Cut off from one another, they face extinction. Monkeys need to mate with other groups to produce healthy young. They also need to move easily to areas where there is food.

Life in a

EASY PICKINGS
A macaque's hands are ideal for foraging among human rubbish. Urban monkeys are clever and adaptable and will eat anything! They watch humans and learn where food supplies may be and how to get at them. They often take just a mouthful before moving on to the next readily available snack.

Monkeys have lived alongside people throughout history. When humans began to farm and settle in communities, monkeys soon learned that there were easy pickings to be had. These monkeys were the generalist (non-specialist) species with a varied diet. Among them were the cheeky capuchins of the New World, the wily macaques of southern Asia and the highly adaptable vervets of Africa.

In many towns and cities there are monkey species that have been urbanized for generations. They have bigger and denser populations than any that exist in the wild.

HUMAN TASTES
Like most urban monkeys, this vervet has developed a taste for human food. In some cities, urban monkeys have become pests and are bold enough to steal from shops and even food cupboards in houses. Tourists from countries where there are no monkeys are fascinated by them and feed them. This encourages the monkeys to scavenge even more.

Concrete Jungle

ALMS FOR THE POOR

In India, one monkey species has become known as the temple macaque, because large numbers of them live around temples. Often, rows of holy men and poor people begging line the roads to the temples. The monkeys have learned to beg, too – for scraps of food from the tourists and worshippers. One of the reasons why macaques and other urban monkeys fit so comfortably into city life is because human society is similar in many ways to how monkey groups live and interact.

ON THE ROCK

Barbary macaques live in Gibraltar, scavenging rubbish and food from tourists. They are the only monkeys to live wild in Europe. Barbary macaques are sometimes confused by the signals from people. They have been known to attack people who smile and stare, because these are aggressive signals in macaque society.

WHERE THERE'S A WILL…

City monkeys, especially vervets, steal food from people's homes. They can open doors, cupboards and boxes containing food. The vervets can be very determined, and will spend a lot of time working out a way to break in, and even brave a shock from an electric fence. People try to protect their homes by putting bars over windows, locking doors, or setting up alarms.

Use and Abuse

Humans have often woven magical stories around monkey characters and even worshipped monkey gods. However, they have also captured monkeys and used them cruelly.

By international law, it is illegal to buy and sell monkeys without a licence. This may be given, say, for the purposes of scientific experiments. Unfortunately, some animals are illegally exported, often travelling long distances in cramped and cruel conditions.

Many monkey and ape species are eaten in Africa and South-east Asia. Most are eaten by local people, for whom monkeys are a cheap source of "bush meat". However, more and more monkey meat is being smuggled around the world, especially into Europe and China, where it is sold illegally for very high prices. Conservationists believe that, if the bush meat trade is not stopped, many monkeys and apes could be wiped out within a few years.

▲ AT YOUR SERVICE
In Malaysia, macaques are trained to climb palm trees to pick coconuts. The agile monkeys easily scramble up to the top of the trees, where the coconuts grow. The macaques have learned to throw the coconuts down to their human owner. Although the monkeys are captive, this task is similar to the behaviour of wild macaques.

◄ PERFORMANCE
In parts of Asia, monkeys, such as this rhesus macaque, are trained to perform tricks to earn money for their human owners. Wild animals in captivity are often treated cruelly by being forced to perform in a way that is completely unnatural to them. Monkeys are not domesticated animals that have been bred to live with humans.

▲ OBEYING ORDERS
This monkey has been put in a cage to stop it escaping. The monkey will behave toward its human owner as a low-ranking male would do to a troop leader in the wild: it will cower to avoid confrontation and will try to do what the owner wants in return for access to food.

▲ FOR THE SAKE OF THE HUMAN RACE
Live monkeys are sometimes used in scientific experiments. This is called vivisection. Some of these monkeys are injured or killed. Many scientists believe that vivisection provides important information that could reduce the suffering of human beings, but others disagree.

◄ TAKE OFF!
A space shuttle blasts off. In the early days of space travel, many monkeys were used in experiments to study weightlessness.

GUIDE MONKEYS ►
Capuchins – the cleverest New World monkeys – have been taught to help disabled people. The monkeys can pick up objects such as telephones and take them to their owners, as well as perform tasks such as operating switches. Although their owners might benefit, not everyone thinks that it is right to use monkeys in this way.

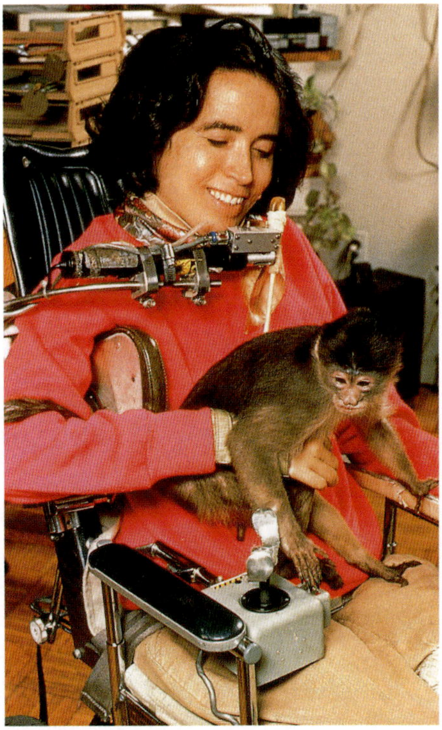

Saving Species

When the last member of a species dies, that species is extinct. It is lost forever. In the last few decades, some people have worked to stop the rarest primate species becoming extinct. Some have started sanctuaries and zoos where animals are helped to breed away from predators. Nearly half of the world's total population of aye-ayes (one of the most threatened lemurs in Madagascar) is in zoos because there is not enough of their natural habitat left to support them.

To save an endangered species, the cause of the threat must be addressed. This might be the destruction of the animal's habitat or the poverty of the people who hunt them. We can all affect the survival of species by making sure that we do not buy products that cause damage to a monkey's habitat.

Did you know? Thanks to conservationists, only one primate has become extinct in 100 years.

▲ **LIFE IN A COLD CLIMATE**
Woolly monkeys come from the steamy jungles of Brazil. This one lives in English woodland. The monkeys' natural habitat in Brazil is being destroyed by humans and the species is disappearing fast. At their cliff-top home in the south-west of England, the monkeys are given special foods to make sure they get the same minerals in their diet as they would in the wild.

◄ **BACK TO THE WILD**
Gerald Durrell, a British conservationist, holds a red-ruffed lemur. Durrell set up a pioneering zoo on the island of Jersey. Many rare primates have been bred there, and the zoo has successfully reintroduced lemurs and tamarins to their natural habitats. The Jersey zoo teaches keepers from other zoos how to raise captive-born animals so they can be released back into the wild.

◄ **HOME FROM HOME**
A rainforest has been created in New York's Bronx Zoo. The temperature, humidity and light are as close as possible to the natural conditions. Captive primates raised in a habitat similar to their wild environment are less likely to become distressed. They will stand a better chance of survival if they are later released into the wild.

KEEPING TRACK ►
These two golden lion tamarins, born in a zoo, were released into a protected area of Brazilian forest. Scientists fitted them with radio transmitters to keep track of them. This species has been saved from extinction by being bred in zoos around the world.

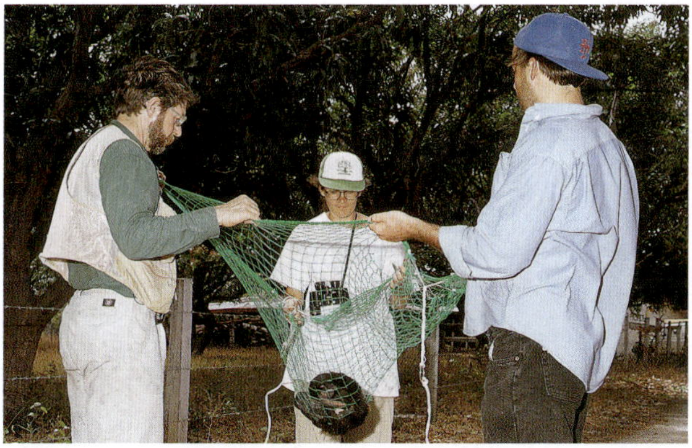

▲ **ONE IN THE BAG**
Scientists have shot a howler monkey with a drugged dart. They will take some measurements then give the monkey another drug to bring it back to consciousness. Their aim is to gain as much information as possible about the howler monkey's population make-up, its diet and the diseases it suffers. This will help them to understand how to conserve the species better.

▲ **SAFE SANCTUARY**
A monkey whose parents were killed by hunters has been rescued and taken to an orphanage. Sometimes, young monkeys that were captured illegally are rescued and sent to zoos to be cared for. They can't be set free as they don't know how to survive in the wild.

GLOSSARY

agile
Moving quickly and easily.

ape
A large, tailless primate with long arms. Chimpanzees, gorillas, gibbons, orangutans, bonobos and humans are all apes.

band
A group of prosimians, such as ring-tailed lemurs, similar to a troop of monkeys.

carcass
The dead body of an animal.

clan
A subgroup of a troop of hamadryas baboons, made up of three of four related male baboons and their harems of females.

diet
The food an animal eats.

digit
A finger, thumb or toe.

dominant
A monkey or prosimian that is in charge of, or is dominant in a troop or band.

ecosystem
A group of living things in a certain area that live together and interact with and depend on each other.

endorphin
A chemical produced by the brain that gives an animal a relaxed, contented feeling.

extinct
When every member of a species of animal or plant is dead.

fermentation
A chemical process that breaks down sugars and starches. Fermentation occurs in the stomachs of animals that eat a lot of plants.

fovea
A patch of highly sensitive cells at the centre of a retina in the eye. Forveas help monkeys to see very clearly.

generalist
An animal that eats all types of food.

gharial
A small type of crocodile found in Asia.

grooming
When monkeys clean each other's fur.

harem
A group of females that is controlled by a single male.

home range
An area of land or territory that generally takes an animal or group of animals several weeks or months to move through.

Jacobson's organ
A small organ at the top of many animals' mouths that helps them to detect smells better. Many prosimians have these organs.

keratin
The protein used to make hair and fingernails.

leader
The monkey or prosimian in charge of the troop, band or harem. Not all primate societies have leaders.

mammal
A hairy animal that feeds its young on milk.

mangrove
A type of tropical tree that grows close to water. Mangrove roots hang down from the plant into the water.

mineral
A type of chemical that is given in food and is essential for good health.

New World
The name given to North and South America. Monkeys from the New World usually have prehensile tails and flat noses with outward-facing nostrils.

niche
The position a species takes in an ecosystem. It may refer to an animal's diet or the time of day it eats.

nutrient
A part of food that gives an animal energy and good health.

Old World
The name given to Europe, Asia and Africa. Monkeys and prosimians from the Old World usually have tails for balance and small noses with downward-facing nostrils.

opposing thumb
A thumb that is 'opposite' the fingers, like a human thumb. Opposing thumbs help a primate to pick up small objects.

orbit
The circular part of an animal's skull that holds the eye.

pollen
The male sex cell of a plant, produced by flowers.

prehensile
A part of the body that is adapted for grasping and gripping. Many monkeys and other mammals have prehensile tails.

primate
A group of mammals, to which prosimians, monkeys and apes belong. All primates have forward-facing eyes and flexible hands.

proboscis
A long, flexible nose, such as that of the proboscis monkey.

prosimian
A primitive type of primate, such as a lemur or bush baby.

protein
A type of nutrient used to make muscles and other structures in an animal's body.

quadruped
An animal that walks on four legs.

rhinarium
The moist area on some animals' noses that joins the nostrils to the lips. It enhances an animal's sense of smell.

savanna
Tropical grassland that has only a few trees. Most savanna is in Africa.

skeleton
A framework of bones inside an animal's body, giving it support and strength.

social animal
An animal that lives in a group.

solitary animal
An animal that lives alone.

species
A group of animals that can breed with each other and share many similar features. Two animals from diffeent species cannot mate with each other.

territory
The area that an animal or group of animals lives in and defends from others.

troop
The name given to a group of monkeys.

venom
Natural poison produced by some reptiles, insects and other animals. these animals use venom to kill their prey and to defend themselves from attackers.

Picture Acknowledgements

Ardea: 49 tr/P Steyn: 5 tr, 15 tr, 22 tl, 56 br/ A and E Bomford: 8 br/M Watson: 9 c / F Gohier: 10 c, 23 tr, 36 br/C Haagner: 21 cl/ J-P Ferrero: 25 tr, 25 b, 32 cl, 42 cl, 42 br/ B Osborne: 22 bl/N Gordon: 28 tl/P Morris: 29 cr, 39 br/K Lucas: 34 bl/Ferrero-Labat: 35 tl/K W Fink: 45 bl /A Warren: 53 cr/D Last: 55 bl/LR Beames: 56 tl/J Cancalosi: 57 cl; **BBC NH Unit**/P Oxford: 7 tl, 11 cl, 30 tl, 47 br, 48 tl, 58 bl/L M Stone: 9 bl/N Gordon: 10 tr, 26 br/D Wechsler: 13 cl/E D Ferrera: 18 tl, 18 bl, 19 cr, 19 b/A Shah: 19 tl, 21 tl, 26 bl, 27 tr, 33 tr, 40 br, 47 tr/S Widstrand: 30 bl/ M Wilkes: 31 tl/T Martin: 31 bl/P Savoie: 32 br/E A Kuttapan: 35 tr/I Arndt: 37 bl/B Walton: 40 bl, 41 tr/T Weald: 49 tl/D Shale: 50 br/B Castelein: 51 tl/B Davidson: 52 bl/ B Britton: 57 bl; **BBC Wild**/ C Taylor: 38 tl/ A Shah: 46 bl, 47 tl; **Biophotos**/B Rogers: 55 tl; **B Coleman**/W Layer: 15 tl/J and P Wegner: 16 tr, 34 tr/B Coleman: 17 tl/S C Kaufman: 24 bl/ G Ziesler: 29 tr/R Magnusson: 36 tr/P van Gaalen: 44 tl; **Corbis Images**/D A Northcott: 44br/ W Kaehler: 55 cr/A Griffiths Belt: 59b/ C and A Purcell 61 br; E/G Neden: 5 bl; Heather Angel: 20 bl, 26 tr, 27 cr; **Hutchinson Library**/HR Dorig: 20 br; **Monkey Sanctuary**/ R Helvesi: 53 tr, 60 tl; **Nat History NZ Ltd**: 21 tr; **Naturepl**/A Shah: 42 cl/R du Toit: 43 cr/ I Arndt: 43 bl/K Bass: 53bl/N Lucas: 55 br; **NHPA**/N Garbutt: 4 tr/D Heuclin: 5 c, 17 br/ P German: 7 br/K Schafer: 13 tr, 37 tl/ T Kitchin and V Hurst: 14 tl/G Lacz: 15 b/ J Warwick: 23 tl/M Bowler: 23 br/A Rouse: 24 br/O Press: 25 tl/M Bowler: 38 br/P Fagot: 51 cl/M Harvey: 61 tl; **OSF**/D Haring: 5 tr, 13 br, 20 tr, 21br, 42 tr, 46 tr/M Deeble and V Stone: 6 tl/M W Richards: 7 c/M Leach: 9 tr/Z Leszczynski: 10 br, 28 bl, 39 cl, 39 c/ C Bromhall: 11 tr/M Colbeck: 11 tr, 18 br/ H Pooley: 11 br/L L T Rhodes: 14 bc/D Curl: 16 bl/S Breeden: 17 cl, 39 tl/K Lucas: 20 cr/ K Wothe: 23 cl/E Hashimoto: 24 tl/J Chellman: 27 bl/P Franklin: 29 tl/B Wright: 31 cr, 39 cr, 43 tr, 50 tl 51 br/R Davies: 32 tl/P Productions Ltd: 33 cl/M Kavanagh: 33 b/S Osolinski: 35 br, 40 tl/P J DeVries: 43 tl/B and B Wells: 45 tl/ B Kenney: 45 cr, 52 tl, 61 cr/J Sierra: 48 bl/ M Powles: 50 bl/D Tipling: 51 tr D J Cox: 53 tl/R Davies: 58a tr/D Cayless: 59/C Tyler: 60 br/M Tandy: 61c bl; P Pics/J Mills: 54 tl/ N Robinson: 54 br; **S American Pictures**/ T Morrison: 47 bl; **SPL**/N Dennis: 8 tl/ T Davis: 12 tr/T McHugh: 14 cr/ M Phillips: 41 cl/G C Kelley: 41 br/A Wolfe: 49 bl/NASA: 59 bl/ R Nannini: 59 br

INDEX

A
adaptability 6, 22, 24–5, 56
agility 12, 13, 18–19
apes 6, 17, 35, 52–3
aye-ayes 9, 60

B
babies 11, 32, 42, 43, 47, 48–9
  macaques 24, 50–1
baboons 13, 17, 28, 29, 31
  chacma 11, 43, 48
  gelada 17, 22
  hamadryas 37, 44–5
  olive 6, 27
birds of prey 34
bodies 12–13, 22
brains 6, 12, 19, 20–1
bush babies 4, 5, 36, 48

C
calls 19, 31, 32–3, 36, 47
captivity 58–61
capuchins 46, 56
chimpanzees 35, 53
colobus monkeys 26, 46
colour 14, 32, 38, 40–1, 49
communication 15, 38–9, 40, 45, 48
  see also calls
communities *see* social groups
co-operation 20, 36, 49
crocodiles 33, 34

D
diet 26–7
  baboons 13, 22, 27
  bush babies 5
  dusky titi monkeys 28
  langurs 18
  lemurs 8, 23, 27
  macaques 21, 24, 25, 26, 51
  mandrills 40
  marmosets 17, 26
  squirrel monkeys 23
digestion 8, 12–13, 27
dominance 5, 37, 41, 42, 44, 46
Durrell, Gerald 60

E
ears 14–15
ecosystems 28, 54
endangered species 60–1
expression 38–9
extinction 8, 9, 28, 54–5, 60–1
eyes 5, 13, 14, 19

F
fathers 36, 48, 49, 50
fighting 30–1, 41
food *see* diet
forest clearance 54–5
fur 11, 22, 24–5, 38, 39, 43

G
galagos *see* bush babies
gibbons 52
gorillas 52
grooming 11, 39, 42, 43, 51
guenons 23, 38, 60
  see also vervets

H
habitats 6, 22–3, 24, 28
  destruction 33, 54–5, 60
hands and feet 4, 7, 9, 10–11, 16, 52
howler monkeys 29, 31
humans *see* people and primates

I
intelligence 13, 20–1, 57, 59

L
langurs 7, 18–19, 31, 39, 48, 49
learning 19, 20–1, 25, 26, 56–7
legs 10, 13, 16
lemurs 4–5, 8–9, 13, 23, 47, 60
  ring-tailed 5, 9, 20, 27, 30, 38
  see also sifakas
leopards 29, 35
lorises 4, 5, 17, 36, 49

M
macaques 11, 17, 21, 22, 47, 50–1
  Japanese 24–5, 26
  and people 56–7, 58
mandrills 10, 17, 27, 39, 40–1, 60
marmosets 26, 37, 39, 46, 48, 49
  pygmy 10, 17
mating 24, 30, 36–7, 40, 43, 46–7
meat-eaters 13, 26, 27, 40
mothers 24, 36, 40, 48–9, 50
movement 5, 10, 12, 16–19, 33

N
New World monkeys 6–7, 10, 15, 16
nocturnal monkeys 14, 29
nocturnal prosimians 8, 20, 21, 30, 36, 38, 48
  see also bush babies; lorises
noses 6–7, 9, 15, 20, 32–3

O
Old World monkeys 6–7, 10, 15, 16

P
people and primates 8, 20, 23, 27, 47, 56–61
  see also habitats, destruction
predators 29, 33, 34–5, 54
primates 4, 52
proboscis monkeys 31, 32–3
prosimians 4–5, 8–9, 14–15, 20, 21, 28
protection 58, 60–1

R
reproduction 24, 36–7, 40, 43, 46–7

S
scent 15, 30, 38, 42
senses 4–5, 9, 12–13, 14–15, 20
  sight 5, 6, 14, 19, 21
sifakas 8, 16, 20, 30, 36, 55
skulls 20
snakes 29, 34, 35
social groups 20–1, 30–1, 36–9, 42–3, 48–51
  baboons 6, 29, 31, 37, 42, 44–5
  langurs 18, 31, 43, 48
  lemurs 5, 9, 31
  macaques 24–5
  mandrills 40–1
  proboscis monkeys 32–3
spider monkeys 10, 16, 38, 43
squirrel monkeys 23, 30, 59
survival 22, 24–5, 28, 36, 56–7, 60–1

T
tails 7, 10, 16, 18, 38
tamarins 7, 13, 26, 46, 48, 61
tarsiers 4, 5, 11, 13, 14
teeth 12, 26, 41, 45
territory 5, 15, 20, 30–1, 33, 40
trees, living in 16, 18–19, 22, 28

U
urban monkeys 56–7

V
vervets 11, 14, 20, 47, 56
vivisection 59

W
woolly monkeys 37, 60

Z
zoos 60–1

NATURE WATCH

ELEPHANTS

BARBARA TAYLOR

Consultant: Dr Adrian Lister,
University College, London

C O N

Published by Anness Publishing Ltd,
Blaby Road, Wigston, Leicestershire LE18 4SE
Email: info@anness.com
Web: www.annesspublishing.com

Anness Publishing has a new picture agency outlet for images for publishing, promotions or advertising. Please visit our website www.practicalpictures.com for more information.

Publisher: Joanna Lorenz
Managing Editor, Children's Books: Sue Grabham
Senior Editor: Nicole Pearson
Editors: Simon Beecroft, Nicky Barber
Designer: Simon Wilder
Picture Researcher: Adrian Bentley
Illustrators: David Webb, Vanessa Card, Julian Baker
Production Controller: Ann Childers
Editorial Reader: Joy Wotton

ETHICAL TRADING POLICY
Because of our ongoing ecological investment programme, you, as our customer, can have the pleasure and reassurance of knowing that a tree is being cultivated on your behalf to naturally replace the materials used to make the book you are holding. For further information about this scheme, go to www.annesspublishing.com/trees

PUBLISHER'S NOTE
Although the advice and information in this book are believed to be accurate and true at the time of going to press, neither the authors nor the publisher can accept any legal responsibility or liability for any errors or omissions that may have been made.

PICTURE CREDITS
b=bottom, t=top, c= centre, l= left, r= right
ABPL/ Daryl Balfour: 9al, 12bl, 23bl, 35bl; ABPL/ Nigel Dennis: 33bl; ABPL/ Paul Funston: 29cl; Clem Haagner: 37al; ABPL/Martin Harvey: 20ar, 23cr, 34b, 36ar &bc, 37bl, 39b, 54b; ABPL/Gerald Hinde: 9bl; ABPL/Luke Hunter: 60br;ABPL/Beverly Joubert: 6al; ABPL/ Philip Kahl: 30cl, 34ar; ABPL/Peter Lillie: 40cl; ABPL/ Philip Richardson: 61al; ABPL/Lorna Stanton: 48ar; ABPL/Gavin Tomson: 9br; Heather Angel: 24bl, 33cr; Ardea: 30bl; BBC Natural History Unit/ Lockwood & Dattari: 5br; BBC Natural History Unit/ Keith Scoley: 13al; CFCL/Image Select: 4ar, 29br; CFCL/Image Select/ Chris Fairclough: 21br; Bios/ Michel Denis Huot: 6bc, 41bl; Bios/ Nicholas Granier: 43br; The Bodleian Library: 15al; Bridgeman: 13ar, 25cl; Bruce Coleman/ Alain Compost: 8ar; Bruce Coleman/ Michael Fogden: 30br; Bruce Coleman/ Paul Van Gaalen: 7bl; Bruce Coleman/M P Kahl: 23ac; Alan G Potts: 21bl; Bruce Coleman/ Rod Williams: 54ar; ET Archive: 11cl, 53al; Mary Evans: 51cr; FLPA/ Peter Davey: 32br; FLPA/ F Hartmann: 11br; FLPA/ David Hosking: 12al, 53b; FLPA/ Terry Whittaker: 14ar; FLPA/ Martin Withers: 28b; Michael Holford: 4bl, 40ar, 55br; Kobal Collection: 49bl; Adrian Lister: 30tl, 51br & bl; Ken Lucas: 10bl; Natural History Museum London: 13br; Natural Science Photo Library/ D Allen Photography: 18al, 21ar; Natural Science Photo Library/C & T Stuart: 52ar; Natural Science Photo Library/S G Neginhal: 35ac; Natural Science Photo Library/Pete Oxford: 19al, 52br, 56cr; Natural Science Photo Library/M W Powles: 46bl; Natural Science Photo Library/M S Price: 59tr; Natural Science Photo Library/P H & S L Ward: 57b; Nature Photographers/ Peter Craig Cooper: 47cl; Nature Photographers/ Hugo Van Lawick: 5ar, Nature Photographers/Paul Sterry: 47al; NHPA/Kevin Aitken: 52bl; NHPA/Daryl Balfour: 19ar & bl, 20bl, 28cl, 31bl, NHPA/Anthony Bannister: 18br, 59bl; NHPA/K Gani: 7al & ar; NHPA/Martin Harvey: 38c, 56al &br, 57ar; NHPA/ T Kitchener/ V Hurst: 60bl; NHPA/ Steve Robinson: 31br; NHPA/Philip Ware: 8c; Oxford Scientific Films: 41a, 43cr; Oxford Scientific Films/ Jen & Des Bartlett: 32ac; Oxford Scientific Films/ G I Bernard: 25al; Oxford Scientific Films/ Martyn Colbeck: 4c, 8bl, 14bl, 22bl, 25br, 29tr, 38b, 39t, 40br, 41br, 42al, 44al, 44bl, 45cr, 45bl, 47bl, 49ar; Oxford Scientific Films/ Cynthia Moss: 45al; Oxford Scientific Films/ Stan Osolinski: 28ar, 32al, 33a, 46ar; Oxford Scientific Films/ Richard Packwood: 55ar; Oxford Scientific Films/Vinay Parelkar/ Dinodia: 55al; Oxford Scientific Films/Anup Shah: 38tl; Oxford Scientific Films/ Gavin Thurston: 14al; Oxford Scientific Films/Steve Turner: 58b; Oxford Scientific Films/Belinda Wright: 25cr; Papillio: 7br; Papillio/ John R Jones: 15ar, 55bl; Planet Earth/ K & K Ammann: 59br; Planet Earth Pictures: 49cl; Planet Earth Pictures/Thomas Dressler: 43al; Planet Earth Pictures/Roger de la Harpe: 61br; Planet Earth Pictures/Frank Krahmer: 21al; Planet Earth Pictures/Doug Perrine: 53ar; Planet Earth Pictures/Jagdeep Rajput: 48br; Planet Earth Pictures/Roger Rogoff: 22al; Planet Earth Pictures/Brendan Ryan: 49br; Planet Earth Pictures/Jonathon Scott: 42c, 43br; Planet Earth Pictures/Anupe Manos Shah: 42br; Ian Redmond: 10br; Ann Ronan: 11ar; Frank Spooner Pictures/ Oliver Blaise: 22br.

INTRODUCING ELEPHANTS

4 What is an Elephant?
6 African or Asian?
8 Elephant Habitats

HOW ELEPHANTS WORK

10 Big Bones
12 Tusks
14 Body Parts
16 Elephant Senses
18 Trunks
20 Keeping Cool
22 Focus on Water Babies

ELEPHANT BEHAVIOUR

24 On the Move
26 Elephant Migrations
28 Feeding
30 Teeth
32 Focus on Daily Life

T E N T S

SOCIAL LIFE

34 Living Together
36 Communication
38 Focus on Bull Elephants
40 Courtship
42 Birth and Babies
44 Focus on Giving Birth
46 Growing Up
48 Life and Death

RELATIVES AND ANCESTORS

50 Ancestors
52 Relatives and Namesakes

ELEPHANT FACTS

54 Life with People
56 Focus on a Working Life
58 Facing Extinction
60 Conservation
62 Glossary
64 Index

What is an Elephant?

Elephants are the largest and heaviest creatures on land. An African male (bull) elephant weighs as much as 80 people, 6 cars, 12 large horses or 1,500 cats! Elephants are extremely strong and can pick up whole trees with their trunks. They are also highly intelligent, gentle animals. Females live together in family groups and look after one another. Like human beings, elephants are mammals. As with all mammals, they can control their body temperature and, like most mammals, they give birth to babies. After humans, elephants are the longest lived of all mammals. Some live to be about 70 years old. Two species of elephant exist today – the African elephant and the Asian elephant. Both have a trunk, large ears and thick, grey skin. Not all elephants have tusks, however. Generally, only African elephants and male Asian elephants have tusks.

▲ **WORKING ELEPHANTS**
In India, domesticated (tamed) elephants are used by farmers to carry heavy loads. In some Asian countries they also move heavy logs by pulling them.

Indra and the Elephant
One of the most celebrated Hindu gods, Indra, rides a mighty white elephant called Airavata. In Hinduism (the main religion of India), elephants are sacred animals. One of Indra's emblems is the ankus, a special pointed stick used to control elephants. Indra is shown holding an ankus in this painting.

Tail, with a brush of thick hair at the end.

▼ UNUSUAL FEATURES

The mighty elephant is a record-breaking beast. Not only is it the largest land animal, it is also the second tallest (only the giraffe is taller). It has larger ears, teeth and tusks than any other animal. The elephant is also one of the few animals to have a nose in the form of a long trunk.

FAMILY LIFE ▲

Adult male and female elephants do not live together in family groups. Instead, adult sisters and daughters live in groups led by an older female. Adult males (bulls) live on their own or in all-male groups.

BABY ELEPHANTS ▲

An elephant baby feels safe between its mother's front legs. It spends most of the first year of its life there. Mother elephants look after their young for longer than any other animal parent except humans. Daughters never leave the family group unless the group becomes too big.

▲ **AFRICAN SHAPE**
You can always tell an African elephant from its large ears and the dip in its back. This species of elephant usually holds its head slightly lower than its shoulders. The male African elephant stands on average 3.3m at the shoulder, while females are about 2.8m tall. The male weighs about 5,000kg (5 tonnes), which is about the same weight as a truck.

African or Asian?

How do you tell an African elephant from an Asian elephant? They seem alike, but they are not identical. The most obvious difference is in the size of the ears – the African elephant's are larger. In Africa, too, elephants have longer legs and a more slender body than their Asian relatives. The back of the Asian elephant is arched, while the African species has a dip in its back. Another difference is tusks – in Asia usually only males have visible tusks, whereas in Africa both male and female elephants normally have them.

▶ **CONTINENTAL EARS**
Some people say that the large, triangular ears of the African elephant (*right*) are shaped like the continent of Africa, while the ears of an Asian elephant are shaped like India. In fact, ears are like fingerprints – they are different on each individual. Scientists look closely at the pattern of veins, notches and tears on the ears to identify individual elephants.

HEAD UP

The Latin name of the Asian elephant, *Elephas maximus*, means huge arch. This refers to the upturned shape of its back. The Asian elephant usually holds its head above its body, so that the top of its head is the highest point of the body.

ASIAN TUSKS

The tusks of a male Asian elephant are shorter and lighter than those of an African elephant. Female Asian elephants have very small tusks or none at all.

Did you know? The heaviest known African elephant weighed 7 tonnes.

SKIN SPOTS

Elephants are usually thought of as grey. However, some Asian elephants have pink patches of skin. These patches are commonly found on the ears, trunk and face. They generally develop as the animal grows older. African elephants rarely have them.

THE HAIRY ONE

Asian elephants have more hair than their African relatives, and baby Asian elephants (*above*) are hairier still. As baby elephants grow up, most of this protective hair is worn away.

Elephant Habitats

The two main species of elephant are divided into smaller groups called subspecies. These subspecies each look a little different from one another and are named after their habitats. Africa has three subspecies – the bush elephant of the open grasslands, the forest elephant of west and central Africa, and the desert elephant of Namibia. The main subspecies in Asia is the Indian and southeast Asian elephant. Asia is also home to the Sri Lankan elephant and the Sumatran elephant, which lives on the islands of Sumatra and Borneo (parts of Indonesia).

▲ **SUMATRAN ELEPHANT**
These elephants wade into swamps to find juicy grasses to feast on. They are the smallest of the three Asian subspecies. Sumatran elephants are also the lightest in colour, and they have fewer pink patches than the other Asian subspecies.

SRI LANKAN ELEPHANT ▶
The rare Sri Lankan elephant is the biggest and darkest of the three Asian subspecies. Many of the 2,500 – 2,700 elephants in Sri Lanka live in protected national parks or nature reserves.

Did you know? The desert elephant is the tallest elephant in the world, at up to 4.21 m high.

◀ **FOREST ELEPHANTS**
The forest elephant (*Loxodonta africana cyclotis*) is the smallest African subspecies. Its size enables it to move easily through the trees. Generally, its ears are small and rounded, and its tusks are less curved than those of other African elephants.

DESERT ELEPHANTS

The hot, dry deserts of Namibia in southwest Africa are home to the rare desert elephant. This subspecies is very closely related to the African bush elephant, but it has longer legs. Desert elephants have to walk long distances to find food and water. Scientists think that this is why they have longer legs than any other subspecies.

ELEPHANT WORLD

African elephants live in a scattered band across central and southern Africa. They became extinct in North Africa around AD300. Today, Asian elephants live in hilly or mountainous areas of India, Sri Lanka, southeast Asia, Malaysia, Indonesia and southern China. In the past, they roamed right across Asia.

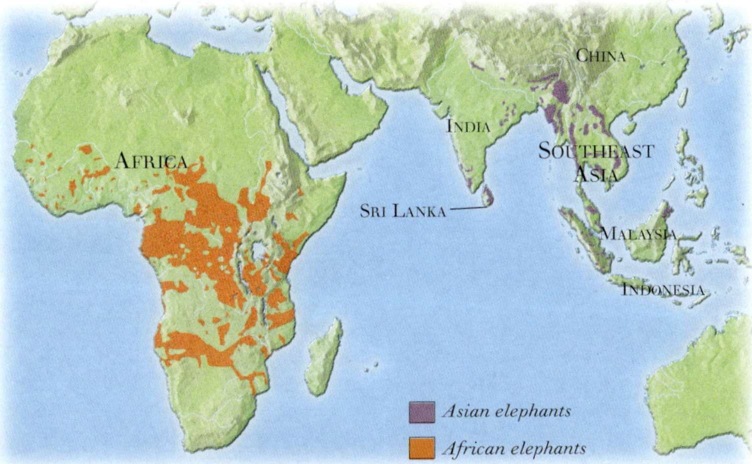

BUSH ELEPHANTS

African bush elephants live in savanna (areas of grassland with scattered trees). However, some live in forests or marshes and even on mountains.

SOLIDLY BUILT

The African bush elephant, *Loxodonta africana africana*, is bulkier and heavier than any other elephant subspecies. Like all elephants, its large size is a useful weapon against lions, tigers and other predators.

Big Bones

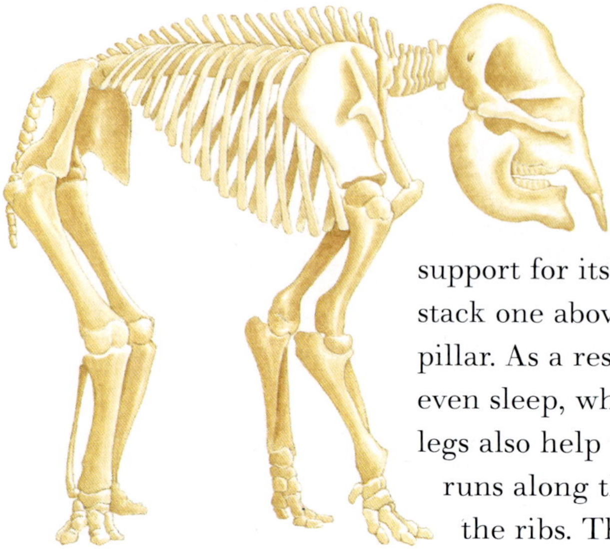

An elephant's legs are placed directly underneath its body, like a table's legs. This arrangement provides a firm support for its great weight. The leg bones stack one above the other to form a strong, tall pillar. As a result, an elephant can rest, and even sleep, while standing up. The pillar-like legs also help to hold up the backbone, which runs along the top of the animal and supports the ribs. The backbones of both African and Asian elephants arch upwards in the middle. Their differently shaped backs are produced by bony spines that stick up from the backbone. The elephant's skeleton is not just built for strength, however. It is also flexible enough to let the elephant kneel and squat easily.

▲ **BONY BACK**
Long spines stick up from the backbone of the Asian elephant's skeleton. The muscles that hold up the head are joined to these spines and to the back of the skull.

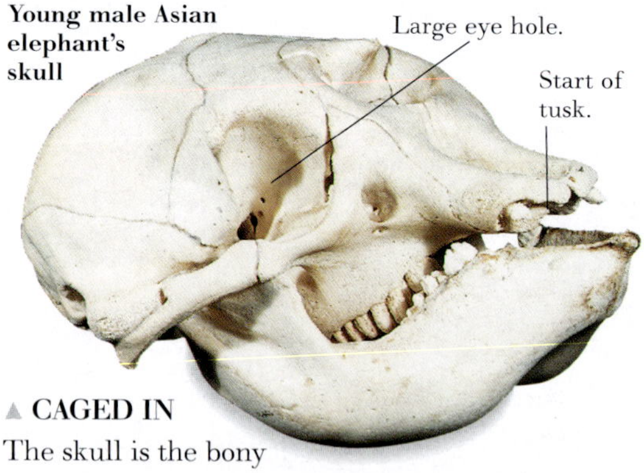

▲ **CAGED IN**
The skull is the bony box that protects the brain and holds the huge teeth and tusks. The skull above is that of a young elephant with undeveloped tusks. On an adult male the upper jaw juts out further than the lower jaw because it contains the roots for the heavy tusks.

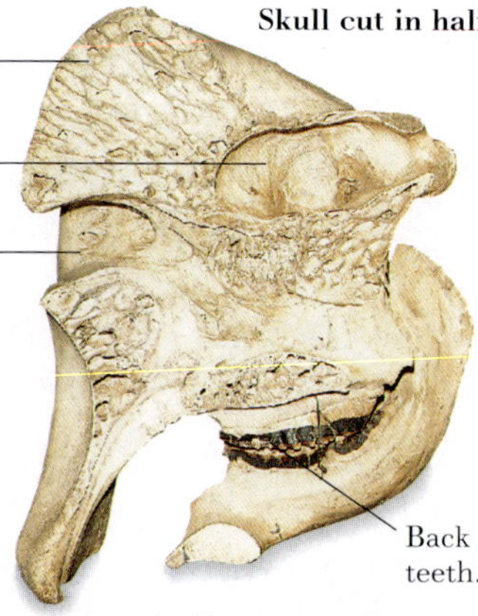

▲ **AIR HOLES**
An elephant has a large skull compared to the size of its body. However, a honeycomb of air holes inside the skull makes it lighter than it looks from the outside.

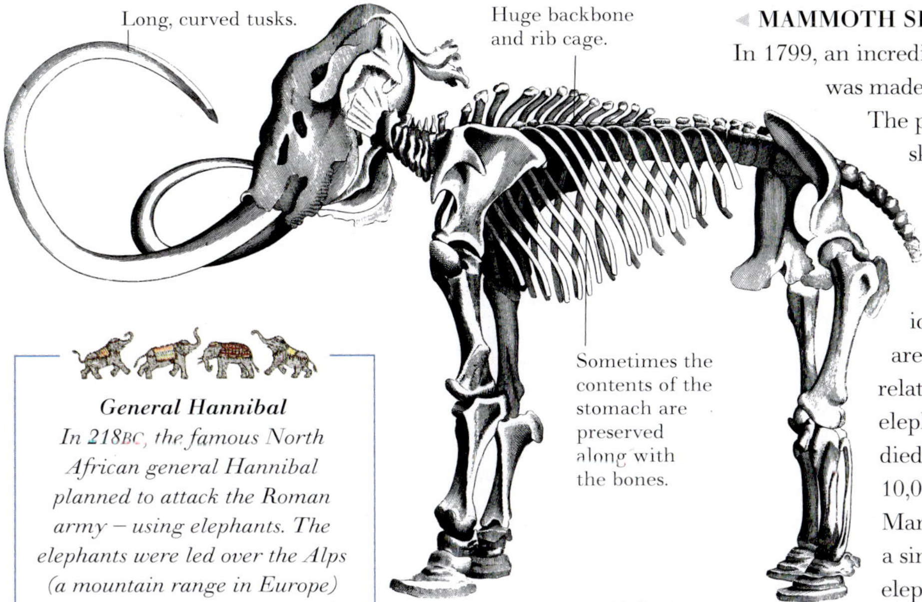

MAMMOTH SKELETON

In 1799, an incredible discovery was made in Siberia. The preserved skeleton of a woolly mammoth was found buried in the ice. Mammoths are prehistoric relatives of the elephant. They died out about 10,000 years ago. Mammoths were a similar size to elephants today, with longer tusks.

General Hannibal

In 218BC, the famous North African general Hannibal planned to attack the Roman army – using elephants. The elephants were led over the Alps (a mountain range in Europe) and taken across rivers by raft. Sadly, most of them died and Hannibal had to admit defeat.

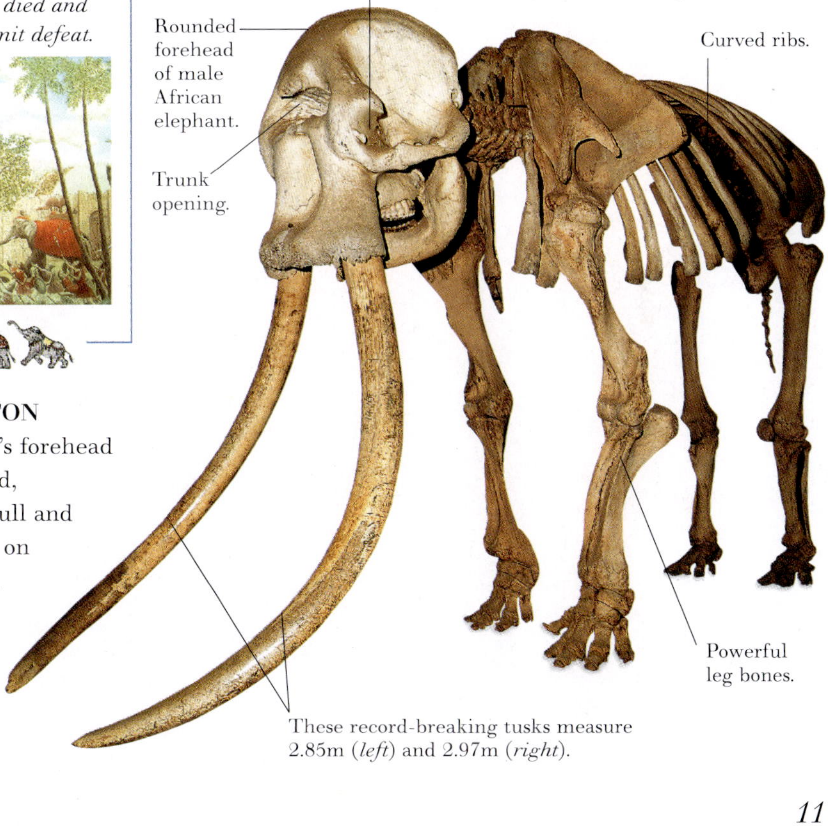

AFRICAN SKELETON

The African elephant's forehead is smooth and rounded, whereas an Asian's skull and head have two domes on top. The large hole in the skull is the trunk opening, and the tusks fit into the skull from the upper jaw.

Tusks

An elephant's tusks are its front teeth. The part that we see is just two-thirds of the tusk's total length. The rest is hidden in the skull. Tusks are made from a very hard material called ivory. Elephants mainly use their tusks for feeding. The sharp ends are ideal for digging up edible roots and stripping bark from trees. Tusks are also used as weapons, and help to protect the trunk, like a car bumper. Elephants are born with milk tusks, which are replaced by permanent tusks when the elephant is between 6-12 months old. They grow continuously at the rate of about 17cm a year. As the tip wears down, more tusk is pushed out from the skull.

▲ **TINY TUSKS**
Some female Asian elephants have small tusks called tushes. They grow so slowly that they hardly stick out of the mouth. Some male Asian elephants have no tusks.

This elephant's left-hand tusk is shorter than the right-hand one because it has been used more.

Male elephant tusks weigh up to 60kg, or about the weight of two children. Female tusks are lighter and weigh 9kg on average.

Did you know? One rare elephant was found with four tusks instead of two.

◄ **TUSK SHAPES**
Elephants tend to be either right-handed or left-handed, just as human beings are. Their tusks are like hands and elephants prefer to use one of them more than the other one. As a result, the favourite one becomes more worn. Tusks also come in different shapes and sizes. Some are slim and straight, while others stick out at different angles or even cross in front.

USEFUL TUSKS
African elephants dig for salt with their tusks. They loosen the soil with the sharp points, just as we use a garden fork. On occasions, elephants also dig for salt in caves under the ground.

FIGHTING ELEPHANTS
In ancient India, humans trained elephants to help them fight battles. The elephant's sharp tusks were deadly when used as weapons. In ancient Rome, elephants were made to battle against gladiators (trained fighters) and other animals for sport. Cruelly, the elephants were usually forced to fight against their own gentle nature.

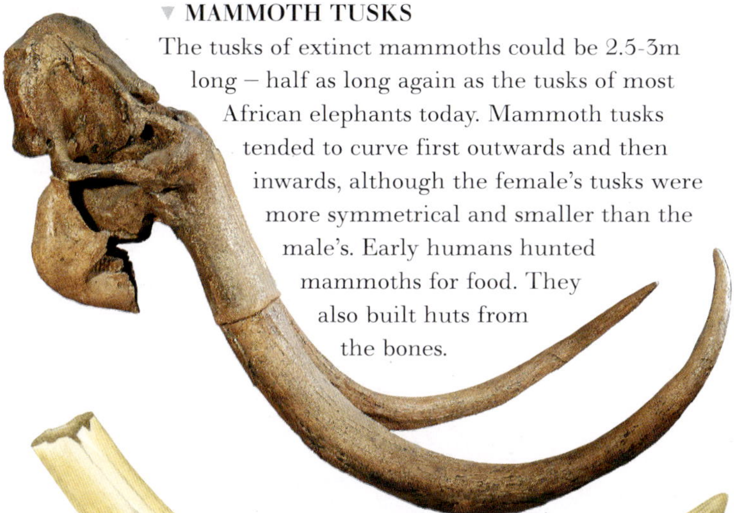

MAMMOTH TUSKS
The tusks of extinct mammoths could be 2.5-3m long – half as long again as the tusks of most African elephants today. Mammoth tusks tended to curve first outwards and then inwards, although the female's tusks were more symmetrical and smaller than the male's. Early humans hunted mammoths for food. They also built huts from the bones.

African bull (male) tusk

African cow (female) tusk

Asian bull (male) tusk

RAW MATERIALS
Tusks are made of solid ivory, which is a type of dentine. This is a hard material that forms the main part of all mammal teeth. Nerves and blood vessels run through the tusks. The blood carries food to feed the growing tusk. Elephants feel pain and pressure in their tusks, just as we can in our teeth.

Body Parts

▲ **THICK SKIN**
An elephant's skin is 2.5cm thick on the back and in some areas of the head. But in other places, such as around the mouth, the skin is paper thin.

An elephant's skin is thick, grey and wrinkly, and surprisingly sensitive. Some insects, including flies and mosquitoes, can bite into it. Often, elephants roll around in the mud to keep flies from biting them (as well as cooling the elephant down). Underneath the skin, the elephant has typical mammal body parts, only much larger. The heart, for example, is about five times bigger than a human heart and weighs up to 21kg – the weight of a small child. Also, an elephant's huge intestines can weigh nearly a tonne, including the contents. The powerful lungs are operated by strong muscles. These let the elephant breathe underwater whilst using its trunk as a snorkel.

▲ **PINK SKIN**
An elephant gets its colour from dots of grey pigment (colouring) in the skin. As it ages, this grey pigment may gradually fade so that the skin looks pink.

Did you know? Some very rare Asian elephants have a white skin.

◄ **ELEPHANT HAIR**
The hairiest part of an elephant is the end of its tail. The tail hairs are many times thicker than human hair and grow into thick tufts. Apart from the end of the tail, the chin and around the eyes and ears, the elephant is not a particularly hairy animal.

▲ **PINK TRUNKS**
Some Asian elephants, particularly the Sri Lankan subspecies, have pink trunks. They may also have pink patches on their ears, face and belly, which are a sign of ageing. This is comparable to the human hair turning grey.

Flying Elephants
According to an Indian folk tale, elephants could once fly. This ability was taken away by a hermit with magical powers when a flock of elephants woke him from a deep trance. The elephants landed in a tree above him, making a lot of noise and causing a branch to fall on his head. The hermit was so furious that he cast a magical spell.

▼ **INSIDE AN ELEPHANT**
If you could look inside the body of an elephant, you would see its huge skeleton supporting the inner organs. The elephant cross-section shown here is of a female elephant.

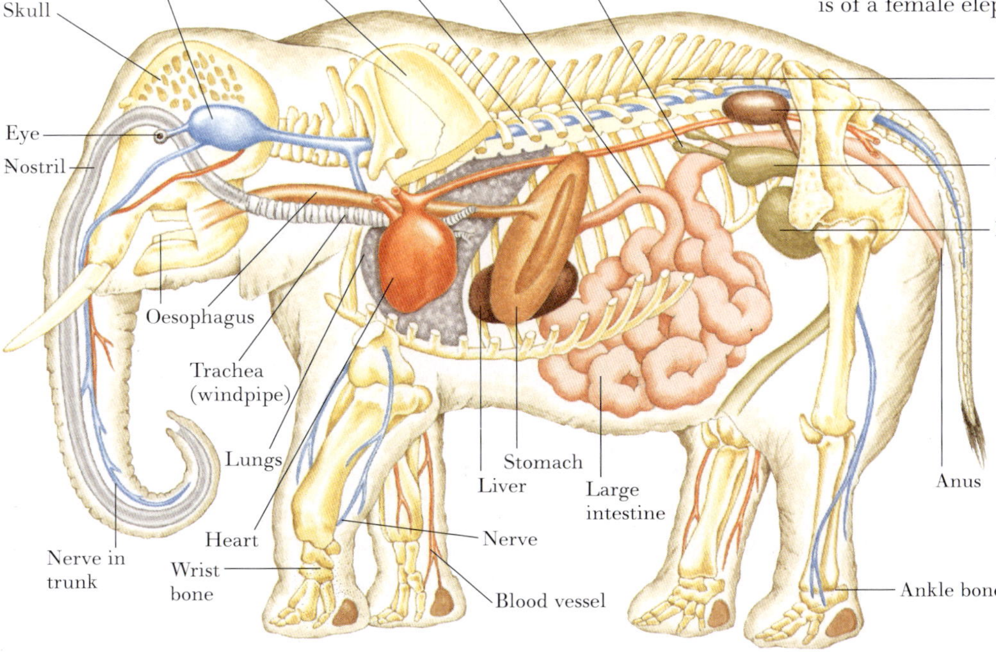

Elephant Senses

Elephants use the five senses to learn about their surroundings – hearing, sight, smell, touch and taste. The most important sense is smell, which they rely on more than any other. Elephants smell through the trunk, using it as a directional nose. The trunk is also particularly sensitive to touch and has short hairs that help the elephant feel things. The tip of the trunk is used to investigate food, water and other objects. It can tell whether something is hot, cold, sharp or smooth. Elephants communicate with each other largely by sound. They make rumbling sounds, most of which are too low for humans to hear. Touch is also crucial for communication. When two elephants meet, each places the tip of its trunk in the other's mouth as a greeting.

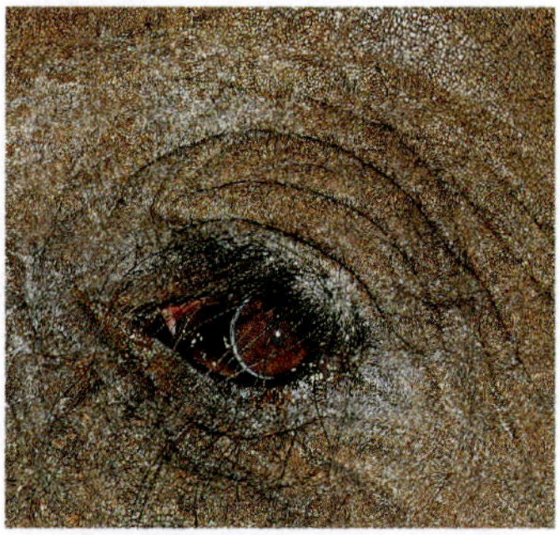

▲ **ELEPHANT EYES**
All elephants' eyes are brown with long lashes. They are small in relation to the huge head. Elephants are colour blind and do not see well in direct, strong sunlight. Their eyesight is better in darker, forest conditions.

Did you know? An African elephant's ear is as big as a single bed sheet and can weigh as much as a person.

◀ **SMELL**
An elephant raises its trunk like a periscope at the slightest scent of danger. It can tell who or what is coming towards it just from the smells picked up by the sensitive trunk. The sense of smell is so powerful that an elephant can pick up the scent of a human being from more than 1.5km away.

USING EARS
An elephant strains to hear a distant noise by putting its ears forward to catch the sound. It also does this when it is curious about a certain noise. Elephants have a well developed sense of hearing. Their enormous ears pick up the rumble of other elephants from up to about 8km away. Male elephants also flap their ears to spread a special scent that lets other elephants know they are there.

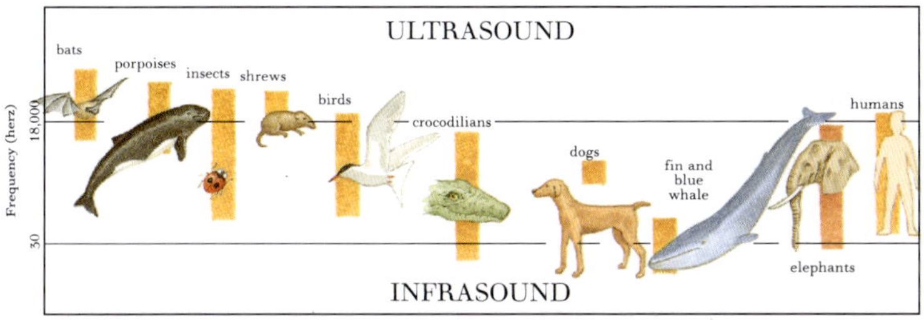

HEARING RANGE
Elephants can hear low sounds called infrasound. Human beings cannot hear infrasound although we can sometimes feel it. Some animals, such as bats and mice, can hear very high sounds called ultrasound.

SENSE OF TOUCH
A young elephant is touched by its mother or another close relative every few seconds. This constant reassurance keeps it from being frightened. Elephants also touch each other when they meet. They often stand resting with their bodies touching.

QUICK LEARNERS
Some young working elephants learn to stop their bells ringing by pushing mud inside them. This allows the clever animals to steal food from farmers' fields without being heard.

Trunks

Imagine what it would be like if your nose and top lip were joined together and stretched into a long, bendy tube hanging down from your face. This is what an elephant's trunk must feel like. It can do everything your nose, lips, hand and arm can do – and more besides. An elephant uses its trunk to breathe, eat, drink, pick things up, throw things, feel, smell, fight and play, squirt water, mud and dust, greet and touch other elephants and make sounds. Not surprisingly, a baby elephant takes a long time to learn all the ways to use its trunk.

DRINKING
An elephant cannot lower its head down to the ground to drink so it sucks up water with its trunk. Baby elephants drink with their mouths until they learn to use their trunks to squirt water into their mouths.

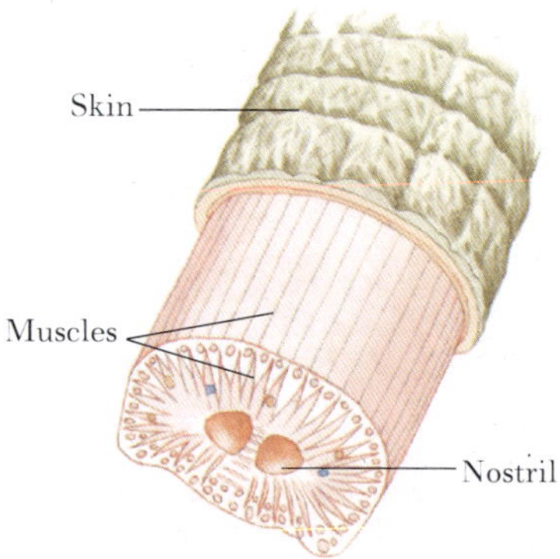

A CLOSER LOOK
The two holes in the centre of the trunk are nostrils, through which the elephant breathes. Thousands of muscles pull against each other in different directions to move the trunk.

TRUNK POWER
An African elephant coils its trunk around a branch to lift it off the ground. Elephants can lift whole tree trunks in this way. The powerful trunk has more than 100,000 muscles, which enable an elephant to lift large, heavy objects easily.

TAKING A SHOWER
An elephant does not need to stand under a shower to be sprayed with water, mud or dust. Its trunk is like a built-in shower, able to cover almost its whole body as it reaches backwards over the head. Showering cools the elephant down and gets rid of insects.

BENDY TRUNK
Elephants sometimes double up their trunks and rest them on their tusks. They can do this because the trunk has no bones inside it, just muscles, which makes it very flexible.

TALL ORDER
African elephants use their long, stretchy trunks to pull leaves off the top branches of tall acacia trees. The highest leaves are the most juicy. The trunk is slightly telescopic, which means that, if necessary, it can be stretched out even longer than usual. The trunk can also be pushed into small holes or gaps between rocks to find hidden pools of water.

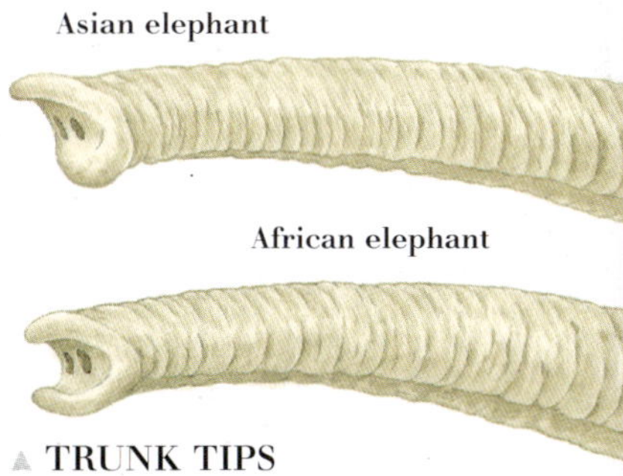

TRUNK TIPS
The trunk of an African elephant has two fingers at the tip, while the Asian elephant has only one. These fingers can pick up an object as small as a leaf or a coin.

Keeping Cool

When we are hot, we lose heat through our skin, especially by sweating. An elephant cannot do this because it does not have sweat glands, so it must cool off in other ways. One way an elephant gets rid of excess body heat is by flapping its enormous ears. This increases the cooling flow of air over the body. The ears also work like a huge car radiator – they prevent overheating by letting heat escape from their vast surfaces. Elephants stay cool by spraying themselves with water, mud or dust. They also love bathing in water and rolling around in mud. The mud dries on the elephant's skin, providing a barrier against the sun's heat.

▲ **RADIATOR EARS**
The African bush elephant has the largest ears of all. It spends more time in open sunlight than the other species, and its large ears increase the amount of skin area through which heat can leave the body.

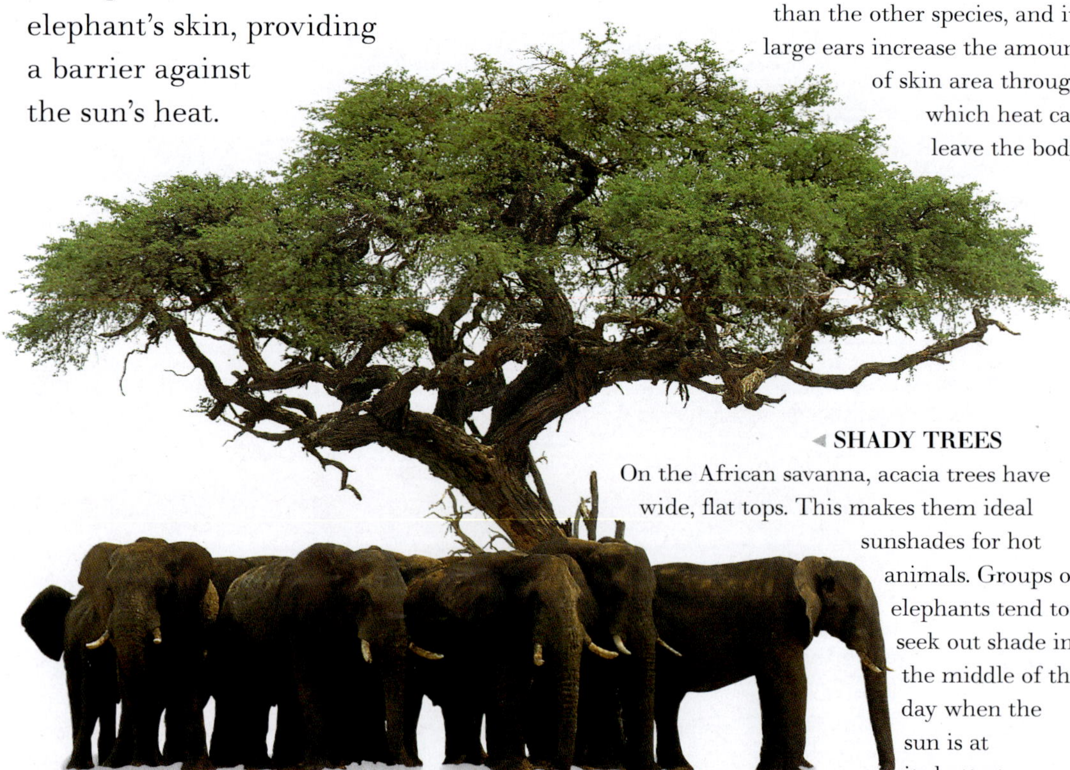

◀ **SHADY TREES**
On the African savanna, acacia trees have wide, flat tops. This makes them ideal sunshades for hot animals. Groups of elephants tend to seek out shade in the middle of the day when the sun is at its hottest.

▼ COOLING MUD

Elephants love to plaster themselves in mud, which changes their ordinary colours to vibrant reds, blacks, browns or yellows. Mud cools the skin, heals cuts, protects against insect bites and helps to stop the skin from getting dry and cracked.

▲ DUST SHOWER

Dust showers are popular with elephants. An earthy coating works like mud to protect the skin from the sun's heat. It also keeps skin in good condition.

▲ BATHTIME

Zoo elephants are often given a helping hand at bathtime. In a zoo, elephants may not have enough space to spray themselves with water or dust so the keepers regularly bathe them and brush their skin to keep it in good condition.

▲ COOL WRINKLES

An elephant's wrinkled skin helps it to keep cool. The wrinkles increase the overall surface area of the skin, so more heat can escape. They also trap cooling moisture.

Focus on

Elephants love water. They drink lots of it and enjoy jumping into lakes and rivers to play and splash around. Elephants are good swimmers and can easily cross rivers or swim out to sea to reach islands with fresh food. They drink at least once a day, or more often when water is available. When water is hard to find in the wild, elephants can be very sneaky, drinking from taps, pipes or water tanks. This usually causes damaged or broken pipes. However, elephants can go without water for up to two weeks.

SPLASHING ABOUT
Elephants spray each other with water, wrestle with their trunks and flop sideways with great splashes. Sometimes they turn upside down and poke the soles of their feet out of the water. All this play strengthens the bonds between individuals and keeps groups together.

TRUNK STRAW
These two elephants are refreshing themselves at a waterhole. Elephants drink by sucking in water through their trunk. They seal off the end with the finger or fingers at the end of the trunk. Then they lift the trunk to the mouth and squirt in the water.

CHAMPION SWIMMERS
Elephants are good at swimming even though they are so big. When an elephant swims underwater, it pokes its trunk above the water and breathes through it like a snorkel.

Water Babies

KEEPING CLEAN
Frequent bathing washes the build-up of mud and dust out of the cracks in an elephant's thick skin. Disease-carrying insects and parasites that feed off the elephant's skin are also washed off in the water.

WATER BABIES
In the water, baby elephants often hold on to the tail of the elephant in front for safety. They can easily be swept away by fast-flowing rivers. Baby elephants are also vulnerable to attack from crocodiles.

THIRST QUENCHER
An elephant needs to drink 70–90 litres of water a day. A full trunk of water holds about 5–10 litres. Incredibly, a very thirsty adult elephant can drink about 100 litres of water in 5 minutes.

On the Move

An elephant looks like a noisy, clumsy animal. In fact, it moves about quietly and is surprisingly agile. Forest elephants can quickly disappear into the trees like silent, grey ghosts. The secret of the elephant's silent movement is the way its foot is made. A fatty pad inside the foot cushions the impact of the foot on the ground. The sole then spreads out to take the weight of each step. Elephants usually walk slowly, at a rate of about 6kmph. They can run at more than 40kmph when angry or frightened, but only for a short distance. Elephants swim well too, and they often reach islands in lakes or off the coast. They are also good climbers, with a ridged sole that grips rough or steep ground well. However, elephants cannot jump. They would crush their legs on impact.

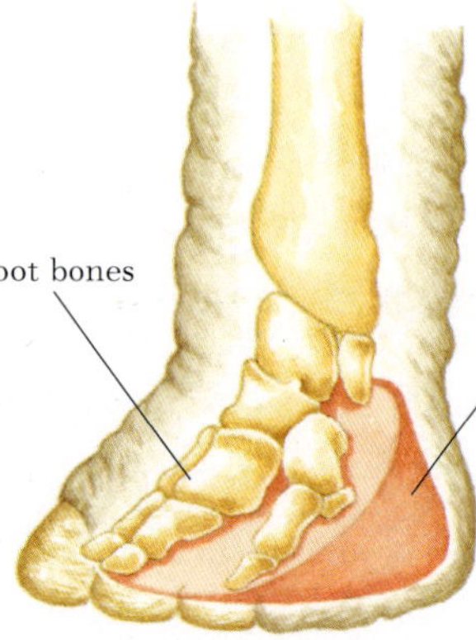

▲ **FATTY FOOT**
An elephant's enormous weight rests on the tips of its toes and on a fatty cushion that works like a giant shock absorber. This fatty shock absorber spreads out as the elephant puts its foot down and contracts as it lifts up its foot. On firm ground, the elephant leaves hardly any track marks.

Did you know? An elephant's foot measures up to 1.5m around.

◄ **ELEPHANT WALK**
An elephant walks and runs with shuffling steps. It cannot trot, canter or gallop. Occasionally, elephants walk backwards. They sometimes find this easier than turning around, which can be a difficult manoeuvre for an elephant.

GRIPPING SOLES

The skin on the sole of an elephant's foot is thick and covered in cracks and deep ridges. These help it to grip rough ground effectively, rather like the treads on tyres or hiking boots.

Did you know? Elephants can swim for six hours without a break.

Royal Hunts

An Indian mythical story tells of King Khusraw, who was killed by his son Shirvieh. In the scene shown here, Shirvieh travels by elephant to the Royal Palace. Here he becomes caught up in a royal hunt. In India, people often rode elephants to hunt.

FOOTCARE

An elephant has a thorn removed from its foot. Elephants in captivity move about less than they do in the wild. As a result, their feet are less tough and need to be looked after. Toenails are not worn away either, so they have to be trimmed.

ON THE MARCH

A line of elephants crosses the savanna in Kenya. Elephants often march along in a row, sometimes with each elephant holding the tail of the one in front. Elephants usually walk about 25km a day, but in the hot deserts of Namibia in southwest Africa journeys of up to 195km a day have been recorded.

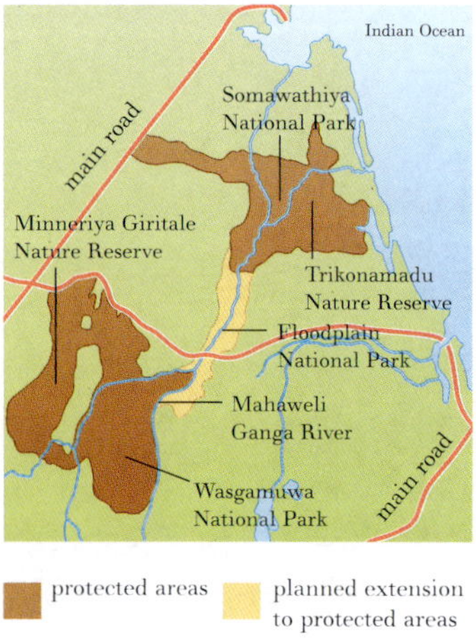

ANIMAL CORRIDORS
Much of the land in Sri Lanka is used for agriculture, so some elephants live in protected areas. They move between the regions along special corridors of land, in the same way that people travel between cities along motorways.

Elephant Migrations

Elephants do not have permanent homes. Every year they make long journeys called migrations to search for food and water. In the dry seasons, elephants gather in large groups for feeding migrations. They follow the same paths when moving from one place to another year after year. These paths are learnt by one generation of elephants after another. Today, elephants have been squeezed into smaller areas as human beings take up more and more land. As a result, elephant migrations are much shorter than they used to be, although they may still cover hundreds of kilometres.

ELEPHANT RAIDERS
Farmers on the island of Sumatra try to chase a herd of elephants away from their maize fields. Migrating elephants can do a lot of damage to crops.

ELEPHANT WELLS
During times of drought, elephants may dig holes in dry stream beds. They use their trunks, tusks and feet to reach water hidden under the ground. Elephants need to drink 70-90 litres of water each day and have been known to travel long distances of up to 30km to reach a tiny patch of rainfall. Elephant wells can be life-savers for other wildlife that come to drink the water after the elephants have gone.

LONG JOURNEYS
African elephants on the savanna grasslands may wander over an area of more than 3,000 square km. The extent of their migrations depends on the weather and other conditions. Asian elephants living in forests migrate over smaller areas of about 100-300 square km.

WATCHING FROM ABOVE
Migrating elephants can be followed by aeroplane in open country. Little or no rain falls during the dry season and the elephants tend to group together in places where water is available. The thirsty animals usually stay around a river valley or a swamp that still has water in it. In rainy seasons, elephants spread out over a wider area.

ELEPHANT BARRIERS
Elephants will try anything to find a way through farmers' fences. They can use their tusks to break electric fence wires and even drop large rocks or logs on top of fences.

KEEPING TRACK
Scientists in Africa fit a radio-collar to an elephant. This device tracks the animal's movements without disturbing its natural behaviour.

Feeding

Elephants are herbivores (plant-eaters). They eat more than 100 different kinds of plants and enjoy almost every part – from the leaves, twigs, bark and roots to the flowers, fruit, seeds and thorns. Plants do not contain much nutrition, however, so an elephant has to eat huge quantities to survive. It spends about 16 hours a day choosing, picking and eating its food. Millions of microscopic organisms live inside an elephant's gut, which help it to digest food. Even with the help of these organisms, half the food eaten by an elephant leaves the body undigested.

◄ **EATING THORNS**
Elephants do not mind swallowing a mouthful of thorns – as long as some tasty leaves are attached to it! Leaves and thorns make up an important part of an elephant's diet as they stay green in the dry season long after the grasses have dried up. This is because trees and bushes have long roots to reach water deep underground.

▲ **STRIPPING BARK**
An elephant munches on tree bark, which provides it with essential minerals and fibre. The elephant pushes its tusks under the bark to pull it away from the tree trunk. Then it peels off a strip of bark by pulling with its trunk.

▼ **GRASSY DIET**
Marshes are packed full of juicy grasses. About 30-60 per cent of an elephant's diet is grass. On dry land, an elephant may even beat grass against its leg to remove the soil before feeding.

Did you know? Elephants can become drunk by eating overripe fruit.

► BABY FOOD

Baby elephants often feed on the dung of adult elephants. They do this to pick up microscopic organisms to live inside their gut and help them digest food. Baby elephants learn what is good to eat by watching their mother and other relatives. They are also curious and like to try new types of food.

◄ DUNG FOOD

Elephant dung provides a feast for dung beetles and thousands of other insects. They lay their eggs in the dung, and the young feed on it when they hatch. Certain seeds only sprout in dung after having first passed through an elephant.

FOOD IN CAPTIVITY ►

People make meals for captive elephants from grasses and molasses (a type of sugar). Zoo elephants eat food such as hay, bread, nuts, fruit, leaves, bark and vegetables. They need a huge pile of food every day – in the wild they eat 100-200kg of plants every day.

Teeth

▲ **EARLY TOOTH**
This fossilized mammoth's tooth is 20cm long. The pattern of ridges on the tooth resembles that of an Asian elephant rather than an African elephant.

An elephant's front teeth are its tusks. They stick out from the mouth. An elephant also has molar teeth inside its mouth. The tusks are only in the top jaw, but the molars are in both the top and the bottom jaw. Two molars sit on each side of the mouth, each one weighing more than a brick. Sharp ridges along the top of the molars grind up tough plants. Tusks grow continuously, but molars gradually wear down and drop out. New teeth develop behind the old ones and slowly move forward along the jaw to take their place. An elephant gets through six sets of molar teeth in a lifetime – that is 24 molar teeth altogether. Each new tooth is bigger than the one before. After the sixth replacement set, the elephant is no longer able to chew its food properly.

▲ **AFRICAN TOOTH**
The ridges on an African elephant's tooth are lozenge (diamond) shaped. The scientific name of this elephant, *Loxodonta*, means lozenge teeth.

▲ **ASIAN ELEPHANT TOOTH**
The ridges on the surface of an elephant's tooth act like scissor blades. They cut up food as the elephant chews.

▲ **CHEWING UP FOOD**
An elephant uses just four huge teeth inside its mouth at any one time. It has two in the top jaw and two in the bottom jaw. Its lower jaws move back and forth to chew. This pulls and pushes the food against the grinding ridges of the molar teeth.

CONVEYOR-BELT TEETH

An elephant's teeth come through at the back of the mouth, move forward and wear out at the front. Once a set of molars has lost its effectiveness, replacement sets of molars are in reserve, ready to push in from behind. The first teeth are replaced at 1-2 years old, the second at 3-4 years, the third at 8-10 years, the fourth around 20-25 years and the fifth at 35-40. The sixth set of molars must last the elephant for the rest of its life.

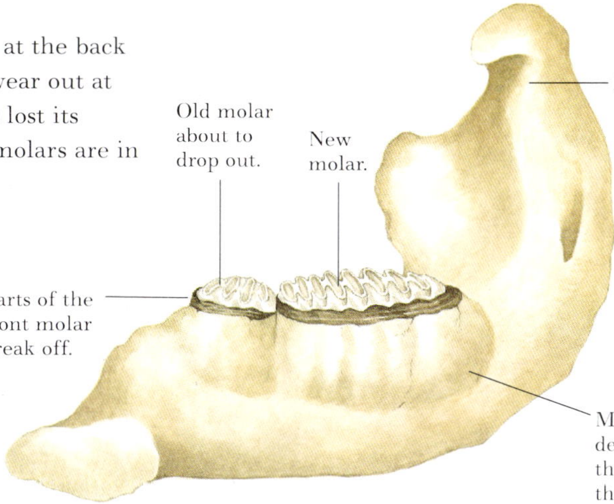

OPEN WIDE

The positions of the molar teeth inside the mouth of an African elephant can be seen by peering down its throat. The mouth is relatively small and delicate and the tongue is fleshy.

Did you know? An adult elephant's tooth can be 30cm long and weigh 4kg.

LONG IN THE TOOTH

Scientists work out the age of an elephant by looking at its teeth. When the animal loses its last tooth, it cannot chew its food properly and soon dies of malnutrition.

Focus on

LIQUID REFRESHMENT
Elephants usually drink only once or twice a day. As they stroll along in their endless search for food, they have to make sure they do not wander too far from the nearest waterhole. Sometimes their migration routes follow rivers.

An elephant's day follows a regular pattern of feeding, sleeping and travelling to new feeding areas. Meeting, greeting and communicating with other elephants is an important part of every day and interrupts other activities from time to time. Feeding takes up most of the time – about 16 hours a day. They also need to drink a lot of water and bathe to cool off. Elephants sleep twice a day, once for a few hours at noon and again in the early hours of the morning. They usually sleep standing up, but sometimes they lie on their sides.

MEAL TIME
When an elephant feeds, it may close its eyes or gaze off into the distance. Seeing its food is less important to the animal than touching and smelling it. The elephant uses its trunk to feel for food.

STICKING TOGETHER
Adult elephants help babies to cross rivers. Young elephants have to keep up with the rest of the herd as it wanders over large areas searching for food and water. However, adult elephants cooperate with each other to protect and guide the young of the herd. When danger is sensed, the group forms a tight circle around the calves. The matriarch (female leader) faces the direction of the threat.

Daily Life

AT PLAY

Elephant calves play together for part of every day. They chase, push and climb over other babies and wrestle with their trunks. These playful calves are testing out their strength and learning to live in the group.

TIME FOR A NAP

Usually, elephants sleep standing up, supported by their firm, rigid legs. They may rest their trunks on the ground while they snooze. At other times, elephants sleep lying on their sides. When they do this, they have been known to make pillows of grass for themselves. Sleeping elephants have also been heard snoring.

NIGHT REFRESHMENT

A group of elephants has a cool night-time drink at a water hole in Namibia, southwest Africa. Elephants use their keen senses of smell, touch and hearing to find their way in the dark. When feeding at night, they also make rumbling noises to let each other know where they are. Family members like to eat at the same time.

Living Together

An elephant family group is made up of related females and their offspring. Each family is led by an older, dominant female known as the matriarch. She makes all the decisions for the group. Her experience, learned over many years, is very important in keeping the family healthy and safe. Female elephants never leave the matriarchal unit unless it becomes too big. Then smaller groups break away and are led by the eldest daughters. Bulls (male elephants) leave their family group when they are between 10 and 16 years old. When they are adults, only the strongest males mate with the females. Bulls spend most of their lives in small, all-male groups (comprising two or three animals) or wander on their own. Each family group has close links with up to five other families in the same area. These linked family units, together with groups of adult bulls, make up a herd.

▲ **LITTLE LEARNERS**
Young elephant calves copy the adults of the family to learn where to find water and food. If they are in trouble, a big sister, a cousin or an aunt is always around to help.

▼ **THE MATRIARCH**
Usually, the oldest and largest adult female becomes the matriarch (female leader). The rest of the group rely on her. She controls when they eat, drink and rest. She also protects them from dangers and controls family members who misbehave.

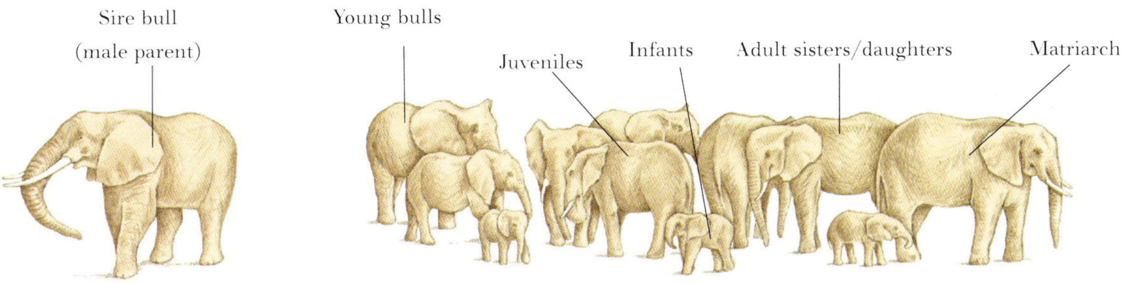

AFRICAN FAMILY

A family of African elephants usually consists of a matriarch, her adult daughters and sisters, their calves and a number of young males and females. Bulls may sometimes join the family for mating but they do not stay with it for long. They soon leave to resume their solitary lives.

ASIAN FAMILIES

Elephants in Asia live in smaller groups than African elephants. Asian families have between four and eight members, although as many as 10-20 individuals may stay in touch.

ALONE

Young bulls have a lot to learn once they leave the safety of their families. Often, they follow older males (patriarchs) around, and sometimes they have mock fights. However, males do not form such strong social bonds with each other as the females in family groups do. As a result, some bulls lead entirely solitary lives.

Communication

Everyone knows the loud trumpeting sound that elephants make. They trumpet when they are excited, surprised, angry or lost. This sound is just one way in which elephants can communicate with one another. They also touch, smell, give off chemical signals and perform visual displays, such as the position of the ears and the trunk. Their sense of smell can even tell them about another elephant's health. Elephants also make a wide range of low, rumbling sounds that carry for many kilometres through forests and grasslands. Different rumbles might mean "Where are you?" or "Let's go" or "I want to play". Females can signal when they are ready to mate, and family members can warn each other of danger.

▲ **ELEPHANT GREETING**
When elephants meet, they touch each other with their trunks, smell each other and rumble greeting sounds. Frightened elephants also touch others for reassurance.

► **BODY LANGUAGE**
Elephants send visual signals by moving their ears and trunk. Spreading the ears wide makes the elephant look bigger. This sends a message to a potential attacker to stay away. The elephant also stands up extra tall to increase the threat. It raises its tusks and shakes its head, making its ears crack like whips against its sides.

Did you know? Humans can only hear about one third of the sounds an elephant makes.

SPECIAL SOUNDS

An elephant makes its familiar high-pitched trumpeting call. Elephants also make a variety of crying, bellowing, screaming, snorting and rumbling sounds. Asian elephants make sounds that African elephants do not, and many of their rumbles last for longer. There are over 20 different kinds of rumble, with females making many more rumbling sounds than males. Females sometimes make rumbling calls when they are together, but male elephants do not do this.

ALARM SIGNALS

This nervous baby elephant is interested in the crocodiles lying on the river bank. It raises its ears, either in alarm or as a threat to the crocodiles. If a baby calls out in distress, its relatives rush to its side, with rumbles of reassurance and comforting touches with their trunks.

TOUCH AND SMELL

An elephant's skin is very sensitive, and touch is an important way of communicating feelings in elephant society. Smells also pass on useful messages, such as when a female or male is ready to mate.

Babar

One of the most famous elephants in children's literature is Babar the elephant king. He was created by French writer Jean de Brunhoff. Babar rules over the Land of the Elephants with the help of his wife Queen Celeste and his old friend General Cornelius. He fights wars with the rhinos, escapes from the circus and has many other exciting adventures.

JUST PRACTISING

Young bulls sometimes have pretend fights. They push and charge at each other, clash their tusks and wrestle with their heads.

Focus on

Male and female elephants live separately most of the time. When young male elephants become adults (at between 10 and 16 years), they are chased away from their family herd. Young bulls then form loose bonds with other males and only visit family groups occasionally. They learn from older males how to fight other males, search for females and mate with them. Bulls compete with each other to become the dominant (top) male elephant in their area. Only dominant bulls mate with the females. This helps to produce healthy, strong babies. Bulls stay near a family herd only when a female is ready to mate, but they will also help to defend the herd.

IN MUSTH

About once every year, adult bulls go through a period of unpredictable, aggressive behaviour called musth. A bull in musth oozes a sticky fluid from the side of his face.

FINDING A MATE

Female elephants prefer to mate with the biggest bulls. Males keep growing until they are aged between 35 and 45 years. Females stop growing at about 25 years.

Bull Elephants

ESTABLISHING CONTROL

Adult bull elephants have mock-fights with each other to decide who is dominant and most likely to mate with the females. The bulls push and wrestle with their heads, trunks and tusks, sometimes even lifting their opponent's front legs off the ground. Eventually, the weaker bull gives up and is chased away. .

BACHELOR HERDS

Young bulls sometimes form small groups for a while, called bachelor herds. These groupings are only temporary, however, and bulls lead a lonely life compared with females.

▲ **FIGHTING BULLS**
In ancient India, people liked to watch bull elephants fighting. They used bulls in musth because the bulls were then in a fighting mood.

Courtship

Male and female animals meet each other and select a mate by a process called courtship. For adult bulls this can happen at any time, but a cow can only mate on a few days every 16 weeks. During this time, she gives off special scents and makes sounds to attract bulls. When the bulls get close, she may begin walking with her head held high while looking back over her shoulder. A bull then follows her around, stroking her with his trunk. At the same time, he tries to chase away other bulls. Large bulls in musth (a period of heightened sexual desire) are best at doing this as they put all their energy into mating. Eventually, the cow may allow the bull to mate with her. The pair stay together for a few hours or a few days.

▲ **MEETING UP**
An African bull elephant meets up with the cows in a group when they are ready to mate. A cow is first ready to mate from about 8-18 years of age, depending on the environment she lives in and the amount of food available.

▼ **THE THRILL OF THE CHASE**
During African elephant courtship, bulls may chase after cows. Depending on her mood, the cow may stop and allow the bull to mate with her, or just keep running. Cows can run faster than the large, heavy bulls.

▲ TESTING THE WATER
When a female is on heat (ready to mate), her urine has a different smell. This musth bull is smelling the female's urine to find out if it has the mating smell. He will curl the tip of his trunk into his mouth to blow the scent over two small openings in the roof of the mouth. These openings detect the scent of a female on heat.

▲ COURTSHIP CONTACT
Elephants often touch each other with their trunks during courtship. African bulls may lay their trunk and tusks along the female's back. In Asia, a bull and a cow elephant face each other and twine their trunks together.

▲ MATING
During the act of mating, the bull elephant stands on his back legs and supports his huge weight on the cow's back. Mating can last for anything up to three minutes and may take place every two or three hours for a few days.

Birth and Babies

BIRTH TIME
When a baby elephant is born, it emerges from the cow in a protective covering called a birth sac. The other females sniff the newcomer and softly touch it all over, while rumbling with excitement.

Female elephants are pregnant for nearly two years. They give birth to their first calf when they are between 10 and 20 years old. After that, they can produce a calf every 4-6 years until the age of about 50, when the birth rate slows down. A female elephant can have between 5 and 12 babies in a lifetime. Female elephants in the same group often have babies at about the same time and look after their young together. In areas where dry seasons make food hard to find, most births take place during rainy seasons. A young calf is highly vulnerable. Its mother feeds it milk for up to six years and protects it from enemies such as lions and tigers, yet about a third of all calves do not survive. Some drown or are crushed by falling trees.

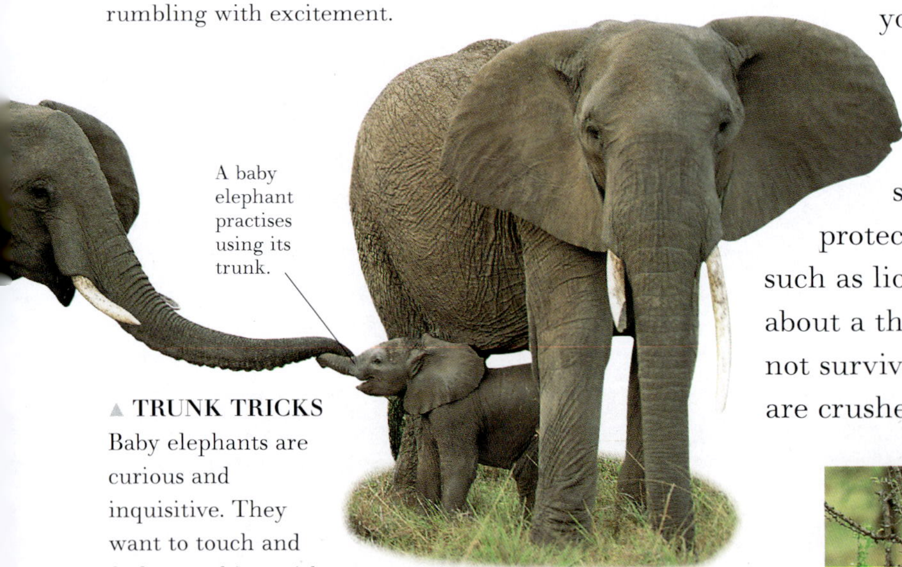

TRUNK TRICKS
Baby elephants are curious and inquisitive. They want to touch and feel everything with their trunk. At first, they cannot control their long, wobbly nose. They trip over it or suck on it – just as human babies suck their thumbs. It takes months of practice to learn how to use the trunk.

MILK BAR
A calf suckles milk from its mother's breasts with its mouth. The milk is thin and watery, but very nourishing. Babies put on weight at a rate of 10-20 kg per month.

LEARNING FAST

A young elephant has to master giving itself a dust bath. It must also learn to pick up and carry things with its trunk, drink, feed and have a mudbath. If a young elephant cannot reach water, the mother sucks up water in her trunk and squirts it down her baby's throat.

Did you know? A newborn baby elephant weighs more than an average-sized adult human.

GUIDING TRUNK

At first, a baby elephant sticks close to its mother night and day. It is always within reach of a comforting touch from her strong, guiding trunk. The mother encourages her baby, helps it to keep up with other members of the herd and often pulls it back if it starts to stray. Baby elephants will die quickly if they are left on their own.

PROPER FOOD

After a few months, calves begin to eat plants. At first, a calf may put its trunk into its mother's mouth to test out which plants are edible.

HELPFUL RELATIVES

The survival of both mothers and calves depends greatly on the support of the family group. Each member takes part in bringing up the babies so that young elephants learn to care for calves, too. This support also gives the mother a break.

Focus on

Giving birth can be quite a dangerous time for elephants because the pregnant female is particularly vulnerable to attack. To protect her, other females may form a large circle around her. More experienced relatives, known as midwives, usually help with the birth. The whole birth process takes only about an hour. Within two hours of birth, the calf usually begins to take its first shaky steps. The mother often takes it to meet the rest of the family group. The other elephants greet the arrival of the new calf with great excitement.

1 Most female elephants squat down to give birth. Occasionally, they give birth lying down. Cows are often restless during the time it takes for the calf to emerge. The other females may trumpet and scream to keep away predators.

2 The female elephant has to push as hard as possible to force the calf out of her body. Finally, the calf begins to emerge from between its mother's back legs. A female relative usually stands nearby to act as a midwife and help the struggling mother at this difficult time.

Giving Birth

3 Sometimes a baby elephant is born feet first. It emerges still partly inside the protective sac in which it has spent the last 22 months within its mother's belly. Amazingly, the delivery of the calf is fairly quick and lasts only about half a minute. After the birth, the midwife encourages the mother to push out the afterbirth (the material that helps support, feed and protect the baby whilst it is in the womb).

4 The mother and the midwife together remove the sticky birth sac from around the newborn baby. The mother then eats the afterbirth, which is full of vitamins and nutritious hormones. They help her to produce milk as quickly as possible in order to feed her newborn calf. She usually recovers her strength after a short time.

5 A calf can usually stand on its feet within about an hour of its birth. Nevertheless, its mother helps support it with her feet and trunk for a while longer. Once the newborn calf gets to its feet, it is guided to its mother's nipple for its first drink of milk. Her nipples are situated on her belly, between her front legs.

Growing Up

A baby elephant is brought up by its family in a fun-loving, easy-going and caring environment – in spite of the dangers. At first, a calf spends a lot of time with its mother. Then, as it grows older and stronger, it is allowed more freedom to explore and it begins to make friends with other calves. Calves spend a lot of time playing together. They can do this because they feed on their mother's milk and do not have to find food all day. Gradually the calf learns all the skills it needs to survive on its own.

BROTHERS AND SISTERS
An older calf may cling to its mother for reassurance even after it has learnt to feed itself. A female elephant may have calves every four to six years. Usually, just when the first calf can feed itself, another one arrives.

Did you know? Female elephants become adult in nine or ten years but males mature a few years later.

MOTHER'S MILK
A calf drinks its mother's milk until it is between four and six years old. By then, the mother usually needs her milk for the next baby. Even so, calves as old as eight have been known to push a younger brother or sister out of the way to steal a drink.

PLAYTIME

A growing elephant learns a lot simply by playing. Male elephants push and shove each other to test their strength. Females play games with lots of chasing, such as tag. Both males and females like to mess about in mud, dust or water.

PROTECTING THE YOUNG

All the adults in a family are protective of the young. They shade calves from the sun and stand guard over them while they sleep. Small calves are vulnerable to attack from many predators, including poisonous snakes.

Ganesh

In the Hindu religion of India, Ganesh is the elephant-headed god of wisdom and the remover of obstacles. Ganesh's father, the god Siva, is said to have cut off Ganesh's head. His mother, the goddess Parvati, was so angry that she forced Siva to give her son a new head. This new head turned out to be that of an elephant. Hindus seek good luck from Ganesh before the start of important business.

LOTS OF MOTHERS

Aunties, also called allomothers, take a special interest in the upbringing of calves. They wake the calf up when it is time to travel, help it if it gets stuck in mud and protect it from danger.

Life and Death

Whilst elephant calves often fall prey to crocodiles and big cats, it is hard to imagine how an animal as big and powerful as an adult elephant can be harmed. Adult elephants have no serious predators apart from people, who kill them for their ivory tusks or destroy their habitats. Yet elephants can also be killed by diseases, accidents, droughts or floods, just like any other animal. When an elephant dies, members of its family try to pick it up. They may throw dust or leaves on the body and stand over it for hours. Elephants also seem to remember where family members have died.

▲ **DROUGHT**
Elephants gather by rivers and swamps during a drought. Food and water is in short supply at this time. For heavy water users like elephants, lack of water can be fatal. Drought is the most likely cause of death in African elephants after being hunted by humans.

Did you know? Elephants can recognize the body of a dead relative.

◀ **OLD AGE**
Older elephants move less quickly than others. The group slows down to wait for them. When an elephant has worn down its sixth and final set of teeth, feeding becomes difficult. It cannot grind up and digest its food properly. Whenever possible, it feeds on soft, watery vegetation at swamps and lakes. Eventually, feeding becomes impossible altogether, and the elephant collapses and dies from malnutrition.

FASCINATING BONES

Elephants appear to be fascinated by the bones of dead elephants. They feel, sniff and scatter them about. Sometimes, the animals pull out the tusks from the skull and smash them. They may even carry bones about in their mouths for a time or bury them. Chewing bones may also be a way of obtaining the mineral calcium, which is contained in bones and teeth.

An elephant gently touches the body of a dying family member to try to rouse it.

RESPECTING THE DEAD

A dying elephant is comforted by another family member, who touches it gently with its trunk. If a family member dies, the group often lingers for days, apparently showing respect for the dead. They comfort any bereaved calves. If a young calf dies, the mother may carry it about on her tusks for a while. When elephants come across the body of a dead elephant, they smell or touch it all over.

ELEPHANT GRAVEYARDS

In stories and films such as *Tarzan of the Apes*, elephants slip away to secret communal places to die. The belief in elephant graveyards arose when people found large collections of bones in one place. Actually, these places were probably the last watering holes of drought-stricken elephants.

KILLING AND CULLING

In the past, people used to kill elephants for sport. Today, this is usually banned or strictly controlled, although poachers still shoot elephants illegally. Elephants may also be killed when there are too many of them in an area and there is nowhere else for them to go. This is called culling. The idea is to leave space for a smaller number of elephants to survive.

Ancestors

The two species of elephant today are the surviving members of a once-huge group of animals. The early members looked nothing like elephants, however. *Moeritherium*, which lived 50-60 million years ago, was a dog-sized animal with no trunk. This creature evolved into a larger animal called *Palaeomastodon*, with tusks, long lips and jaws for feeding. The first true elephant ancestor, *Stegodon*, evolved about 12 million years ago. It gave rise to the African and Asian elephant, as well as the now-extinct mammoth.

▲ **EARLIEST ANCESTOR**
Moeritherium was a small animal that lived in and around lakes and swamps about 50-60 million years ago. It had nostrils instead of a trunk, which were high on its face for breathing whilst submerged. Its eyes were also set quite high, similar to those of a hippo, to allow it to see easily whilst in the water. *Moeritherium* had sturdy legs, like an elephant, and small tusks.

GOMPHOTHERIUM ▶
This distant elephant ancestor was up to 3m tall. Its skull was longer than that of a modern elephant, and it had tusks in both the top and bottom jaws. The lower jaw was longer so that the short lower tusks reached out as far as the upper ones.

◀ *STEGODON GANESA*
Stegodon ganesa lived in India and is named after the Hindu elephant-headed god Ganesh. It probably lived in forests and ate bamboo shoots and leaves. The long tusks of this animal were parallel along most of their length and grew very close together. The family to which this elephant belonged was similar to, but distinct from, true elephants. It died out about one million years ago.

▲ *MASTODONS*
Creatures known as mastodons were more common than mammoths in the woodlands of North America. They lived more than 10,000 years ago and were about the same size as an Asian elephant, with long black-brown hair. Some mastodons had two small tusks in the lower jaw as well as the big, curved tusks in the top jaw.

► WOOLLY MAMMOTHS
These elephant relatives lived in cold areas. They had long, woolly coats to protect them from freezing temperatures. Mammoths lived at the same time as early humans and died out about 10,000 years ago.

The Cyclops
In Greek legend, Cyclops were fierce giants with only one eye. The legend may have been based on the fossilized skull of a mastodon found in Greece. The huge nose opening could have been mistaken for a large, central eye socket.

▼ PRESERVED SKULL
Mammoth skeletons, including skulls, have been found in Asia. From their preserved flesh, we can work out that mammoths had small ears. They would have helped to stop the animal from losing valuable body heat in very cold areas.

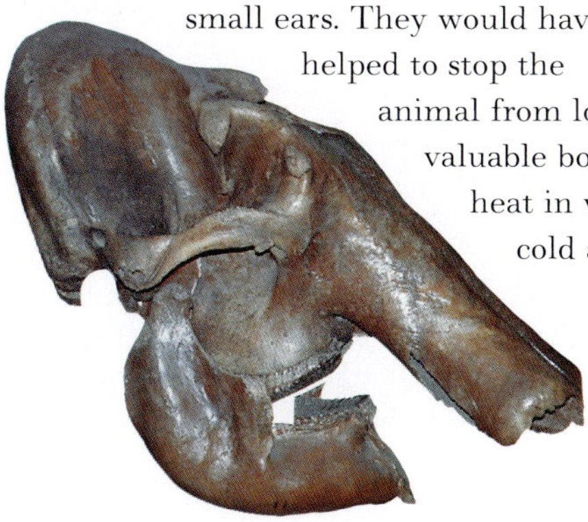

▲ CAVE PAINTING
Stone-age people painted pictures of mammoths on cave walls, many of which can still be seen today. These early humans may have dug pits to trap mammoths or driven them over cliffs.

Relatives and Namesakes

Elephants do not look like any other mammals alive today, yet they have some unlikely-looking relatives. In fact their closest living relatives are manatees and dugongs, which live in rivers, swamps and seas. Another animal that is closely related to elephants is about the size of a squirrel. It is called a hyrax and lives among rocks or in trees. The tiny elephant shrews of Africa may also be distant relatives. Evidence for these relationships comes from studying fossils and from comparing the bones and internal organs of living animals. Scientists also compare the DNA of different animals. DNA is a chemical that is passed on from parent to offspring. It is very similar in closely related animal groups.

▲ **ELEPHANT SHREWS** This tiny mammal is named after its trunk-like snout, which it uses to find insects to eat. The DNA of elephant shrews is similar to the DNA of elephants, which suggests that they are related.

▼ **HYRAXES** The hyrax has several features in common with the elephant. For example, both have ridges on their teeth and similar foot bones.

◄ **DUGONGS** These elephant relations live in the western Pacific and the Indian oceans. They are the only vegetarian sea mammals. Dugongs, like elephants, are social animals and live in groups. Male dugongs also have tusk-like front teeth.

Mermaids

In seafaring legend, mermaids and mermen were strange creatures with human bodies and fish tails. They caused shipwrecks, floods and other disasters. These legends may have begun when sailors first saw manatees or dugongs. These animals suckle their young in a similar way to humans, yet the fish-like tail of a manatee or a dugong looks like that of a mermaid or merman.

MANATEES

The manatee, like its relative the dugong, is a plump, slow-moving, vegetarian. Manatees and dugongs are sometimes called sea cows. This is because they graze on sea plants just as cows graze on land plants. Sea cows swim as whales do, using their flippers and tail.

Did you know? Some giant prehistoric hyraxes were as big as a small horse.

ELEPHANT SEAL

Despite its name, the elephant seal is not related to the elephant. It gets its name from the long, trunk-like nose of the male. Female elephant seals do not have the huge, swollen noses. During the breeding season, male elephant seals roar at their rivals and the nose works like a loudspeaker.

Life with People

For thousands of years, people have hunted elephants for their meat or ivory. They have also captured and trained elephants to carry out tasks such as fighting in battles, carrying loads and pulling heavy logs out of forests. Elephants are also used in cultural and religious ceremonies, while others are trained to entertain people in circuses or zoos. Many parts of dead elephants are used in traditional Chinese medicines, including eyeballs to treat eye diseases. Today, both scientists and tourists ride on the backs of elephants in order to study or watch other animals in the wild.

▲ **ZOO ELEPHANTS**
Zoos and wildlife parks can help elephant conservation through education, research, fund raising and careful breeding. The best zoos keep small herds in large grassy paddocks to imitate the elephants' own habitats.

◄ **ON SAFARI**
In Africa, some elephants are used to take tourists on wildlife trips called safaris. The elephants can easily walk through tall grass and over rough ground where vehicles cannot go. Also, dangerous animals such as tigers and rhinos respect the elephant's great size and strength.

CEREMONIAL ELEPHANTS

These elephants are taking part in an elephant march in South India. This is a competition to decide which of the elephant riders have the most colourful and striking umbrellas. Another Indian festival features a parade of colourfully painted elephants draped with rich cloth.

TOURIST ELEPHANTS

In the past, hunters with guns shot animals from the backs of elephants. Now the shooting is done with cameras by tourists on organized elephant safaris. In Africa, fewer elephants are used for this purpose than in Asia. This is because there is no tradition in Africa of training elephants for work and ceremonial purposes.

CIRCUS ELEPHANTS

Elephants have appeared in circuses since Roman times. They can be trained to perform tricks, although circuses with animals are becoming increasingly rare today.

ELEPHANTS IN WAR

Ancient Indian war elephants wore armour and were used to trample on the enemy. Soldiers rode on top of the elephants in a wooden castle known as a howdah.

Focus on a

Elephants were first tamed and made to work about 5,000 years ago. In Asia today between 14,000 and 17,000 elephants (about a third of all elephants in Asia) are put into service. In the past, they were caught in the wild and tamed before being trained. Now, most working elephants are born in captivity. Training usually begins when the elephant is about ten years old. It finishes when the animal is about 25. The process can be cruel. The elephant may be tied to a tree without food or water for days in order to break its spirit.

SADDLE UP
This elephant is being trained in Thailand, southeast Asia. It has to learn to accept a rider on its back and must wear a wooden saddle with chains for dragging logs. Trained elephants may be used to help train others.

BATHTIME
Working elephants enjoy a daily bath. It helps the keeper and the elephant build up a friendship and learn to trust each other.

ELEPHANT-BACK
Elephants are still the best form of transport in many areas. They move through hilly forests and cross rivers more easily than trucks. Elephants do not burn costly fuel or pollute the air, and they have longer working lives than many machines.

Working Life

HEAVY LOAD

Some elephants today are used to carry loads in hilly areas where it is difficult for vehicles to move easily. An adult elephant can carry a load weighing up to 500kg on its back. Elephants are not well suited to heavy carrying, however. They have weak shoulders and high ridged backbones.

MAHOUTS

A working elephant usually has one trainer or handler for its whole working life. In India trainers are called mahouts. In Myanmar (Burma) they are called oozies. The trainers control their elephants with up to 60 words of command. In addition, they touch the elephant with their feet or hands. A spiked stick called an ankus is used during training to help reinforce commands.

LOGGING

In southeast Asia, elephants are used to transport heavy logs. Many people are concerned that the logging industry is destroying the forest homes of many animals including elephants themselves.

Facing Extinction

Today, elephants are in great danger. Wildlife experts warn that the species faces extinction and must be protected if it is to survive in the future. Asian elephants are most at risk, with only between 36,000 and 44,000 individuals left in the wild. In Asia, the main cause of the decline is humans taking the elephants' land. Activities such as building houses, mining, growing crops and constructing dams take up a lot of space. In Africa, the biggest threat is the ivory trade. African elephants have bigger tusks than Asian elephants and are therefore more sought after. Although the ivory trade was banned in 1989, it will take a long time for elephant numbers to recover.

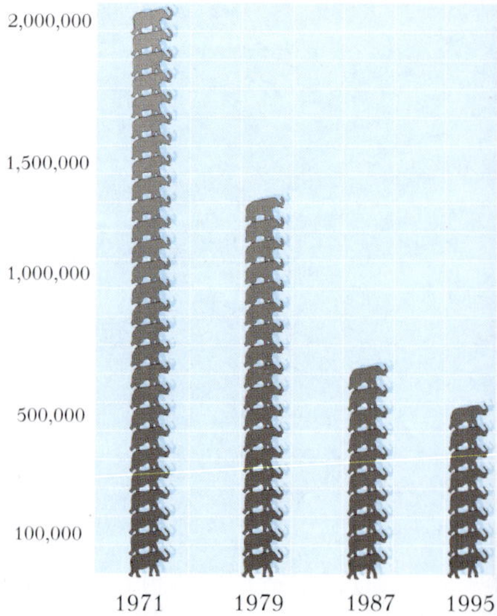

ON THE DECLINE ▲
Between 1971 and 1989, the number of elephants in Africa more than halved, from 2 million to 609,000. Up to 300 were killed each day. Over 90% of these elephants were killed illegally by poachers.

Did you know? Elephant meat is sold as food in certain Asian countries.

◄ IVORY BONFIRE
In 1989, the Kenyan government burned US $3.6 million worth of ivory on this huge bonfire. They did this in order to support the worldwide ban on the trade in ivory. However, poachers will go on killing elephants illegally as long as people are prepared to pay huge sums of money for the tusks.

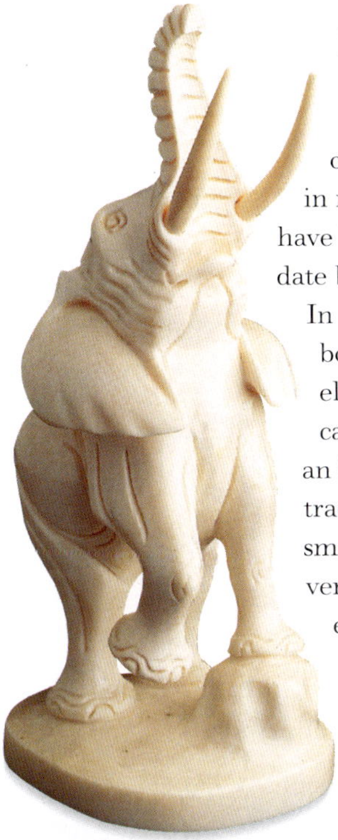

WHITE GOLD

People have carved ivory for tens of thousands of years. Carvings in mammoth ivory have been found that date back 27,000 years. In ancient Egypt, both hippo and elephant ivory was carved. Ivory makes an ideal material for trade because even a small amount has a very high value. It is easy to carve but also hard enough to last, with a smooth, cool surface.

POACHING

When poachers kill elephants, they only want the tusks. They leave the rest of the elephant to rot. Before the ban on ivory trading, an African poacher could earn hundreds of dollars for just one pair of tusks. It would take the poacher a year to earn this much money in an ordinary job.

TOO MANY ELEPHANTS

African elephant feet are sometimes sold to raise money for conservation. These animals were legally killed in a National Park where elephant numbers had grown too high.

ELEPHANTS OR PEOPLE?

These elephants are invading a farmer's home and fields in Kenya, east Africa. The human population of Kenya is expected to double by the year 2020, putting enormous pressure on the land. Finding enough land for both people and elephants will be a problem. Elephants will not be able to roam freely, as they have been able to do in the past, but will be confined to special areas.

Conservation

CAPTIVE BREEDING
A zoo elephant has its foot cleaned with a hoof knife. Zoos play a role in conserving animals. However, elephants are not often bred in zoos because bulls are difficult to handle and can be dangerous.

Elephants need to be conserved if the species is to survive. Many African and Asian countries have set aside areas of land called national parks or nature reserves. Here, elephants are protected from the threat of poachers. This is not a perfect solution, however. Elephants in reserves are so well protected that their numbers steadily increase. Eventually, they eat themselves out of food. The rangers are then forced to kill some elephants to let others live. Other conservation efforts include keeping elephants on game ranches and banning trade in ivory. Alternatives to ivory, such as plastics, resins and the nuts of a South American palm tree, are less damaging to wildlife.

WALRUS WORRIES
When the trade in elephant ivory was banned in 1989, poachers turned their attention to walrus ivory instead. During 1989, poachers in speedboats shot at least 12,000 Alaskan walruses, whose tusks can grow almost a metre long.

POACHING PATROLS
Guards in a national park in southern Africa hold wire traps left by poachers. Protecting elephants from poachers is dangerous work. As well as removing traps left for elephants, guards may become involved in gun battles as poachers try to kill elephants.

ELEPHANT TRAVELS
Sometimes, elephants are moved to areas where they have better chances of survival. Moving elephants is not an easy thing to do, however. Getting these enormous, heavy animals into a truck or an aeroplane can be a tricky business. Sometimes a whole family is moved, which helps these sensitive animals get over the trauma. They then settle into their new home more easily.

BORN FREE FOUNDATION

ELEFRIENDS

SUPPORT GROUPS
Conservation groups such as Elefriends in the UK raise money to help conservation work in the wild. They also persuade people not to buy and trade in ivory.

ORPHAN BABIES
With a great deal of patience, care and understanding, orphaned baby elephants can be raised in National Parks in Africa and returned to the wild. Raising a baby elephant is just as hard work as raising a baby human.

TOURIST TRAPS
Tourists pay to watch elephants in national parks. This money is used to help run the parks and look after the elephants. It is also used to improve the lives of people living nearby. Many of these people have given up their land to save the elephants.

INDEX

A
afterbirth 45
ageing 15, 31, 48
alarm signals 36-7
allomothers 47
ancestors 50-1
ankus 4, 57

B
Babar 37
bachelor herds 39
backbone 10, 15
bathing 20-1, 22-3, 32, 56
birth 4, 42-5
birth sac 42, 45
body language 36
bones 10-11, 15, 48-9
brain 10, 15
bull elephants 4, 5, 6, 34-5, 38-41, 60
bush elephant 8-9

C
calves 4, 5, 12, 18, 23, 29, 32-5, 37, 42-7, 61
chewing 30-1
circuses 54, 55
colour 4, 7, 14, 15
communication 16, 18, 32-3, 36-7
conservation 54, 60-1
courtship 40-1
cow elephants 5, 6, 15, 34-5, 40-5
culling 49, 59
Cyclops 51

D
danger 32, 36-7, 42, 47
death 48-9
desert elephant 8-9
digestion 28-9
digging 5, 12, 13, 26
DNA 52
drinking 18-19, 22-3, 26, 32-3, 43
drought 26, 48-9
dugong 52, 53
dung, elephant 29
dust showers 18, 20, 21, 43

E
ears 5, 6, 16-17, 20, 36
elephant seal 53
elephant shrew 52
endangered species 58-9
eyes 5, 10-11, 15, 16

F
family groups 4, 5, 22, 29, 32-7, 42-7, 49
feeding 12, 13, 18-19, 28-31, 32-3, 42-3, 46
feet 5, 24-5, 59, 60
fighting 5, 12, 13, 18, 35, 38-9, 40, 54
forest elephant 8

G
Ganesh 47, 50
Gomphotherium 50
graveyards, elephant 49

H
habitats 8-9
Hannibal 11
hearing 16-17, 33, 36-7
herds 34-5, 38-9
Hinduism 4, 47
howdahs 55
hunting 54, 58-9, 60
hyrax 52, 58

I
Indra 4
infrasound 16-17, 36-7
intelligence 4, 17
ivory 13, 54, 58-9, 60-1

L
legs 5, 10-11
life expectancy 4, 42, 48-9

M
mahouts 57
mammoth 11, 13, 30, 50-1
manatee 53
Mastodon 51
mating 37, 38, 40-1
matriarchs 32, 34-5
mermaids 53
midwives 44-5
migrations 26-7, 32
milk 42, 45, 46
Moeritherium 50
movement 24-5
mud rolling 14, 20-1
musth 38, 40-1

N
nature reserves 60-1
nose 5, 16, 18
nostrils 10, 15, 18

P
Palaeomastodon 50
patriarchs 35
play 18, 22, 33, 46-7
poaching 49, 58-9, 60
predators 23, 42, 48, 54

R
rumbling 16-17, 33, 36-7, 42

S
salt, digging for 13
senses 16-17, 32-3, 36-7
sight 16, 32
skeleton 10-11, 15
skin 4, 5, 7, 14, 20-1, 37
skull 10-11, 12, 15
sleep 10, 32-3, 47
smell, sense of 16-17, 18, 32-3, 36-7, 40-1
Stegodon 50
swimming 22-3, 25

T
taste, sense of 16
teeth 5, 10, 12-13, 30-1, 48
temperature control 14, 19, 20-1, 32
touch, sense of 16-17, 18, 32-3, 36-7, 42-3
tourism 54-5, 61
training 4, 56-7
trunk 4, 5, 10-12, 15-19, 22, 32, 36, 42-3
tushes 12
tusks 4, 5, 6-7, 10-11, 12-13, 30-1, 58-9

Z
zoos 21, 29, 54, 60

NATURE WATCH

SNAKES

BARBARA TAYLOR

Consultant: Michael Chinery

Published by Anness Publishing Ltd,
Blaby Road, Wigston,
Leicestershire LE18 4SE
Email: info@anness.com
Web: www.annesspublishing.com

Publisher: **Joanna Lorenz**
Managing Editor: **Sue Grabham**
Senior Editor: **Nicole Pearson**
Editor: **Nicky Barber**
Designer: **Simon Wilder**
Special Photography: **Kim Taylor**
Illustrations: **Julian Baker, David Webb**

ETHICAL TRADING POLICY
Because of our ongoing ecological investment programme, you, as our customer, can have the pleasure and reassurance of knowing that a tree is being cultivated on your behalf to naturally replace the materials used to make the book you are holding. For further information about this scheme, go to www.annesspublishing.com/trees

PUBLISHER'S NOTE
Although the advice and information in this book are believed to be accurate and true at the time of going to press, neither the authors nor the publisher can accept any legal responsibility or liability for any errors or omissions that may have been made.

PICTURE CREDITS
b=bottom, t=top, c=centre, l=left, r=right
Jane Burton/Warren Photographic: pages 29c, 40-41 and 48tr; Bruce Coleman Ltd: pages 6bl, 8bl, 14br, 15r, 16tl, 17tl, 18tl, 19cr, 20cl, 20tr, 20br, 21tr, 21bl, 22br, 27br, 28t, 29b, 30br, 31t, 33cr, 33bl, 33r, 34bl, 36c, 37bl, 39br, 44cl, 44br, 46bl, 46br, 48tr, 48cr, 48b, 50cr, 50b, 53tr, 54bl, 56bl, 56cr, 59br, 60tr and 61b; Ecoscene: pages 17cr and 59tl; Mary Evans Picture Library: pages 45tr and 53br; FLPA: pages 11cr, 11cl, 16c, 18bl, 23tl, 23tr, 26tl, 26c, 31cr, 32-33, 35tr, 35bl, 38cr, 39t, 39bl, 40-41, 43tr, 45cl, 45br, 48cl, 51bl, 50-51, 52b, 52r, 55tl, 57bl and 60br; Holt Studios International: pages 22tl, 46tr and 58tl; Nature Photographers: page 30bc; NHPA: pages 7bl, 8tl, 9tr, 9bl, 11tl, 14tl, 18br, 19tl, 19br, 22bl, 23bl, 23br, 31cl, 36, 37, 37tr, 37cr, 38bl, 42cl, 47tl, 47cl, 47bl, 49cl, 49r, 50tl, 51t and 53c; Oxford Scientific Films: pages 6tc, 12-13, 24-25, 42t, 43b and 56br; Planet Earth Pictures: pages 28b, 32cl, 32br,38tr, 42br, 43c, 54tr, 54br, 55cr, 57cr, 60cl, 61tl and 61tr; Visual Arts Library: pages 5br, 11br and 27bl; Zefa Pictures: page 51br.

Special photography: Kim Taylor/Warren Photographic: pages 1, 2, 3, 4-5, 7tr, 10bl, 10bc, 11tr, 12-13, 14-15, 16cl, 16bl, 24-25, 27tr, 29t, 31bl, 34tl, 34-35, 35cr, 36bl, 47r, 52cl, 56tr, 57t, 58b, 59c, 59bl, 62, 63 and 64.

No animals were harmed during the making of this book.

C O N

4–5 Snake Life
6–7 Shapes and Sizes
8–9 How Snakes Work
10–11 A Scaly Skin
12–13 A New Skin
14–15 Snakes on the Move
16–17 Snake Senses
18–19 Food and Hunting
20–21 Teeth and Jaws
22–23 Stranglers and Poisoners
24–25 Let's Do Lunch!
26–27 Poisonous Snakes
28–29 Vipers
30–31 Defence
32–33 The Cobra and its Relatives

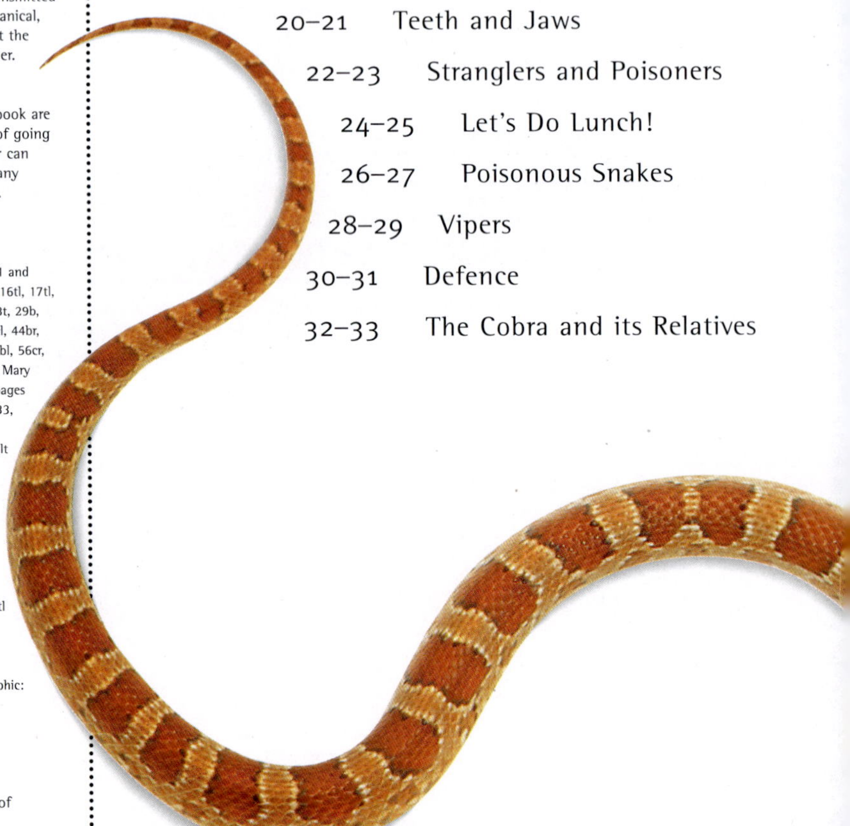

TENTS

34–35 Colour and Camouflage
36–37 Reproduction
38–39 Eggs
40–41 Hatching Out
42–43 Giving Birth
44–45 Growth and Development
46–47 Where Snakes Live
48–49 Tree Snakes
50–51 Emerald Tree Boa
52–53 Desert Snakes
54–55 Water Snakes
56–57 Snake Families
58–59 Snake Relatives
60–61 Conservation
62–63 Glossary
64 Index

Snake Life

Snakes are beautiful and secretive animals, and they are not nearly as dangerous as people think. They are a kind of reptile related to lizards, crocodiles and turtles. Altogether, there are about 2,700 different kinds of snake, but out of this number only 300 or so are able to kill people. In Europe or North America, you are more likely to be struck by lightning than to be bitten by a poisonous snake. All snakes have several things in common. They are long, thin animals with no legs, eyelids or outside ears. Their bodies are covered with tough, waterproof scales. Snakes have forked tongues for smelling and tasting the air. They are all flesh eaters and swallow their prey whole. Snakes have always had a special place in myths and legends, being used as symbols of both good and evil.

A Snake's Tail

It is often hard to tell where a snake's body ends and where its tail begins. The tail is the part behind a small opening called the cloaca, where the body wastes pass out. The snake narrows slightly at this point. The tail of a male snake is generally longer than that of the female.

Snake Heads

Like this grass snake, most snakes have a definite head and neck. But in some snakes, one end of the body looks very much like the other end, except that the head has eyes, nostrils and a mouth, and the tail does not!

rattlesnake

Forked Tongues

Snakes and some lizards are the only animals with forked tongues. A snake flicks its tongue in and out of its mouth to taste and smell the air. This gives the snake a picture of what is around it. A snake does this every few seconds if it is hunting or if there is danger nearby. The tongue will not sting if it touches your skin. The snake is just collecting information about you.

Colombian rainbow boa

Scaly Armour

A snake's skin is hidden under a covering of tough scales. The scales grow out of the skin and usually hide the skin from view. However, after a big meal, the scaly skin stretches so that the skin becomes visible between the scales. A snake's scales protect its body while allowing it to stretch, coil and bend. If you touch a snake, you will feel that the scales are dry. Some snakes have rough scales, whilst others have smooth ones.

Did you know? Snakes never feel slimy to the touch.

Did you know? A boa squeezes its prey to death in its coils.

red-tailed boa

eye has no eyelid

forked tongue

Did you know? Most snakes lay eggs but some give birth to baby snakes.

Medusa

An ancient Greek myth tells the story of the Medusa, a monster with snakes for hair. Anyone who looked at her was turned to stone. Perseus managed to avoid this fate by using his polished shield to look only at the monster's reflection. He cut off the Medusa's head with his sword and carried it home, dripping with blood. As each drop touched the earth, it turned into a snake called a viper.

Shapes and Sizes

Can you imagine a snake so big that it could stretch all the way from the ground to the roof of a three-storey house? The reticulated python is this big. The biggest snakes have thick, powerful bodies that measure nearly a metre round. Other snakes are as thin as a pencil and small enough to fit into the palm of your hand. Snakes also have different shapes to suit their different environments. Sea snakes have flat bodies and tails like oars to help them push against the water and move forwards. Burrowing snakes have rounded, tube-shaped bodies to help them slide easily through the soil. Pythons and vipers have fairly short tails, while the tails of some tree snakes are longer than their bodies.

Thick and Thin

Vipers mostly have thick bodies with much thinner, short tails. They are not very long snakes. The bags of poison on either side of a viper's head take up a lot of space, so the head is large and broad. It is often a triangular shape.

rhinoceros viper

Long and Thin

Tree snakes are often long and thin in shape, to help them slide through leaves and along branches. Even their heads are long and pointed. The head of a tree snake is also very light so that it does not weigh the snake down as it reaches out for the next branch. The long tail of a tree snake coils around branches to help it grip tightly as it climbs.

green tree snake

red-tailed boa

Being Big

This picture shows the head of a red-tailed boa at its actual size. The head measures about 15cm. The red-tailed boa usually grows to an overall length of 3.5m. The longest snake in the world is the reticulated python, which can grow up to 10m. Other giant snakes include the anaconda, other boas and the pythons.

Did you know? The shortest snake in the world, Leptotyphlops bilineata, is only 10.8cm long.

Slender Snakes

A garter snake has a narrow head because it is not poisonous and does not need space in its head for bags of poison. The body of a garter snake is slender and small - from 0.3m to 1.2m long. Garter snakes get their name from their stripes. These stripes look like the pattern on old-fashioned elastic garters, which people once used to hold up stockings.

garter snake

blind snake

Smallest Snake

The world's smallest snakes are the blind snakes and the thread snakes. These tiny snakes are less than 40cm long. They have rigid skulls for burrowing underground. Their bodies are shaped like a long cylinder and covered with smooth, shiny scales so that they can slide through the soil easily.

Snake Shapes

Snakes have four general body shapes and lengths - large and round (like a python), short and thick (like a viper), small and thin (like a burrowing snake), and long and thin with a pointed head (like a tree snake).

egg-eating snake

Stretchy Stomach

The big lump inside this snake is a whole egg. Luckily, the throat and gut of the snake are so elastic that the thin body of a snake can stretch wide enough to eat its large meal. The throat and first part of the gut are also very muscular, to help force food right down into the snake's stomach. The egg is broken on the way down.

How Snakes Work

A snake has a stretched-out inside to match its long, thin outside. The backbone extends along the whole body with hundreds of ribs joined to it. There is not much room for organs, such as the heart, lungs, kidneys and liver, so these organs are thin shapes to fit inside the snake's body. Many snakes have only one lung. The stomach and gut are very stretchy so that they can hold large meals. When a snake is swallowing large animals, it pushes the opening of the windpipe up from the floor of the mouth in order to keep breathing. Snakes are cold-blooded, which means that their body temperature is the same as their surroundings. This is why most snakes live in warm places where the temperature is high enough for them to stay active both day and night.

banded rattlesnake

Warming Up and Cooling Down

Like all snakes, the banded rattlesnake is an ectotherm. This means that it is cold-blooded. In order to warm up or cool down, it has to move from sun to shade. The word ectothermic means outside heat because the main source of a snake's body-heat is outside its body.

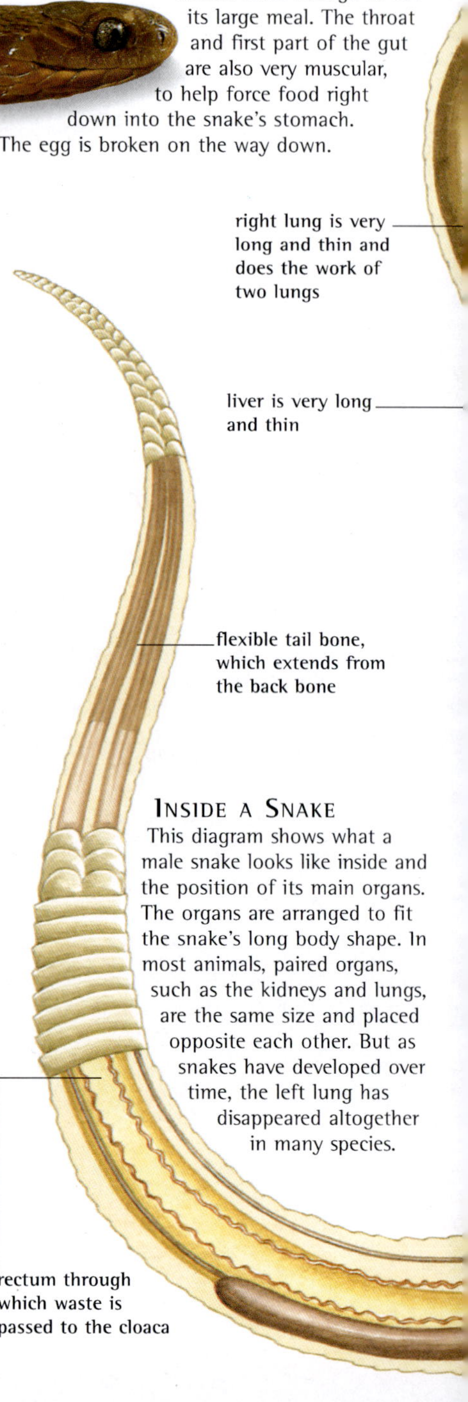

Inside a Snake

This diagram shows what a male snake looks like inside and the position of its main organs. The organs are arranged to fit the snake's long body shape. In most animals, paired organs, such as the kidneys and lungs, are the same size and placed opposite each other. But as snakes have developed over time, the left lung has disappeared altogether in many species.

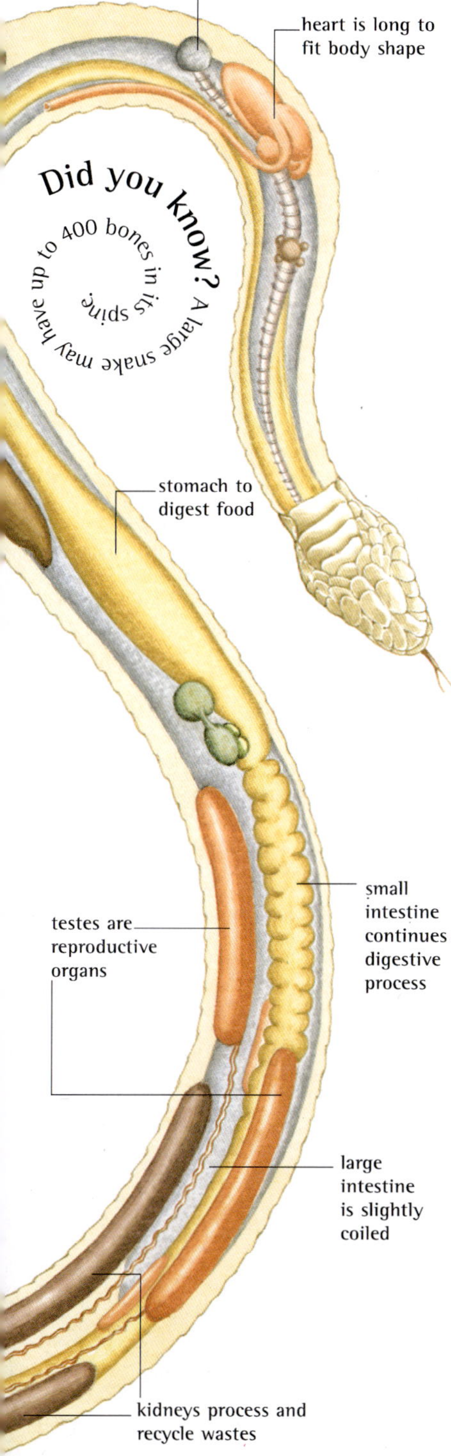

Snake Bones

Look at this X-ray of a grass snake and you will see the bones that make up its skeleton. There are no arm, leg, shoulder or hip bones. A few other snakes do have very tiny hip and back leg bones, but these are not used for walking. If you look carefully, you will see that the ribs do not extend into the tail. Snake bones are delicate and easily damaged.

viper skeleton

Skeleton

A snake's skeleton is made up of a skull and a long backbone with ribs arching out from it. The free ends of the ribs are linked by muscles, forming a tube of bone and muscle that protects the snake's organs. The loose joints in a snake's backbone allow it to bend and coil its body.

A Scaly Skin

A snake's scales are extra-thick pieces of skin. They work like a suit of armour to protect the snake from bumps, knocks and scrapes as it moves about. They also allow the skin to stretch when the snake moves or feeds. Scales are usually made of a horny substance, called keratin, which forms your hair and nails, as well as the claws, hoofs and feathers of other animals. Every part of a snake's body is covered by a layer of scales, including the eyes. The clear, bubble-like scale that protects each eye is called a brille or spectacle. The scales of most snakes also contain the pigments that give snakes their many colours. Every so often a snake grows a shiny new skin underneath its old one. Then it wriggles out of the dead skin.

Horned Snake

As its name suggests, the European nose-horned viper has a strange horn on its nose. The horn is made up of small scales that lie over a bony or fleshy lump sticking out at the end of the nose. No one knows what the horn is used for. Other snakes, such as shovel-nosed snakes, have large scales on the ends of their noses, which they use for burrowing in the ground.

nose-horned viper

Scutes

Most snakes have a row of broad scales, called scutes, underneath their bodies. The scutes go across a snake's body from side to side, and end where the tail starts. Scutes help snakes to grip the ground, like the tyres of a caterpillar tractor. Many burrowing and water snakes do not have scutes.

corn snake's scutes

Warning Rattle

The rattlesnake has a number of hollow tail-tips that make a dry, buzzing sound when shaken. The snake uses this sound to warn enemies to keep away. Every time a rattlesnake sheds its skin a section at the end of the tail is left behind, adding another piece to the rattle.

rattlesnake's rattle

corn snake's scales

Skin Scales

When a snake's skin is stretched, the scales pull apart so that you can see the skin between the scales. The scales grow out of the top layer of the skin, called the epidermis. There are different kinds of scales. Keeled scales may help snakes to grip surfaces, or break up a snake's outline for camouflage. Smooth scales make it easier for the snake to squeeze through tight spaces.

Look closely at the rough scales of the puff adder (left) and you will see a raised ridge, or keel, sticking up in the middle of each one.

Did you know? Most snakes get their colours from pigments in the scales.

The wart snake (right) uses its scales to hold on to its food. Its rough scales help the snake to keep a firm grip on slippery fish until it can swallow them. The wart snake's scales do not overlap.

The green scales and stretched blue skin (left) belong to a boa. These smooth scales help the boa to slide easily through leafy branches. Burrowing snakes also have smooth scales so that they can slip through the soil.

Did you know? The hairy bush viper has pointed scales with curled tips, making it look hairy

Eternal Youth

A poem written in the Middle East about 3,700 years ago tells a story about why snakes can shed their skins. The hero of the poem is Gilgamesh (shown here holding a captured lion). He finds a magic plant that will make a person young again. While he is washing at a pool, a snake eats the plant. Since then, according to the poem, snakes have been able to shed their skins and become young again. But people have never found the plant of eternal youth – which is why they always grow old and die.

1 Ready to Shed
A snake sheds its outer layer of scaly skin when a new layer of skin and scales has grown underneath the old one. In the days before it peels, a snake is sluggish and bad-tempered, and its colours become dull. Its eyes turn cloudy or blue as the eye coverings gradually become loose. About 24 hours before moulting, the eyes clear again.

A New Skin

One of the most amazing things about snakes is the almost magical way they wriggle out of their old, tight skin to reveal a bright, new, shiny skin underneath. Unlike humans, who shed little flakes of old skin and grow new skin all the time, snakes shed their worn-out skin and scales in one piece. This process is called moulting or sloughing. Snakes only moult when a new layer of skin and scales has grown underneath the old skin. Adult snakes slough their skin up to about six times a year.

2 The First Split
The paper-thin layer of outer skin and scales first starts to peel away around the mouth. The snake rubs its jaws and chin against rocks or rough bark, and crawls through plants. This helps to push off the loose layer of skin.

Did you know?
A baby snake may shed its skin when it is only a few days old

3 Peeling Off

The outer layer of skin gradually peels back from the head over the rest of the body. The snake slides out of its old skin, which comes off inside-out. It is rather like taking hold of a long sock at the top and peeling it down over your leg and foot!

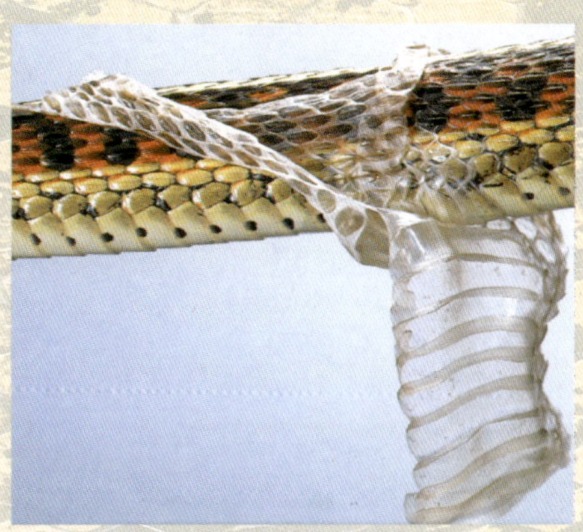

4 Shedding Skin

A snake usually takes several hours to shed its whole skin. The old skin is moist and supple soon after shedding, but gradually dries out to become crinkly and rather brittle. The moulted skin is an exact copy of the snake's scale pattern. It is very delicate. If you hold a moulted snake skin up to the light, you will notice that it is almost see-through.

Did you know? Female snakes often shed their skin just before giving birth.

5 The Old Skin

A shed skin is longer than the snake itself. This is because the skin stretches as the snake wriggles free. It is not coloured but may have traces of the snake's pattern left on it. The snake's new skin is brightly coloured and shiny, with a well-marked pattern.

Snakes on the Move

For animals without legs, snakes move around very well. They can glide over or under the ground, climb trees and swim gracefully through water. A few snakes can even parachute through the air. Snakes are not speedy animals, however. Most move at about 3kph. With their bendy backbones, snakes have a flexible, wavy movement. They push themselves along using muscles joined to their ribs. The scales on their skin also grip surfaces to help with movement. Snakes move in four main ways – wriggling from side to side, crawling straight forwards, moving like a concertina and sidewinding. The type of movement depends on the size of the snake and, most importantly, on its surroundings.

Did you know? A person can walk faster than most snakes can move.

corn snake

S-shaped Mover

Most snakes move in an S-shaped path. They push the side curves of their bodies backwards against the surface they are travelling on or through. The muscular waves of the snake's body hit surrounding objects and the whole body is pushed forward from there. A snake track in sand has small heaps on the outside edges of the curves where the snake has pushed hard against the sand.

banded sea snake

Swimming Snake

The startling stripes of the banded sea snake stand out as it glides effortlessly through the water. Snakes always swim using S-shaped movements. Instead of pushing against the ground, a sea snake pushes against the water. A sea snake's tail is flattened from side to side to give it extra power, like the oar on a rowing boat.

Concertina Snake

The green whip snake moves with an action rather like a concertina. The concertina is played by squeezing forwards and backwards. The green whip snake moves by gripping on to a surface with the back part of its body while stretching the front part forwards. Then the front half holds on while the back half is drawn up to meet it.

green whip snake

Peringuey's viper

Sidewinding

This way of moving is used by snakes that live on loose sand. The snake anchors its head and tail in the sand and moves the middle part of its body sideways. Then it grips with the middle part of the body and moves its head and tail sideways, winding along at an angle of 45° to the direction of travel.

Did you know? The fastest land snake is the black mamba, moving at up to 11 kph.

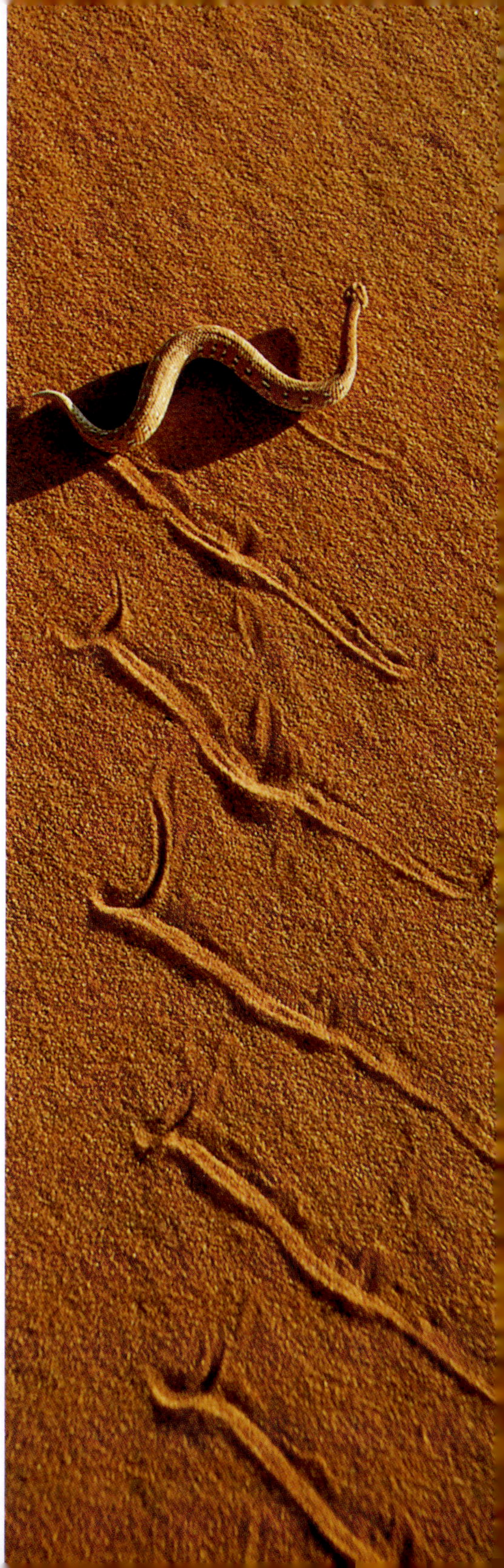

How Snakes Move

Most land snakes move in four different ways, depending on the type of terrain they are crossing and the type of snake.

1 S-shaped movement: the snake wriggles from side to side.

2 Concertina movement: the snake pulls one half of its body along first, then the other half.

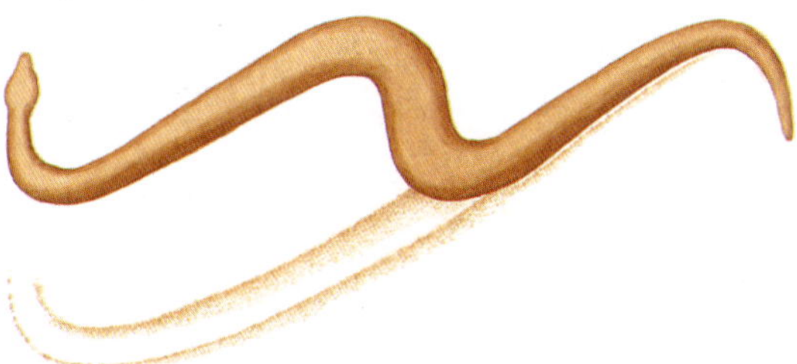

3 Sidewinding movement: the snake throws the middle part of its body forwards, keeping the head and tail on the ground.

4 Caterpillar movement: the snake uses its belly scutes to pull itself along in a straight line.

Snake Senses

To find prey and avoid enemies, snakes rely more on their senses of smell, taste and touch than on sight and hearing. Some snakes can see well – tree snakes for example – but others, such as burrowing snakes, are almost blind. Snakes have no outside ears or eardrums, but they do have one earbone joined to the jaw. They can pick up sound vibrations through the ground and probably some low sounds travelling through the air. As well as their ordinary senses, snakes also have some special ones. They are almost the only animals to taste and smell with their tongues. There are also some snakes, such as pit vipers, that can sense the heat given off by their prey through pits on the sides of their faces. No other animal can do this.

Eyesight

This snake seems to have a glassy stare because it has no eyelids to cover its eyes. Eyesight varies in snakes, but even those with good sight probably detect movement rather than the shape or form of objects. The snakes with the best eyesight are tree snakes, such as this green mamba, and day hunters such as garter snakes.

Night Hunters

At night, the eyes of the horned viper open wide to let in as much light as possible (*above*). The horned viper is a night hunter. During the day, its pupils close down to narrow slits to protect the eyes from bright light (*below*).

green tree python

heat pits

Sensing Heat

The green tree python tracks other animals by detecting the heat given off from their bodies. Both boas and pythons have heat holes along the lips. These heat sensors are lined with nerves. They allow a snake to find and kill prey in total darkness.

black-tailed rattlesnake

The Forked Tongue

When snakes want to investigate their surroundings, they flick out their tongues to taste the air. The forked tongue picks up tiny chemical particles of scent from the air or off the ground.

Hearing

The cobra cannot hear the music played by the snake charmer. It follows the movements of the pipe and rises up as it prepares to defend itself.

Snakes are not deaf, though. The lower jaw picks up vibrations carried through the ground and passes them on to the small ear bone. From these vibrations, a snake can tell if an animal is coming towards it or going away, and if the animal is big or small.

Jacobson's Organ

As a snake draws its tongue back into its mouth, it presses the forked tip into the two openings of the Jacobson's organ. This organ is in the roof of the mouth and it analyses tastes and smells. Information from the Jacobson's organ tells the snake about food, predators, mates and other important things.

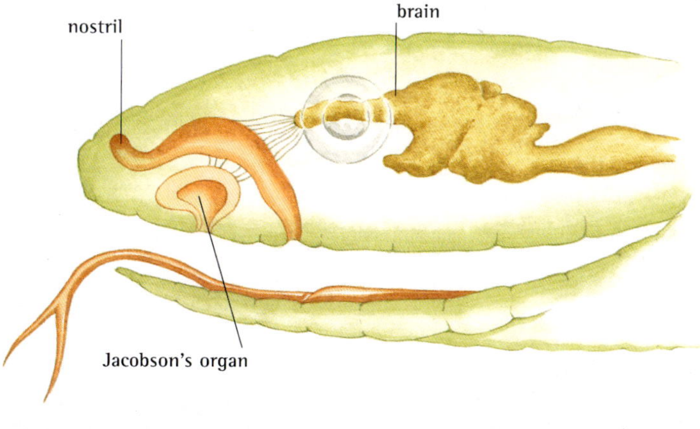

Food and Hunting

All snakes are hunters, catching and killing other animals. However, snakes eat different foods and hunt in different ways depending on their size, their species and where they live. Some snakes eat a wide variety of food, while others have a more specialized diet. Thread snakes eat only slugs and snails, and egg-eating snakes eat only the eggs of birds and reptiles. A few snakes, such as the king snake, hunt and eat other snakes. A snake has to make the most of each meal because it moves fairly slowly and does not get the chance to catch prey very often. Luckily, a snake's body works at a slow rate so it can go for months without eating. A snake uses its senses to detect other animals and, instead of hunting, many snakes simply lie in wait for their prey to come to them.

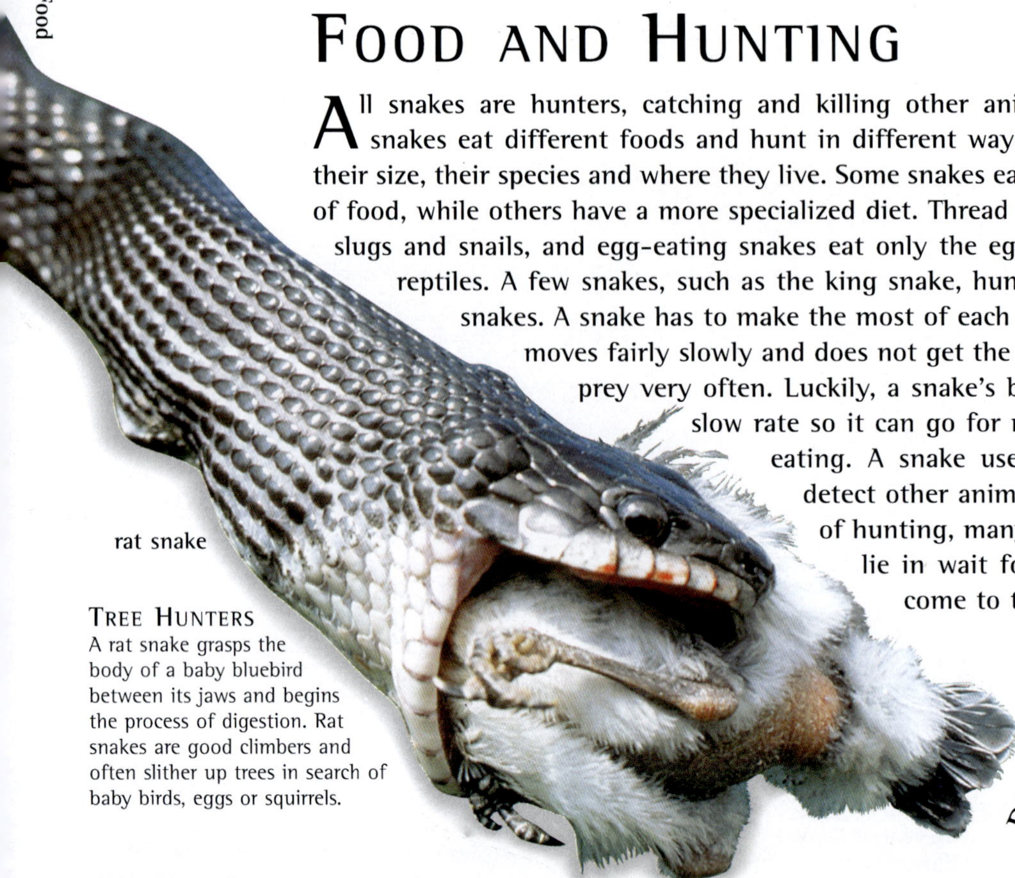

rat snake

Tree Hunters

A rat snake grasps the body of a baby bluebird between its jaws and begins the process of digestion. Rat snakes are good climbers and often slither up trees in search of baby birds, eggs or squirrels.

Did you know? Some Cuban boas feed on bats.

tentacled snake

Fishy Food

Like many snakes, the tentacled snake lives on fish. Tentacled snakes probably hide among plants in the water and grab fish as they swim past. Sea snakes have narrow heads and necks to search out fish that are hidden among rocks and coral, or down mud burrows. These snakes catch mainly slow-moving fish, or fish that are resting at night.

Tricky Lure

The Australasian death adder has a brightly coloured tail tip that looks just like a worm. It uses its tail tip as a lure to trick prey. The adder wriggles the 'worm' to lure lizards, birds and small mammals to come within range of its deadly, poisonous fangs.

death adder

African egg-eater

Egg-eaters

Snakes eat eggs because they are a rich source of food – and they do not fight back! The snake first checks the egg with its tongue to make sure it is fresh. Then it swallows the egg whole. It uses the pointed ends of the bones in its backbone to crack the eggshell. The snake swallows the yolk and white of the egg, then coughs up the crushed eggshell.

Surprise Attack

Lunch for this gaboon viper is a mouse! The gaboon viper hides among dry leaves on the forest floor. Its colouring and markings make it very difficult to spot. It waits for a small animal to pass by, then grabs hold of its prey in a surprise attack. Many other snakes that hunt by day also ambush their prey.

Did you know? Sometimes a snake coughs up its prey – alive!

gaboon viper

smooth snake

Snake Eats Snake

This smooth snake is eating an asp viper. The viper fits neatly inside the body of the smooth snake. This makes it easier to swallow than animals that are different shapes. Although the viper is poisonous, the smooth snake is not affected by the viper's poison. King snakes can eat poisonous rattlesnakes without being harmed, too.

Teeth and Jaws

Most snakes have short, sharp teeth that curve backwards. Snakes' teeth are good for gripping and holding prey, but not for chopping or chewing it into smaller pieces. The teeth are not very strong, and often get broken, so they are continually being replaced. Poisonous snakes have some teeth that are much larger than the others. These teeth are called fangs and they may be in the front or the back of the mouth. When the snake bites, poison flows down the fangs to paralyse the prey and break down its body. All snakes swallow their prey head-first and whole, even when the victim's body is bigger than the snake's. Special loose jaws allow the snake to open its mouth wider than any other animal. The jaws 'walk' over the prey to force it down the snake's throat.

Back Fangs

A few poisonous snakes, such as this African boomslang, have fangs at the back of their mouths. Poison runs down grooves in the fangs. The boomslang is digging its fangs hard into the flesh of a chameleon to get enough poison inside.

boomslang

Open Wide

An eyelash viper opens its mouth as wide as possible to scare off an enemy. You can see that its fangs are folded back against the roof of the mouth. If the viper were to attack its enemy, the fangs would swing quickly forwards.

Folding Fangs

There are two main kinds of snake with fangs at the front of the mouth – vipers and elapid snakes. Snakes in the viper family have long fangs that can be folded back. When a viper strikes at its prey, the fangs swing forward so that they stick out in front of the mouth. The fangs inject poison deep into the victim.

viper skull

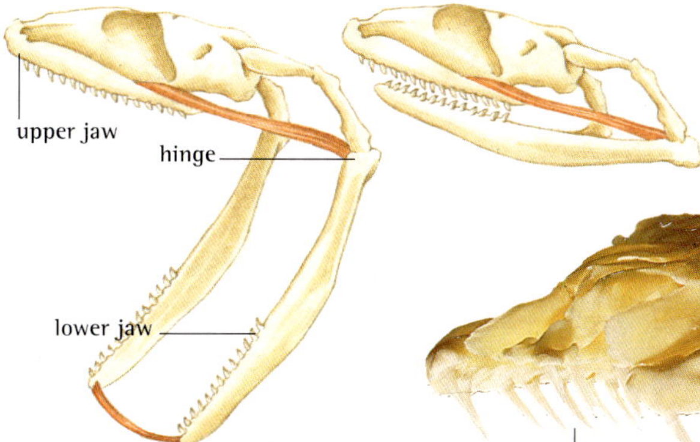

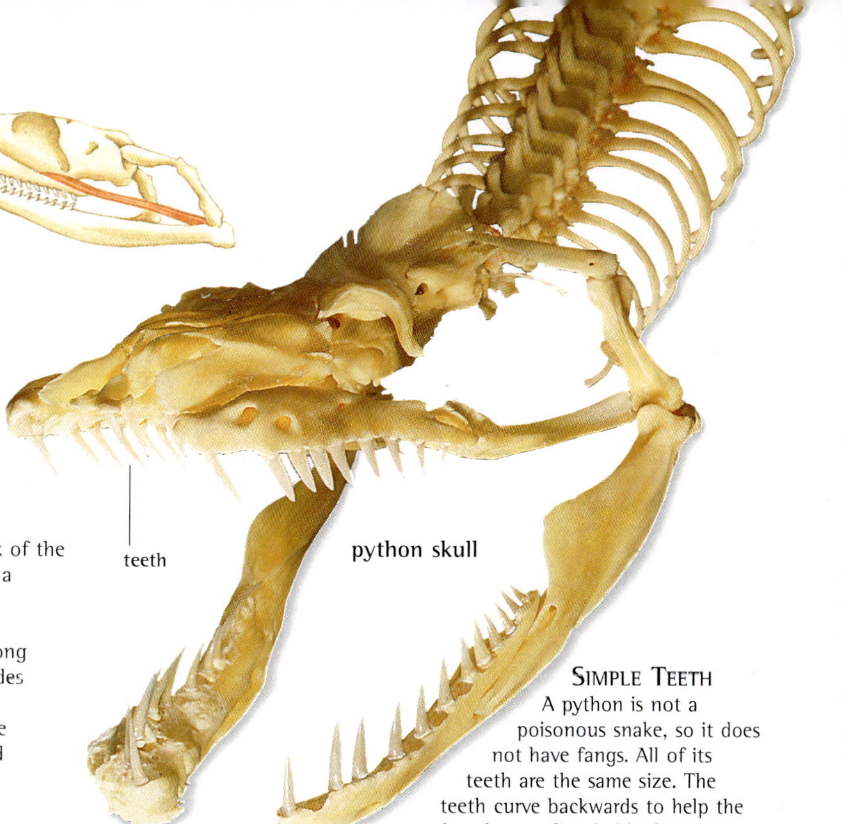

python skull

Stretchy Jaws

When a snake eats, a hinge at the back of the lower jaw lets the jaw swing wide, like a gate. The lower jaw is in two halves connected at the front by a stretchy ligament. This means it can stretch a long way sideways. It also means the two sides of the jaw can move independently of each other. One side can hold on to the prey, while the other side slides forward to get a new grip. The jaws walk from side to side over the food. A fully grown python can swallow prey that is twice as big as its head.

Simple Teeth

A python is not a poisonous snake, so it does not have fangs. All of its teeth are the same size. The teeth curve backwards to help the python keep a firm hold of its prey. The teeth of a snake are attached to the inner edges of the jawbones rather than on top of them. Replacement teeth develop under existing ones.

Did you know? The gaboon viper has the longest fangs of any snake.

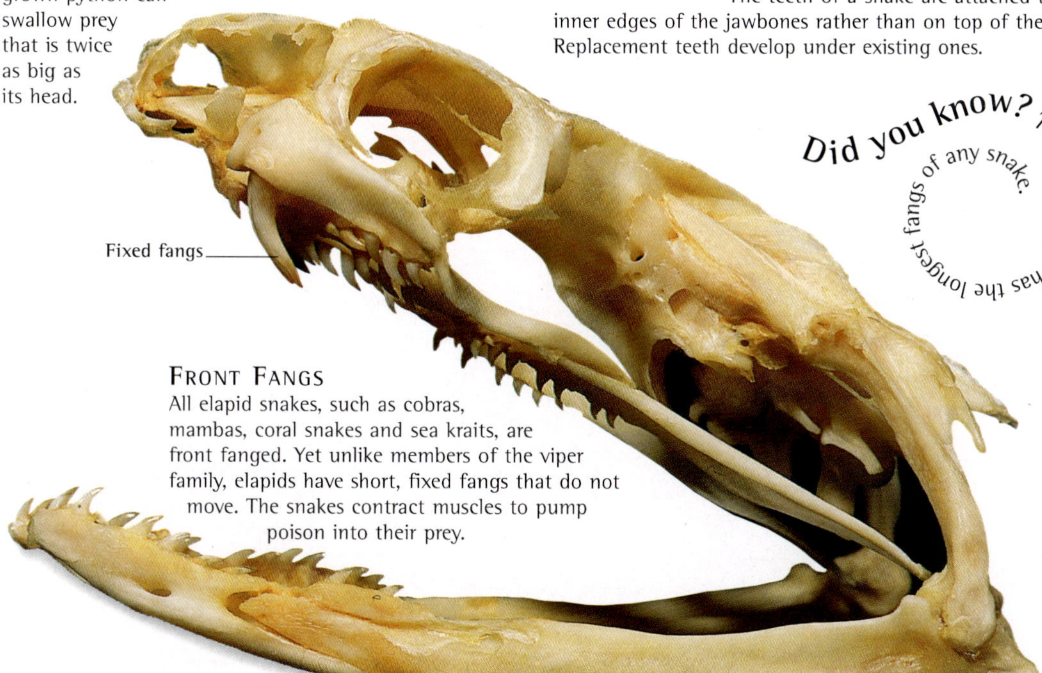

cobra skull

Front Fangs

All elapid snakes, such as cobras, mambas, coral snakes and sea kraits, are front fanged. Yet unlike members of the viper family, elapids have short, fixed fangs that do not move. The snakes contract muscles to pump poison into their prey.

American racer

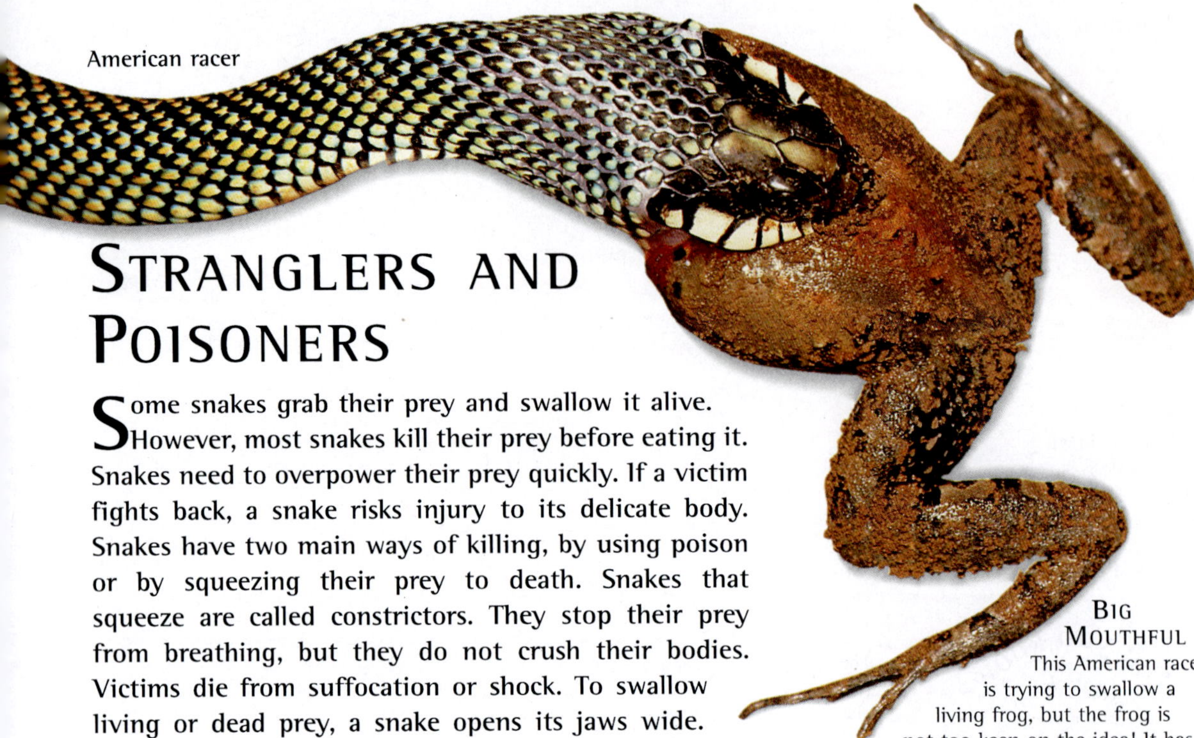

STRANGLERS AND POISONERS

Some snakes grab their prey and swallow it alive. However, most snakes kill their prey before eating it. Snakes need to overpower their prey quickly. If a victim fights back, a snake risks injury to its delicate body. Snakes have two main ways of killing, by using poison or by squeezing their prey to death. Snakes that squeeze are called constrictors. They stop their prey from breathing, but they do not crush their bodies. Victims die from suffocation or shock. To swallow living or dead prey, a snake opens its jaws wide. Lots of slimy saliva helps the meal to slide down. After eating, a snake yawns widely to put its jaws back into place. Digestion may take several days, or even weeks, but happens faster when the snake's body is warm.

BIG MOUTHFUL

This American racer is trying to swallow a living frog, but the frog is not too keen on the idea! It has puffed up its body with air to make it more difficult for the snake to swallow. Sometimes, a snake eats a frog feet first and squeezes the air out through the frog's mouth.

fer-de-lance

AT FULL STRETCH

This fer-de-lance snake is at full stretch to swallow its huge meal – a whole agouti. The fer-de-lance is probably the most dangerous snake in South America. It is a large pit viper that kills with poison.

copperhead

SWALLOWING A MEAL

A copperhead holds on to its dead prey – a mouse. Copperheads are the most widespread poisonous snakes in eastern North America. They often live near homes, but they have weak poison and rarely bite people. As well as mice, they eat frogs, toads and salamanders.

rock python

KILLING TIME

A crocodile is slowly squeezed to death by a rock python. The time it takes for a constricting snake to kill its prey depends on the size of the prey and how strong it is. Small animals may die in just a few seconds. Large prey may struggle for much longer.

Did you know? Pythons can kill and eat leopards.

spotted python

Did you know? King cobras sometimes kill Indian elephants by biting them on the trunk.

COILED KILLER

The spotted python sinks its sharp teeth into its victim. Once it has a firm grip it throws two or more coils around the victim's body. Each time the animal breathes out, the snake tightens its grip. Eventually, the animal cannot breathe in at all.

carpet python

HEAD-FIRST

The hind legs of a whiptail wallaby are the last thing to disappear inside this carpet python's body. Snakes usually try to swallow their prey head-first so that legs, wings or scales fold back. This helps the victim to slide into the snake's stomach more easily.

BREATHING TUBE

An African python displays its breathing tube. Swallowing prey can take a long time. With a mouth full of food, a snake could have a problem breathing. So, the windpipe moves to the front of the mouth. This means that air can get to and from the lungs as the snake eats.

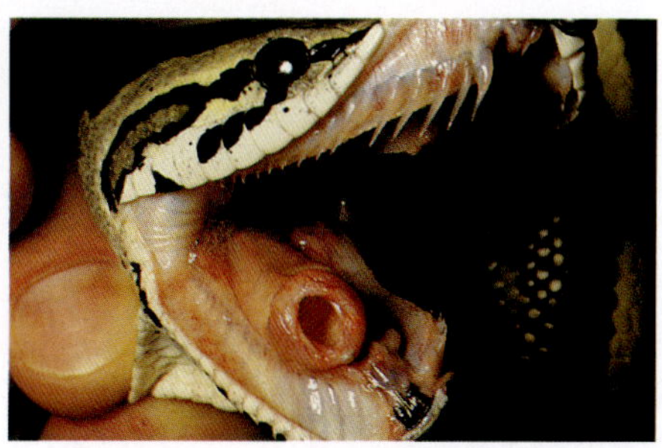

1 Looking for Food
Rat snakes feed on rats, mice, voles, lizards, birds and eggs. Many of them hunt at night. They are good climbers and can even go up trees with smooth bark and no branches. Rat snakes find their prey by following a scent trail or waiting to ambush an animal.

Let's Do Lunch!

This rat snake is using its strong coils to kill a vole for its supper. Rodents, such as voles and rats, are a rat snake's favourite food. After using its senses to find the vole, the snake slides near enough for a quick strike. With the vole held tightly in its sharp teeth, the rat snake coils around its body. It squeezes hard to stop the vole breathing. When the vole is dead, the rat snake opens its jaws very wide and swallows its furry meal head-first.

2 Teeth and Coils
When the rat snake is near enough to its prey, it strikes quickly. Its sharp teeth sink into the victim's body to stop it running or flying away. The snake then loops its coils around the victim as fast as possible, before the animal can bite or scratch to defend itself.

3 A Tight Squeeze
Each time the victim breathes out, the rat snake squeezes a little harder around its rib cage. This stops the animal from breathing in again. Breathing becomes more and more difficult as the rat snake tightens its coils. Soon the prey cannot breathe at all and dies from suffocation.

4 Open Wide

Once the victim is dead, the rat snake stops squeezing. It loosens its coils and begins the slow process of swallowing. The snake may flick its tongue over its meal to work out which end is the head. Then it unhinges its jaws and walks its mouth over its meal. The loose lower jaw stretches sideways to fit around the shape of the dead prey.

5 Stretchy Throat

The meal is swallowed head-first. As the vole moves down the snake's throat, its legs fold back against the sides of its body. The way the fur lies makes the vole easier to swallow. The snake's skin stretches as the meal moves down its body. The scales pull apart, showing the skin in between them.

6 Disappearing Meal

As the vole moves further down inside the snake's body, the skin stretches more and more. The ribs move apart at the front to make space for the vole's body. The snake pushes its windpipe to the front of its mouth, so that it can use it like a snorkel for breathing. It may take only one or two gulps for a snake to swallow a small animal, but it takes an hour or more for larger prey to disappear completely.

Poisonous Snakes

Less than a quarter of all snakes – about 700 species – are poisonous, and only about half of these poisonous species can kill people. Snake poison is called venom. Venom is very useful for snakes because it allows them to kill without having to fight a long battle against their prey. This means that poisonous snakes are less likely to be wounded by bigger animals with sharp teeth and claws. Some snake venom also works on the body of the prey, softening it and making it easier for the snake to digest. There are two main kinds of snake venom. One type attacks the blood and muscles. The other attacks the nervous system, stopping the heart and lungs from working. A poisonous snake stores venom in a bag in its head and injects it into prey through large teeth, called fangs.

Poisonous Bite

A copperhead gets ready to strike. Poisonous snakes use their sharp fangs to inject a lethal cocktail of chemicals into their prey. The death of victims often occurs in seconds or minutes, depending on the size of the prey and where it was bitten.

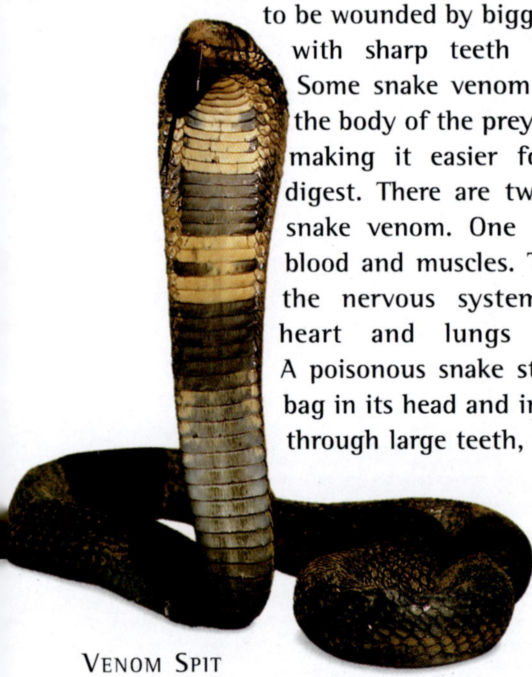

spitting cobra

Venom Spit

Spitting cobras use their venom for attack and defence. They have an opening in their fangs to squirt venom into an enemy's face. They aim at the eyes, and the venom can cause pain and blindness.

Warning Colours

The brightly coloured stripes of coral snakes warn predators that these snakes are very poisonous. Birds and other predators soon learn to leave them alone. There are more than 50 species of coral snake, all with similar patterns. But predators remember the basic pattern and avoid all coral snakes.

coral snake

Did you know? The most poisonous snake in the world is the black-headed sea snake.

Fangs Forward

The copperhead opens its mouth wide and displays its fangs. This snake is a type of viper, so its fangs swing down from the roof of the mouth ready for action. It stabs rather than bites at its prey. The muscles around the venom glands squeeze poison through the fangs and into the wound.

copperhead

Bible Snake

At the beginning of the Bible, a snake is the cause of problems in the Garden of Eden. God told Adam and Eve never to eat fruit from the tree of knowledge of good and evil. However, the snake persuaded Eve to eat the fruit. It told Eve that the fruit would make her as clever as God. Eve gave some fruit to Adam too. As a punishment, Adam and Eve had to leave the garden of Eden and lose the gift of eternal life.

black mamba

Milking Venom

Venom is collected by making a black mamba bite through the top of a jar. The venom is used to make a medicine called antivenin. It can stop people dying from snake bites.

green bush viper

VIPERS

From adders and asps to rattlesnakes and copperheads, vipers are the most efficient poisonous snakes in the world. Their long fangs can inject venom deep into a victim. The venom acts mainly on the blood and muscles of the prey. Vipers usually have short, thick bodies and triangular heads covered with many small scales. All the scales are ridged. There are two main groups of vipers. Typical vipers and adders live in Europe, Asia and Africa. Pit vipers, such as rattlesnakes, live in the Americas and southern Asia. The main difference between the two groups is that pit vipers have large heat pits on the face and other vipers do not. Most vipers give birth to fully developed, or live, young.

TREE VIPER

The green bush viper lives in tropical forests, mainly in the trees. Its colouring means that it is well camouflaged against the green leaves. It lies in wait for its prey and then kills it with a quick bite. Once the prey has been caught, the snake must hold tight to stop it falling out of the tree.

puff adder

BALLOON SNAKE

When threatened, the puff adder swells up like a long balloon. It does this by taking a lot of air into its lungs. Being larger makes it look more dangerous. Many of the true vipers can swell up in this way. Puff adders also hiss with a loud sound that can be heard some distance away.

rattlesnake

Fearsome Fangs

This rattlesnake is exploring its surroundings with its forked tongue. When the rattlesnake strikes at its prey, its mouth opens wide. The long, hinged fangs swing forward and lock into place. The viper stabs its prey to give it a quick injection of powerful venom, then lets go. The prey soon dies, so there is no need for the snake to hold on to it. If the prey does manage to move away, the snake can easily track it down.

Heat Detectors

This beautiful Sumatran pit viper has a large heat pit on each side of its head, between the nostril and the eye. The heat pit is larger than the nostril. It can detect the heat given off by warm-blooded prey. By turning the head from side to side, a pit viper can work out the direction of its prey. When both pits receive the same amount of warmth, the prey is straight ahead. Pit vipers can also work out how far away their prey is, which helps them to strike accurately.

Slow snake

Asp vipers are slow-moving snakes. They are active both by day and by night. Their main sources of food are mice, lizards and baby birds in their nests. The asp viper does not usually kill people, but its bite is painful.

Sumatran pit viper

DEFENCE

Most snakes, especially baby snakes and snake eggs, are eaten by a wide range of animals. Snakes provide food for birds of prey, such as eagles. They are also hunted by predators such as foxes, racoons, mongooses, baboons, crocodiles, frogs and even other snakes. If they are in danger, snakes usually prefer to hide or escape. Many come out to hunt at night, when it is more difficult for predators to catch them. If they cannot escape, snakes often make themselves look big and fierce, hiss loudly or strike at their enemies. Some pretend to be dead. Giving off a horrible smell is another good way of getting rid of an enemy! Many snakes will twist into a ball or figure of eight, smearing the strong-smelling substance over their body. Poisonous snakes defend themselves with their venom, but they try to avoid fighting in case they get hurt.

SMELLY SNAKE

The poisonous cottonmouth is named after the white colour of the inside of its mouth. It opens its mouth to threaten enemies. If a cottonmouth is attacked, it can also give off a strong-smelling liquid from near the tail. By swishing its tail from side to side, the cottonmouth can spray this smelly liquid up to 1m away.

EAGLE ENEMY

The short-toed eagle uses its powerful toes to catch snakes on open hillsides. It tears apart a large snake with its sharp beak and eats it on the ground. It can carry small snakes back to the nest to feed its chicks.

SCARY MOUTH

Like many snakes, this vine snake opens its mouth very wide to startle predators. The inside of the mouth is a bright red colour. If the predator does not go away, the snake will give a poisonous bite with the fangs at the back of the mouth. The open mouth is a warning message saying 'Go away or I will bite you!'

vine snake

cottonmouth

hognose snake

Playing Dead

This grass snake knows that most predators prefer healthy, living prey. So it protects itself by pretending to be dead. Many other snakes also do this. They roll on to their backs, open their mouths and keep quite still. When the danger has passed, they come back to life again.

grass snake

Dramatic Display

The hognose snake of North America is harmless, but makes itself look dangerous in a dramatic display. First, it flattens its neck to make a hood. Then it hisses loudly and strikes towards the enemy. If that does not work, the hognose wriggles and smears itself with smelly scent. Finally, it rolls over and pretends to be dead.

Hidden Snake

The best defence for many snakes is to avoid being seen at all. The horned viper buries itelf. Its scales keep the sand from piling up and covering the eyes when the snake is buried.

horned viper

cobra

Looking Larger

This cobra is rearing up and spreading its hood wide to make itself look as big and frightening as possible. This helps to scare away predators and also makes the cobra look too big for other animals to swallow.

Did you know? In some cultures cobra venom is used as a pain reliever.

Hood Variety

Like spitting cobras, the king cobra and the water cobra, this Egyptian cobra has a narrow hood. The Indian cobra and the Cape cobra of Southern Africa have much wider hoods. The Egyptian cobra ranges over much of Africa and into Arabia.

The Hood

The hood of this Indian cobra is made from flaps of skin supported by long ribs. Most of the time the skin rests flat against the body and the cobra looks much like any other snake. But when it is alarmed the cobra spreads its neck ribs, stretching and tightening the neck skin to form a wide hood.

Hood Patterns

Some cobras have eyespots, horseshoes or bands on the back of their hoods to make them look more frightening to predators. The Indian cobra has a particularly wide hood with large eyespot markings on the back.

The Cobra and its Relatives

Cobras are very dramatic and very poisonous snakes. They have short, fixed fangs at the front of their mouths. Some cobras can squirt their deadly venom at their enemies. Cobra venom works mainly on the nervous system, causing breathing or heart problems. Cobras are members of the elapid snake family, which includes the mambas of Africa, the coral snakes of the Americas and all the poisonous snakes of Australia, such as tiger snakes.

Large Cobra

The king cobra is the largest of all venomous snakes, growing to a length of 5.5 m. It is shy and prefers to keep well away from people. King cobras are the only snakes known to build a nest. The female lays 20 to 40 eggs and guards the eggs and hatchlings until they leave the nest.

Mind the Mamba

The green mamba lives in trees. Other mambas, such as the black mamba, live mostly on the ground. Mambas are slim, long snakes that can grow up to 4m long. Their venom is very powerful and can kill a person in only ten minutes!

Do Not Disturb!

The Australian mainland tiger snake is the fourth most poisonous snake in the world. It has a powerful body and smooth, shiny scales. If the tiger snake is disturbed, it puffs up its body, flattens its neck and hisses loudly. Australian tiger snakes eat almost anything, from fish and frogs to birds and small mammals.

rainbow boa

Changing Colours

The rainbow boa is named after its iridescent colours, which change each time the snake moves. Light is made up of all the colours of the rainbow. When light hits the thin, see-through, outer layer of the snake's scales, it is split up into the different colours. The colours we see depend on the type of scales and the way light bounces off the scales.

Colour and Camouflage

milk snake

The colours of snakes come from the pigments in the scales and from the way light reflects off the scales. Dull colours and patterns help to camouflage a snake and hide it from its enemies. The patches of colour on a snake's skin are often arranged in such a mixed-up way that the overall shape of the snake is hard to pick out. Bright colours startle and confuse predators or warn them that a snake is poisonous. Harmless snakes sometimes copy the warning colours of poisonous snakes. Dark colours may help snakes to absorb heat during cooler weather. Some female snakes become darker when they are pregnant. The extra warmth they soak up speeds up the development of the young inside their bodies. Young snakes are sometimes a different colour from their parents, but no one knows the exact reason for this.

Did you know? Milk snakes always have black bands between the red and yellow – coral snakes have the red and yellow touching.

ring-necked snake

Startling Colours

If ring-necked snakes are threatened, they show off their brightly coloured undersides to startle predators. The red tail draws attention away from the head – the most important part of the snake.

Colour Copies

The bright red, yellow and black bands of this milk snake copy the colouring of the poisonous coral snake. The milk snake is not poisonous, but predators leave it alone – just in case! This colouring is found in milk snakes in the southeast of the USA. In the northeast of the USA, milk snakes are greyish with rusty blotches. There are no coral snakes in this region, so there is no point copying their warning colours.

No Colour

White snakes, with no colour at all, are called albinos. People have bred some albino and other very pale snakes in captivity, out of interest. In the wild, these snakes stand out against background colours and are usually killed by predators before they can reproduce.

Snake Markings

Like the red-tailed boa, many snakes have markings made up of patches of different colours or shades, as well as spots and stripes. These markings are usually caused by groups of different pigments in the scales.

red-tailed boa

Clever Camouflage

Among the dead leaves of the rainforest floor, the gaboon viper becomes almost invisible. Many snakes have colours and patterns that match their surroundings. They usually have markings that break up their outline, so it is difficult to see where the snake ends and the background begins.

Reproduction

Snakes do not live as families, and parents do not look after their young. Males and females come together to mate, and pairs may stay together for the breeding season. Most snakes are ready to mate when they are between two and five years old. In cooler climates, snakes usually mate in spring so that their young have time to feed and grow before the winter starts. Some snakes, such as garter snakes, mate after they come out of hibernation in spring, when there are a lot of snakes around. In tropical climates, snakes often mate before the rainy season, when there is plenty of food for their young. Male and female snakes usually look alike, although females are often larger than males. Male snakes find females by following their scent trails. Sometimes the males compete in a test of strength to decide who will mate with the females.

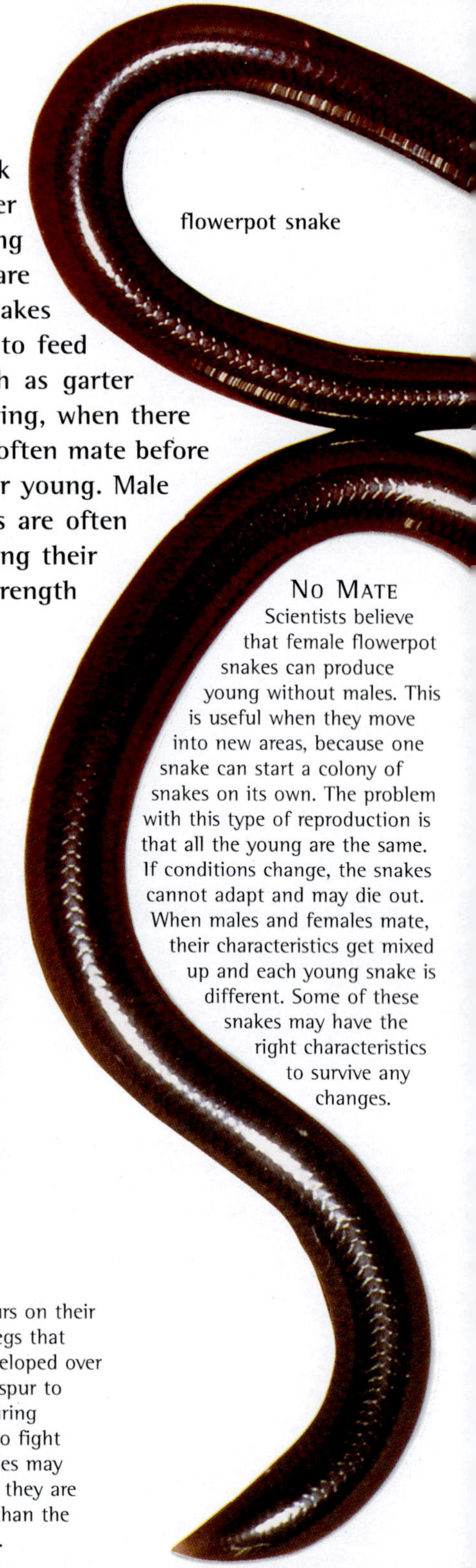
flowerpot snake

No Mate

Scientists believe that female flowerpot snakes can produce young without males. This is useful when they move into new areas, because one snake can start a colony of snakes on its own. The problem with this type of reproduction is that all the young are the same. If conditions change, the snakes cannot adapt and may die out. When males and females mate, their characteristics get mixed up and each young snake is different. Some of these snakes may have the right characteristics to survive any changes.

Fighting for Females

Rival male adders fight to test which one is the stronger. Rattlesnakes and adders often fight like this. The two males rear up and face each other, then twist their necks together. Each snake tries to push the other to the ground. In the end, one of the males gives up and crawls away.

adders

spur

Snake Spurs

Both boa and python males have small spurs on their bodies. These are the remains of back legs that have disappeared as snakes have developed over millions of years. A male uses its spur to scratch or tickle the female during courtship, and sometimes to fight with other males. Females may also have spurs, but they are usually smaller than the males' spurs.

Did you know? The female Javan wart snake can store sperm in her body for up to 7 years.

Wrestling Match

These male Indian rat snakes are fighting to see which is the stronger. The winner will stand a better chance of mating with the females. The fight is rather like people in a wrestling match. The snakes hiss and strike out, but they seldom get hurt in these tests of strength.

Did you know? Some male cobras have longer fangs than the female.

Similarities and Differences

No one knows why the male and female of the snake shown here have such different head shapes. In fact, male and female snakes of the same species usually look similar because snakes rely on scent rather than sight to find a mate. However, differences between males and females can include eye colour, length of fangs and head shape as seen here.

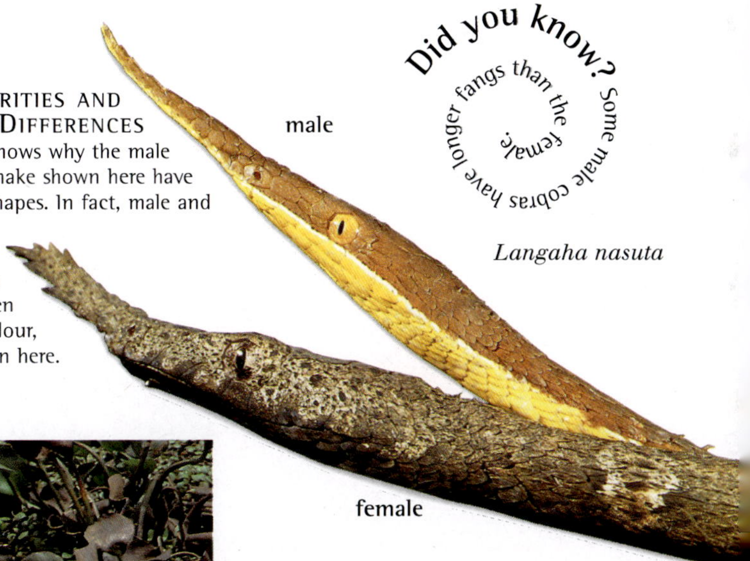

Langaha nasuta

anacondas

Mating

Two anacondas are in the act of mating. When a female is ready to mate, she allows the male to lift her tail and coil his tail around hers. The male has to place his sperm inside the female's body to fertilize her eggs. The eggs can then develop into baby snakes. Mating lasts from a few minutes to several hours, depending on the species of the snakes.

Eggs

Some snakes lay eggs and some give birth to fully developed, or live, young. Egg-laying snakes include rat snakes, milk snakes, cobras, pythons and hognose snakes. A few weeks after mating, a female egg-laying snake looks for a safe, warm, moist place to lay her eggs. This may be under a rotting log, in sandy soil, under a rock or in a compost heap. The king cobra is the only snake that builds a nest. The average number of eggs laid at one time is between 6 and 30. Most snakes cover their eggs and leave them to hatch by themselves. A few snakes, such as the bushmaster, some cobras and most pythons, stay with their eggs to protect them from predators and the weather. However, once the eggs hatch, all snakes abandon their young and leave them to look after themselves.

BEACH BIRTH
A sea krait moves across the sand near an unsuspecting holidaymaker. Sea kraits are the only sea snakes to lay eggs. They often do this in caves near the sea, above the level of the salt water.

EGG CARE
This female python has piled her eggs into a pyramid shape and coiled herself around them. She does this to protect the eggs from predators. All female pythons protect their eggs. The female Indian python even twitches her muscles – rather like shivering – to warm up her body. The extra heat helps the young to develop. Snake eggs need to be kept at a certain temperature if they are to develop properly.

LAYING EGGS
The Oenpellis python lays several small eggs that have a rounded shape. Smaller snakes than pythons tend to lay fewer eggs at a time. The eggs are usually long and thin to fit inside the small body of the snake. Some snakes lay long, thin eggs when they are young, but more rounded eggs when they grow larger.

Did you know? The mud snake lays over 100 eggs at a time.

Children's Python Mass Hatching

As they hatch, these children's pythons dent and flatten their egg shells. The shell of a snake's egg is tough and leathery, not hard and brittle like the shell of a bird's egg. Birds' eggs would break into pieces if they were squashed like this. A snake's egg is not as watertight as a bird's egg, so it has to be laid in a moist place to stop it drying up.

children's pythons

Did you know? Some snakes eat birds' eggs.

Where Eggs are Laid

Most snake eggs are white in colour. They have to be hidden in the soil, or under rocks and logs so that predators will not spot them. However, female snakes never completely bury their eggs, because the baby snakes need to breathe air that flows through the outer shell.

rat snake

Hot Spots

This female grass snake has laid her eggs in a pile of rotting plants. Both grass snakes and rat snakes do this. They often use compost and manure heaps. The heat given off by these heaps speeds up the development of the eggs.

Hatching Out

About two to four months after the adult snakes mate, the baby snakes hatch out of their eggs. Inside the egg, the baby snake feeds on the yolk, which is full of goodness. Once the snake has fully developed and the yolk has been used up, the snake is ready to hatch. Development depends on the temperature. A baby snake develops more quickly in warm climates than in colder places. All the eggs in a clutch tend to hatch at the same time. A baby snake makes slits in its shell with a sharp egg tooth. This egg tooth drops off a few hours after hatching. A few days later, the baby snake wriggles away to start a life on its own. It has to survive without any help from its parents.

1 Snake Eggs
Eight weeks after being laid, these rat snake eggs are beginning to hatch. While they were developing inside the egg, each baby rat snake fed on its yolk. A day or so before hatching, the yolk sac was drawn inside the snake's body. A small scar, rather like a tummy button, will show where the baby snake was once joined to the yolk.

2 Starting to Hatch
The baby snake has become restless, twisting inside its shell. It is now fully developed and cannot get enough oxygen through its shell. A snake's egg has an almost watertight shell, but water and gases, such as oxygen, pass in and out of it through tiny holes (pores). As the baby snake prepares to hatch, it cuts a slit in the shell with a sharp egg tooth on its snout. This egg tooth will drop off a few hours after hatching.

3 Having a Rest

After it has broken through the stretchy shell, the baby snake has a rest. It pokes its nose through the slit in the egg to breathe the air and take a first look at the strange and exciting world outside.

4 Waiting Until it is Safe

All the eggs in this clutch have hatched at the same time (a clutch is a set of eggs laid by a snake). After making the first slits in their leathery shells, the baby snakes will not crawl out straight away. They poke their heads out of their eggs to taste the air with their forked tongues. If they are disturbed, they will slide back inside the shell where they feel safe. They may stay inside the shell for a few days.

Did you know? Some snakes lay as many as 100 eggs in one clutch.

5 Breaking Free

Eventually, the baby snake slithers out of the egg. It may be as much as seven times longer than the egg because it was coiled up inside.

Pope's tree viper

TREE BIRTH
Like most vipers, this Pope's tree viper gives birth to live young. Tree snakes often give birth high up among the branches. The clear membrane around the baby snakes sticks to the leaves and helps to prevent the baby snakes from falling out of the branches to the ground.

GIVING BIRTH

Some snakes give birth to fully developed or live young instead of laying eggs. Snakes that do this include boas, rattlesnakes, garter snakes, adders and most sea snakes. The eggs develop inside the mother's body surrounded by see-through bags, called membranes, instead of shells. While the baby snake is developing inside the mother, it gets its food from the yolk of the egg. In a few snakes, the babies take food from their mother and also pass wastes into her blood. The babies are warm and protected inside the mother, but their weight makes the mother snake heavy and slow-moving. The babies are born after a labour that may last for hours. Anything from 6 to 50 babies are born at a time. At birth, they are still inside their membranes.

BIRTH PLACE
This female sand viper has chosen a quiet, remote spot to give birth to her young. Snakes usually give birth in a safe, hidden place. They hide in underground tunnels, among rocks or logs, or in between tree roots. There, the young are safe from enemies. Water snakes and sea snakes always give birth in the water.

BABY BAGS
These red-tailed boas have just been born. They are still inside their see-through bags. The bags are made of a clear, thin, tough membrane, rather like the one inside the shell of a hen's egg.

Did you know? Newborn anacondas are only 6cm long.

Breaking Free

This baby rainbow boa has just pushed its head through its surrounding membrane. Snakes have to break free of their baby bags on their own. The mother does not help the babies to escape. Each baby has an egg tooth to cut a slit in the membrane and wriggle out. The babies usually do this a few seconds after birth.

Did you know? Baby boa Constrictors are 30cm long when they are born.

New Baby

A red-tailed boa has just succeeded in breaking free of its baby bag, or egg sac, which is in the front of the picture. The baby's colours are bright and shiny. Some newborn babies crawl off straight away, while others stay with their mother for a few days.

Did you know? Timber rattlesnake mothers defend their newborn babies for a few days.

Colour Change

Believe it or not, this baby is an emerald tree boa. When young, these boas are a vivid red colour. As they grow up, they turn green. Although boas and pythons are very similar snakes in some ways, one of the principal differences between them is that boas give birth to live young while pythons lay eggs. Mother boas do not look after their young.

young emerald tree boa

Growth and Development

The rate at which snakes grow depends on many different things. The size when baby snakes are born or when they hatch from their eggs, how much they eat and the climate around them all affect their development. In warm climates, snakes may double or triple their length in just one year. Some snakes are mature and almost fully grown after three to five years, but slow growth may continue throughout their lives. Young snakes shed their skin more often than adults because they are growing quickly. While they are growing and developing, young snakes are easy prey for animals such as birds, racoons, toads and rats. Many rely on their camouflage colours to hide them from enemies, although poisonous snakes can defend themselves with their venom from the moment they are born.

Fast Food

Like all young snakes, this Burmese python must eat as much as possible in order to grow quickly. Young snakes eat smaller prey than their parents, such as ants, earthworms, flies and grasshoppers.

Deadly Baby

This baby European adder may look tiny and harmless, but it is not! Poisonous snakes hatch out with their venom glands in full working order. They may be small, but they can give a dangerous bite soon after hatching. Luckily, the venom of the European adder is not very strong and rarely kills people.

mother European adder

baby European adder

Mother and Baby

Female European adders give birth towards the end of the summer. The young adders need to grow quickly so that they will be big enough to survive the winter in hibernation. Growth stops during the hibernation period.

Burmese python

Did you know? Snakes can live for as long as 30 or 40 years.

Heracles the Strong

Heracles (or Hercules) was the son of Zeus, king of the ancient Greek gods. His mother was Alcmene, an ordinary human being. Zeus's wife, Hera, was jealous of Alcmene's baby. She sent two poisonous snakes to kill Heracles as he slept. But Heracles was strong and killed the snakes by strangling them with his bare hands.

Diet Change

Many young Amazon tree boas live on islands in the West Indies. They change their diet as they grow up. Young Amazon tree boas start by feeding on lizards but, later on, they switch to feeding on birds and mammals.

Did you know? Female snakes reach maturity later than male snakes.

Rattle Age

It is not possible to work out the age of a rattlesnake by counting the sections of its rattle. This is because several sections may be added each year and pieces of the rattle may break off.

Where Snakes Live

Snakes live on every continent except Antarctica. They are found in most types of habitat, but are most common in deserts and rainforests. Snakes live in the fresh water of rivers, ponds and streams as well as the salty water of the sea. They cannot survive in very cold places, such as the icy polar lands or very high mountains, because they use the heat around them to make their bodies work. Snakes become slow and sluggish when it gets cold and they may die if the temperature drops too low. This is why most snakes live in warm places where the temperature is high enough for them to stay active day and night. In cooler places, snakes may sleep through the colder winter months.

grass snake

Grasslands

The European grass snake is one of the few snakes to live on grasslands. There is not much food for snakes in grasslands and not many places to hide from predators, either.

Mountains

The Pacific rattlesnake is sometimes found in the mountains of the western USA, often on the lower slopes covered with loose rock. In general, though, mountains are problem places for snakes because of their cold climates. However, some snakes have adapted to mountain life. Vipers are the most common mountain snakes.

Pacific rattlesnake

Winter Sleep

Thousands of garter snakes emerge after their winter sleep. In colder parts of the world, many snakes spend the winter asleep. This is called hibernation. Snakes often hibernate in caves or underground, where it is warmer. In spring, the snakes slowly come back to life.

Living in Trees

The eyelash viper lives in the Central American rainforest. The climate in rainforests is warm all year round, so snakes can stay active all the time. There are also plenty of places to live – in trees, on the forest floor, in soil and in rivers.

Tropical Rainforests

The greatest variety of snakes lives in tropical rainforests, including the Brazilian rainbow boa shown here. There is plenty to eat, from insects, birds and bats to frogs. Many rainforest snakes feed on smaller snakes and some have specialised diets, for example feeding only on snails.

Did you know? The Himalayan pit viper lives at an altitude of 4,900m.

Brazilian rainbow boa

Burrowing Snakes

The yellow-headed worm snake lives under the bark of a tree. Many worm, or thread, snakes spend most of their lives under the ground, where the soil is warm enough for them to stay active.

Deserts

This African puff adder lives in the Kalahari desert of southern Africa. Large numbers of snakes live in deserts because they can survive with little food and water. Desert snakes have to cope with huge temperature changes, from boiling hot days to freezing cold nights.

Tree Snakes

With their long, thin, flat bodies and pointed heads, tree snakes slide easily through the branches of tropical forests. Some can even glide from tree to tree. Most tree snakes are lightweight. Tree boas may look heavy, but they weigh less than their relatives on the ground. Tree boas and pythons have ridges on their belly scales to give them extra gripping power. Many tree snakes also have long, thin tails that coil tightly around branches. To catch prey, tree snakes use their sharp teeth or poisonous fangs. Green or brown camouflage colours keep tree snakes well hidden among the leaves and branches. They can hide from predators or lie in wait for their prey without being seen.

Amazon tree boa

Colour and Pattern

This Amazon tree boa is coloured and patterned for camouflage. Many tree snakes are green or brown with patterns that break up the outline of their body shape. Some even have patterns that look like mosses and lichens.

green tree python

Tree Twins

The green tree python lives in the rainforests of New Guinea. It looks very similar to the emerald tree boa and even behaves in a similar way. But these snakes are not closely related. The emerald tree boa lives thousands of kilometres away in South America. The two snakes are similar because they both live in rainforests and both survive in the same way.

Catching Prey

High in the rainforests of Costa Rica, a blunt-headed tree snake has caught a lizard. Tree snakes have to grasp their prey firmly so that it does not fall out of the tree. They usually have long, sharp teeth, which are very good at piercing skin.

blunt-headed tree snake

long-nosed whip snake

Did you know? Many snakes have long tails to grip tree trunks and branches.

Heads and Eyes

The long-nosed whip snake opens its brightly coloured mouth to scare away a predator. Like most tree snakes, it has a long head, with a pointed snout. This shape is ideal for sliding through branches. Tree snakes usually have large eyes for night-time hunting.

eyelash viper

Viper Reflexes

The green eyelash viper, with its mossy camouflage, has such speedy reflexes that it can catch unsuspecting birds as they fly through the trees. This viper is poisonous, but it has to hold on to its prey while its deadly venom takes effect.

Cook's tree boa

Body Shape

A Cook's tree boa winds its long body around the branch of a tree. Tree snakes have long, thin bodies that are light in weight. This helps them to crawl along small branches without breaking them. They can also stretch easily from one branch to another.

Emerald Tree Boa

With their bright green coils looped around branches, emerald tree boas lurk among the leaves in the rainforests of South America. The creamy coloured bands along the back of the snake break up its outline. These tree boas are good climbers, hanging head-first from branches to seize fast-moving prey in their sharp teeth. To rest, they lie with their coils encircling a narrow branch, and their head lying on top.

Upside-Down Meals

To catch a meal, emerald tree boas drape their coils over a horizontal branch and hang their heads down. Once the snake has a firm hold on its prey with its teeth, it coils around its victim. It slowly squeezes with its coils to stop the animal breathing. When the animal is dead, the emerald tree boa swallows it head-first, so that it slides down easily.

Climbing Skills

Emerald tree boas are longer and more slender than boas that live on the ground. This helps them to slide easily through the leafy branches.

Long Tail

The emerald tree boa's long tail is strongly prehensile, which means it grips and holds the branches tightly. As an emerald tree boa climbs, it reaches up with its front end and coils itself around a branch, then pulls up the rest of its stretchy body.

Colour Change

Emerald tree boas mate while they are coiled around branches. They give birth to live young. The young snakes are orange, pink or yellow when they are born. They gradually change to a green colour in their first year by producing new colour pigments in their skin. No one is quite sure why the young snakes are a different colour from the adults. They may live in different places in the forest from the adults and so need a different colour for effective camouflage.

Hot Lips

All along their lips, emerald tree boas have large pits between the scales. These pits can sense the heat given off by prey animals. From these heat messages, the snake can tell exactly where its prey is.

Lethal Teeth

The emerald tree boa has long, sharp, backwards-pointing teeth. The teeth help the snake to grab hold of its prey and stop it escaping. The snake can also open its mouth very wide to fit more of its prey inside. This is why the snake can feed on animals that move quickly, such as birds.

Desert Snakes

Deserts are full of snakes. This is partly because snakes can survive for a long time without food. They don't need to use energy from their food to produce body heat because they get heat energy from their surroundings. It is also because their waterproof skins stop them losing too much water. A snake can push its long thin body between rocks or down rodent burrows to escape the fierce heat of the Sun and the bitter cold of the night. Using their special senses, such as heat pits and forked tongues, snakes can find prey at dawn or dusk, when it is cooler. Some snakes rest quietly through very hot, dry periods. They burrow deep underground, where the sand is cooler than at the surface. Then they stay there until the worst of the heat is over. This is called aestivation.

horned viper

Scale Sounds

If threatened, the desert horned viper makes a loud rasping sound by rubbing together jagged scales along the sides of its body. This warns predators to keep away. If the viper hissed a warning, it would lose water into the air. Water is scarce in a desert, so a snake must keep as much water inside its body as possible.

Did you know? Most rattlesnakes live in the Southwest and Mexico.

rattlesnake

Tail Rattle

Rattlesnakes shake their rattle to warn enemies to keep away. The rattlesnake raises its tail to shake it and often lifts its head off the ground. The rattlesnake cannot hear the buzzing noise it makes – but its enemies can.

sand boa

Sand Shuffle

The desert horned viper shuffles under the sand by rocking its body to and fro. It spreads its ribs to flatten the body and pushes its way down until it almost disappears. Only the eyes and horns show above the surface. The snake suddenly strikes out at its prey from this hidden position.

Peringuey's viper

Sidewinding

A Peringuey's viper leaves behind a line of bar-shaped tracks as it moves across the desert sand. Many desert snakes travel in this way. It is called sidewinding. As the snake moves, only a small part of its body touches the hot sand at any one time. Sidewinding also helps to stop the snake sinking down into the loose sand.

Hidden Boa

The brown and orange colours of this sand boa make it hard to spot amongst the desert rocks and sand. This helps it to hide from predators and sneak up on its prey. Sand boas have a long, round body. Their body shape helps them to burrow down into the sand.

The Hopi Indians

This Native North American was a Hopi snake chief. The Hopi people used living snakes in their rain dances. They hoped the snakes would carry their prayers to the rain gods to make rain fall on their desert lands.

Water Snakes

Some snakes live in marshy areas or at the edge of freshwater lakes and rivers, where there is plenty of food. Two groups of snakes live in the salty water of the sea. They breathe air, but they can stay underwater for a long time. Glands on their heads get rid of some of the salt from the water. Most sea snakes give birth to their young in the water, while some have to come on to land to lay eggs. Sea snakes tend to swim in the shallow seas near the coast. They have hollow front fangs and are very poisonous. This is because a sea snake has to subdue its prey quickly in order to avoid losing it in the depths of the sea. We know very little about how sea snakes live.

sea snake

Sea Snake Senses

A sea snake's eyes and nostrils are positioned towards the top of the head. This means it can take a breath without lifting its head right out of the water, and the eyes can watch out for predators about to attack from above.

Champion Swimmer

Northern water snakes are good swimmers, rarely found far from fresh water. They feed mainly on fish, frogs, salamanders and toads. At the first sign of danger, they dive under the water. Northern water snakes often bite when they are captured. Luckily, they are not poisonous.

Northern water snake

Breath Control

Most sea snakes never leave the water. When a sea snake is underwater, it can close off its nostrils to stop the water flooding in. It has a large lung that allows it to stay underwater for a few hours at a time. Only then does it have to come to the surface to breathe.

anaconda

Heavy Weight

The green anaconda lurks in swamps and slow-moving rivers, waiting for birds, turtles and caimans to come within reach of its strong coils. This one has caught a bird called an ibis. Green anacondas are the heaviest snakes in the world. They weigh up to 227 kg!

Did you know? Anacondas sometimes eat people!

sea krait

Half and Half

This sea krait is crossing a sandy beach. Sea kraits are less well adapted to life in the sea than other sea snakes. Instead of giving birth in the water, they lay their eggs on land. All sea kraits have black or dark brown bands and feed on fish, particularly eels.

Snake Families

Scientists have divided the 2,700 different kinds of snake into about ten groups, called families. These are the colubrids, the elapids, the vipers, the boas and pythons, the sea snakes, the sunbeam snakes, the blind snakes and worm snakes, the thread snakes, the shieldtail snakes and the false coral snake. The snakes in each family have features in common. The biggest family is the colubrid family, which has over 1,800 different species of snake. Each species of snake has a Latin name that is used by scientists all over the world. This name can be used no matter what language a person speaks. A snake may have several common names, which can be confusing. But it only ever has one Latin name.

Indian cobra

Did you know? The viper family includes rattlesnakes, adders, asps and pit vipers.

Elapids

Elapids are poisonous snakes that live in hot countries around the world. They include cobras in Asia and Africa, mambas in Africa and coral snakes in the Americas. All the poisonous snakes of Australia are elapids, such as the tiger snake, king brown snake and the taipan. Elapids have short, fixed fangs at the front of their mouths.

milk snake

Colubrids

About three-quarters of all the snakes in the world, including this milk snake, belong to the colubrid family. They come in almost every shape, size or colour. Most colubrids are not poisonous. Colubrids have no left lung or hip bones, and their heads are covered with large, symmetrical scales.

sand viper

Vipers

Snakes in this family, such as the sand viper, have long, hollow fangs that can be folded back inside the mouth when they are not needed. Vipers are usually short, with thick bodies and wide heads. Their scales have heavy ridges, or keels.

Classification Chart

Kingdom	Animalia
Phylum	Chordata
Class	Reptilia
Order	Squamata
Suborder	Ophidia
Family	Boidae
Genus	Boa
Species	Boa constrictor

This chart shows how a boa constrictor is classified within the animal kingdom.

Colombian rainbow boa

Boas and Pythons

This family includes snakes that kill by constriction rather than poisoning. They have curved teeth, hip bones and tiny back leg bones. Most boas live in Central and South America, whereas pythons live in Africa, Southeast Asia and Australia.

Sea Snakes

The family of sea snakes includes some snakes that are born in the sea and spend all their lives there, and others that spend part of their time on land. Sea snakes have flattened tails for swimming and nostrils that can be closed off under the water. Most live in warm waters, from the Red Sea to New Zealand and Japan. No sea snakes live in the Atlantic Ocean. Sea snakes are sometimes grouped together with the elapid family.

sea snake

sunbeam snake

Sunbeam Snakes

The sunbeam snake family has only two members. The name of the sunbeam snake comes from the way its scales reflect the light. Sunbeam snakes are burrowing snakes that live in Southeast Asia and southern China. Unlike most other snakes, they have two working lungs, although the left lung is about half the size of the right lung.

Did you know? Some scientists put pythons and boas into different families.

sand lizard

Snake Relatives

Snakes are part of a large group of animals called reptiles. There are about 6,000 different kinds of reptiles alive today, nearly half of which are snakes. Other reptiles include lizards, crocodiles and alligators, turtles and tortoises. Reptiles have bony skeletons with a backbone and bodies covered in scales. They lay eggs with waterproof shells or give birth to live young. Young reptiles look like tiny copies of their parents. Reptiles are cold-blooded and rely on their surroundings for heat, so they live mostly in warm places. Snakes appeared on the Earth more recently than other reptiles. The first snakes lived between 100 and 150 million years ago, alongside the most famous reptiles of all – the dinosaurs.

Lizards

This sand lizard is threatening an enemy by making itself look big and frightening. Lizards usually have four legs, feet with sharp claws and a long tail. They also have movable eyelids and good eyesight. Most lizards have pointed tongues, but monitor lizards have a forked tongue.

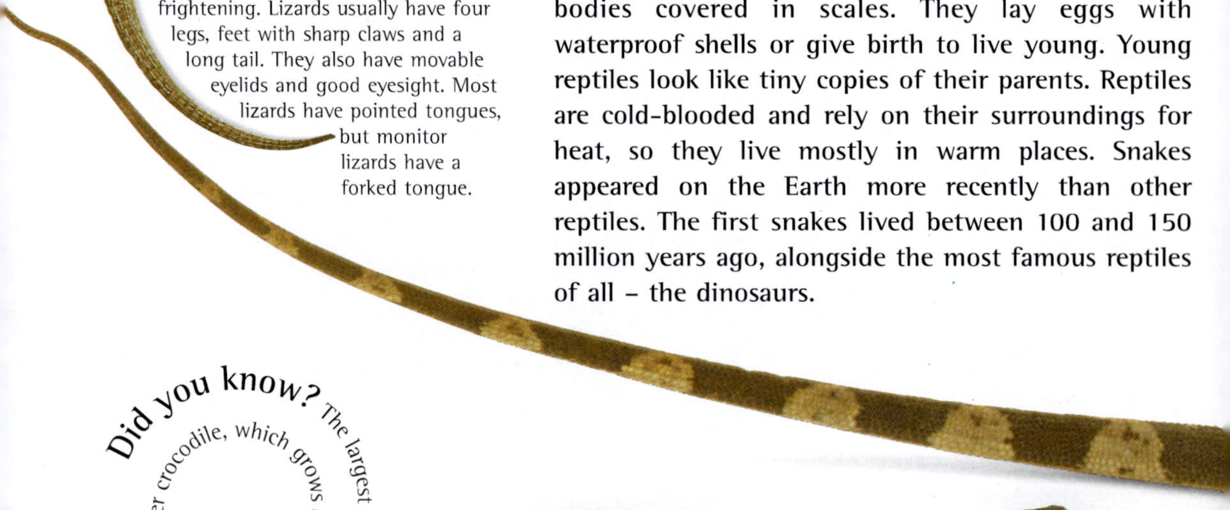

Did you know? The largest reptile in the world is the saltwater crocodile, which grows over 7m long.

baby crocodile

Caring Crocodiles

Crocodiles are some of the largest and most dangerous living reptiles. Yet female crocodiles make doting mothers. They stand guard over their eggs and protect their young until they are old enough to fend for themselves.

legless lizard

Legless Lizards

Some burrowing lizards, including skinks, have long, snake-like bodies and very short legs – or no legs at all. Snakes possibly developed from burrowing lizards, which did not need legs for sliding through the soil or leaf litter.

Did you know? The only two poisonous lizards are the gila monster and the Mexican beaded lizard.

Lizard Tails

Lizards, like this water dragon, generally have long tails and shed their skin in several pieces. Some lizards can grow a new tail if their tail breaks off in a fight with a predator.

water dragon

Tortoises

A tortoise has a shell as well as a skeleton inside its body. The shell is made from bony plates fused to the ribs, with an outer covering of horny plates. It is useful for protection, but it is also very heavy. The tortoise's long neck can be drawn right back inside its shell.

tortoise

Worm Lizards

These strange reptiles are fierce predators that look rather like earthworms. Worm lizards live underground, digging burrows with their strong, hard heads. Their nostrils point backwards and they close

worm lizard

Conservation

The biggest threat to snakes comes from people. Some snakes are killed because people are afraid of them. Farmers often kill snakes to protect their farm animals and workers, although many snakes actually help farmers by eating pests. In some countries, people catch snakes for food or use parts of their bodies to make medicines. Snake skins are also used to cover expensive jewellery, shoes, belts and bags. Snakes are sometimes captured and sold as exotic pets. Also, when people move into an area, snakes' habitats are often destroyed. To help snakes survive, people need to learn more about them and take action to preserve their habitats, so that they have safe places in which to live.

Finding out More
These scientists are tracking a black-tailed rattlesnake in Arizona, USA. They use an antenna to pick up radio signals from a transmitter fitted to the snake. This allows them to follow the snake's movements even when they cannot see it. There are still many gaps in our knowledge about snakes. The more we know about how they live, the easier it is to protect them.

Trophy
Although snakes are killed for food in some parts of the world, there are unfortunately still those who shoot them for recreation. As a trophy demonstrating their sporting achievements, the hunters put the rattles or head of the snake on display.

Snakes in Danger
Snakes such as this Dumeril's boa are in danger of dying out altogether if people do not do something to save them. Laws can be passed to stop collectors taking snakes from the wild, but it is difficult to make sure people obey the law. Other threats to snakes are more difficult to solve, such as people building roads or towns in places where snakes live.

Using Snake Skins

Snake skins have been used for many years to make bracelets and earrings such as these. Certain snake skins are more popular than others, and some species have declined as a result of intensive killing for skins in some areas. Recently, countries such as Sri Lanka and India have banned the export of snake skins.

Round-up

This show has been put on for tourists to show the skill needed to capture a rattlesnake. Early settlers in North America used to set aside a day each year to kill as many rattlesnakes as possible. Today, such hunts are not so common.

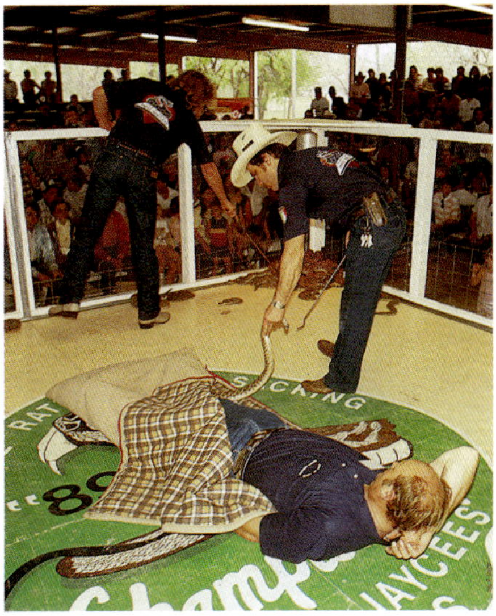

Did you know? Legend says St Patrick banished snakes from Ireland to rid the country of evil.

Pet Snakes

Some people like to keep pet snakes. However, they do not do very much and are not happy in captivity. Snakes can lose the ability to hunt and dislike being kept in a confined space.

GLOSSARY

A

aestivation
A period of rest during heat and drought, similar to hibernation.

albino
An animal that has no colour on all or part of its body but belongs to a species that is usually coloured.

anaconda
A type of boa.

antivenin
A substance made from the blood of mammals and/or snake venom that is used to treat snakebite.

B

boas
A group of snakes that live mainly in North and South America. They kill by constriction and give birth to live young.

brille
A transparent scale covering a snake's eye. It is also called a spectacle.

C

camouflage
Colours or patterns that allow a snake to blend in with its surroundings.

classification
Grouping of animals according to their similarities and differences in order to study them. This also suggests how they may have developed over time.

cloaca
Combined opening of the end of the gut, the reproductive system and the urinary system in reptiles, amphibians and birds.

clutch
The number of eggs laid by a female at one time.

cobras
Poisonous snakes in the elapid family, with short, fixed fangs at the front of the mouth.

cold-blooded
An animal whose temperature varies with that of its surroundings.

colubrids
Mostly harmless snakes. These snakes make up the biggest group –nearly three-quarters of the world's snakes.

compost heap
A pile of layers of garden plants, leaves and soil. Compost gives off heat as it rots down and can eventually be dug into the soil to make it rich. This helps plants to grow.

constrictor
A snake that kills by coiling its body tightly around its prey to suffocate it.

D

digestion
The process of absorbing food into the stomach and bowels.

E

ectotherm
A cold-blooded animal.

egg tooth
A small tooth sticking out of the mouth of snake hatchlings, which is used to slit open the egg.

elapids
A group of poisonous snakes that includes the cobras, mambas and the coral snakes. Elapids live in hot countries.

epidermis
The outer layer of the skin.

F

fang
A long, pointed tooth that may be used to deliver venom.

H

hibernation
A period of rest during the winter when body processes slow down.

J

Jacobson's organ
Nerve pits in the roof of a snake's mouth into which the tongue places scent particles.

K

keratin
A horny substance that makes up a snake's scales.

M

mature
Developed enough to be capable of reproduction.

membrane
A thin film, skin or layer.

moulting
The process by which a snake sheds its skin.

P

pigment
Colouring

pits
Heat sensors located on either side of a snake's head.

predator
A living thing that catches and kills other living things for food.

prey
An animal that is hunted by animals or by people for food.

python
A group of snakes that lives mainly in Australia, Africa and Asia. Pythons kill their prey by constriction. They lay eggs.

R

rainforest
A tropical forest where it is hot and wet all year round.

rattlesnakes
Snakes that live mainly in the south-west United States and Mexico. They have a warning rattle made of empty tail sections at the end of the tail.

S

saliva
A colourless liquid produced by glands in the mouth. Saliva helps to slide food from the mouth to the throat. In some snakes, saliva also aids digestion.

spurs
Leg bones attached to the hip bones, which are found in boas and pythons and used during courtship displays.

V

venom
Poisonous fluid produced in the glands of some snakes to kill their prey.

vipers
A group of very poisonous snakes with fangs that fold. Some vipers have heat pits on their faces. Most vipers give birth to live young.

W

warm-blooded
An animal (such as a mouse or a human being) that is able to maintain its body at roughly the same temperature all the time.

warning colours
Bright colours that show others that an animal is poisonous. Bright colours also warn predators to keep away.

windpipe
The tube leading from the mouth to the lungs. It is through this that an animal takes in fresh air (containing oxygen) and gives out used air (containing carbon dioxide).

Y

yolk
Food material that is rich in protein and fats. It nourishes a developing embryo inside an egg.

INDEX

A
adders 36, 42, 44, 47
African pythons 23
albino snakes 35
Amazon tree boas 45, 48
anacondas 7, 37, 55
antivenin 27
asp vipers 19, 29

B
baby snakes 34, 40-5, 51
backbones 8, 9, 14, 58
birth 42-3, 54, 58
blind snakes 7, 56
boas 7, 11, 16, 36, 50-1, 56, 57
bones 8-9, 58
breathing 8, 23, 25, 54
Burmese pythons 7, 44
burrowing snakes 6, 7, 10, 11, 16, 52

C
camouflage 34-5, 44, 48, 49
carpet pythons 23
children's pythons 39
cobras 17, 21, 26, 31, 32-3, 38, 56
cold-blooded snakes 8, 46, 58
colours 26, 34-5, 43, 44, 48, 51
conservation 60-1
constrictors 22, 23, 24-5, 50, 57
Cook's tree boas 49
copperheads 22, 26-7
coral snakes 21, 26, 33, 35, 56
corn snakes 10, 11, 14
cottonmouths 30-1
crocodiles 23, 58

D
defences 30-1, 34
desert snakes 47, 52-3
digestion 8-9, 22, 26
Dumeril's boas 60

E
ears 16, 17
egg-eating snakes 8, 18, 19
egg teeth 40, 43
eggs 38-41, 54, 58
elapid snakes 20, 21, 33, 56, 57
emerald tree boas 43, 48, 50-1
eyelash vipers 20, 47, 49
eyes 5, 10, 16, 31, 49, 53, 54, 58

F
families 56-7
fangs 20-1, 26, 27, 29, 30, 33, 48, 54, 56
fer-de-lance 22
fighting 36, 37
food 18-19, 22-5, 44, 45

G
gaboon vipers 19, 35
garter snakes 7, 36, 42, 46
grass snakes 4, 9, 31, 46
green tree pythons 6, 16, 48
growth 44-5

H
habitats 46-7
heads 4, 37, 49
heat pits 16, 28, 29, 51, 52
hibernation 44, 46
hognose snakes 31
hoods, cobras 31, 32
Hopi Indians 53
horned vipers 10, 16, 31, 52, 53
hunting 18-19, 22-3

J
Jacobson's organ 17
jaws 20-1, 22, 24, 25

L
lizards 58, 59
lungs 8, 9, 23, 28, 54, 57

M
mambas 16, 21, 27, 33, 56
markings 32, 35, 48
mating 36-7, 38, 51
milk snakes 34-5, 56
moulting 12-13, 44
mouths 20-1, 30
movement 14-15

O
Oenpellis pythons 38
organs 8-9

P
Peringuey's vipers 15, 53
pit vipers 16, 22, 28, 29
poisonous snakes 4, 6, 7, 19, 20, 22, 26-9, 30, 33, 34, 35, 44, 54, 56
Pope's tree vipers 42
prey 4, 18-19, 20, 22-5, 26, 29, 44, 48, 50
puff adders 11, 28, 47
pythons 6, 7, 21, 23, 36, 38, 39, 48, 56, 57

R
rainbow boas 5, 34, 47, 57
rainforests 47, 48, 50
rat snakes 18, 24-5, 37, 38, 40-1
rattlesnakes 5, 8, 10, 19, 28, 29, 36, 42, 45, 46, 52, 60, 61
red-tailed boas 4-5, 35, 42, 43
reproduction 36-41
rhinoceros vipers 6
ribs 8, 9, 25
ring-necked snakes 34
rock pythons 23

S
sand boas 53
sand vipers 42, 56
scales 4, 5, 10-11, 14, 25, 28, 34, 48, 52, 56, 57, 58
scutes 10
sea kraits 38, 55
sea snakes 6, 14, 18, 38, 42, 54-5, 56, 57
senses 16-17, 18
sidewinding 15, 53
skeletons 9, 58
skin 5, 10-13, 25, 32, 34, 44, 52, 60, 61
sloughing 12-13
smell, sense of 16, 17
smooth snakes 19
sounds 30, 31, 52
spotted pythons 23
spurs 36
stomach 8, 9, 23
sunbeam snakes 56, 57
swimming 14, 54

T
tails 4, 6, 48, 50, 59
taste, sense of 16, 17
teeth 20-1, 23, 24, 26, 27, 40, 43, 48, 50, 51
tentacled snakes 18
thread snakes 7, 18, 56
tiger snakes 33
tongues 4, 5, 16, 17, 19, 25, 29, 52, 58
tree snakes 6, 7, 16, 42, 47, 48-9

V
venom 26-7, 28, 29, 30, 33, 44, 49
vine snakes 30
vipers 6, 7, 9, 27, 28-9, 46, 56

W
water snakes 54-5
whip snakes 14, 49
worm snakes 56

NATURE WATCH

BIG CATS

RHONDA KLEVANSKY

Consultant: Dr. Nigel Dunstone

C O N

Published by Anness Publishing Ltd,
Blaby Road, Wigston, Leicestershire LE18 4SE
Email: info@anness.com
Web: www.annesspublishing.com

Anness Publishing has a new picture agency outlet for images for publishing, promotions or advertising. Please visit our website www.practicalpictures.com for more information.

Publisher: Joanna Lorenz
Managing Editor, Children's Books: Sue Grabham
Project Manager: Charlotte Hurdman
Designers: Traffika Publishing Ltd, Mirjana Nociar
Picture Researchers: Elizabeth Walsh, Kay Rowley
Illustrators: David Webb, Vanessa Card, Peter Bull
Production Controller: Ann Childers
Editorial Reader: Felicity Forster
With thanks to Nicole Pearson for her work on this title.

ETHICAL TRADING POLICY
Because of our ongoing ecological investment programme, you, as our customer, can have the pleasure and reassurance of knowing that a tree is being cultivated on your behalf to naturally replace the materials used to make the book you are holding. For further information about this scheme, go to www.annesspublishing.com/trees

PICTURE CREDITS
b=bottom, t=top, c= centre, l= left, r= right
ABPL: 7tl, 17cbl&cbr, 23c; /Daryl Balfour: 43t; /Peter Chadwick: 19t; /Nigel Dennis: 6b, 18t, 20t, 53t; /Clem Haagner: 4–5c, 10b, 26t, 41t, 51cr, 52b; /Dave Hamman: 25tr, 51t; /Roger de la Harpe: 17ctl; /Lex Hes: 28t&b; /Gerald Hinde: 12b, 29t, 33tl; /Luke Hunter: 32–33c, 39tl, 43b; /Beverly Joubert: 24c, 38t, 41c; /Peter Lillie: 21t, 24b; /Anup Shah:20c, 40b, 41b; /Giselle Wurfsohn:57tl. Anness/Jane Burton: 59t. BBC Natural History Unit/Keith Scholey: 21b; /Lynn Stone: 6c; /Tom Vezo: 13tl. Bridgeman Art Library: 24t, 31tr, 53br, 56b. Bruce Coleman Ltd/Erwin & Peggy Bauer: 19b, 26bl, 27b, 37c, 46b; /Alain Compost: 17t, 47tl&cl; /Gerald Cubitt: 60bl; /Peter Evans: 33b; /Christer Fredriksson: 25b; Paul van Gaalen: 13tr; /HPH Photography: 9bl; /Leonard Lee Rue: 15t, 27c; /Joe McDonald: 27t, 31br, 34b, 51b; /Antonio Manzanares: 25tl; /Rita Meyer: 15b; /Dieter & Mary Plage: 37br, 45c; /Michael Price: 61c; /Hans Reinhard: 19c, 44b; /John Shaw: 5b, 30b, 59b; /Kim Taylor: 9t; /Norman Tomalin: 22t, 33tr; /Rod Williams: 35b, 45b, 47tr, 53c, 59cl; /Gunter Ziesler: 26br, 44t. C M Dixon: 54t, 55t, 60t. Mary Evans Picture Library: 51cl, 54b, 57tr. FLPA: 43cl; /E&D Hosking: 61b; /Leonard Lee Rue: 50b; /Mark Newman: 40t; /Philip Perry: 14t, 23tr, 35t, 52t, 53bl; /Fritz Pölking: 18b, 20b, 21c; /Terry Whittaker: 5tl, 11t, 13b, 17b, 23b, 36t, 49t&b, 59cr; /E Woods: 37bl. Grant Museum/A Lister: 9br. Griffith Institute, Ashmolean Museum: 56t. Michael Holford: 10t. Niall McInerney: 57br. Natural History Museum: 58. Natural Science Photos/D Allen Photography: 6t, 22b, 29c, 42t, 43cr, 50t; /Ken Cole: 7c, 8b, 48t; /Carlo Dani & Ingrid Jeske: 7b, 37t, 61t; /Lex Hes: 17b, 30–31t; /C Jones: 34t; /David Lawson: 30c; /Michael Kock: 57bl; /M W Powles: 31bl; /Keren Su: 48br; /John Warden: 36b, 42b, 45t. NHPA/Martin Harvey: 55br; /Yves Lanceau: 12t; /Christophe Ratier: 60br /Andy Rouse: 46t. Papilio Photographic: 4t, 7tr, 15c, 16t, 17ctr, 23tl, 47bl, 48bl. Planet Earth Pictures: 35c, 39b, 47br; /Anup Shah: 39tr; /Manoj Shah: 38bl&br. Tony Stone: 4b; /Geoff Johnson: 14b; /Nicholas Parfitt: 32b; /Bryan Parsley: 49c; /Mark Petersen: 29b; /Manoj Shah: 32t. Visual Arts Library: 5tr. Werner Forman Archive: 55bl. Photo of Jim Corbett p55 by permission of the British Library (neg B19186).

INTRODUCING BIG CATS

4 What is a Cat?
6 The Big Cats

HOW BIG CATS WORK

8 Bones and Teeth
10 Muscles and Claws
12 Sight and Sound
14 Touching, Tasting, Smelling
16 Spots and Stripes
18 On the Move
20 Focus on the Hunting Cheetah
22 Communication

CATCHING PREY

24 Hunting Prey
26 Killing Prey
28 Focus on the Lone Leopard

T E N T S

HOW BIG CATS BEHAVE

30 Living Together
32 Focus on a Pride
34 Finding a Mate
36 Giving Birth
38 Focus on Cute Cubs
40 Growing Up
42 Enemies of Big Cats

WHERE BIG CATS LIVE

44 Mountain Cats
46 Forest Dwellers
48 Focus on Siberian Tigers
50 On the Savanna
52 Desert Cats

FACTS ABOUT BIG CATS

54 Killer Cats
56 Cats in Danger
58 Big Cat Relatives
60 Protecting Cats
62 Glossary
64 Index

What is a Cat?

Cats are native to every continent except Australia and Antarctica. All cats are mammals with fine fur that is often beautifully marked. They are skilled hunters and killers with strong agile bodies, acute senses and sharp teeth and claws. Cats are stealthy and intelligent animals and many are solitary and very secretive. Although cats vary in size from the domestic (house) cat to the huge Siberian tiger, they all belong to the same family, the Felidae. This means that both wild and domestic cats look alike and behave in very similar ways. In all, there are 38 different species of cat.

LONG TAIL
A cat's long tail helps it to balance as it runs. Cats also use their tails to signal their feelings to other cats.

All cats have short, rounded heads.

Whiskers help a cat to feel its surroundings.

The body of a cat is muscular and supple, with a broad, powerful chest.

BIG CATS
Cats are very specialized flesh eaters. They are the perfect carnivore, with excellent hearing and eyesight. Their curved, razor-sharp claws, used for catching and holding prey, are retractable. This means they can be pulled into the paws to protect them when running. The hair covering a cat's paws and pads helps it to move silently.

BIG BITE
As this tiger yawns it reveals its sharp teeth and strong jaws that can give a lethal bite. Cats use their long, curved canine teeth for killing prey.

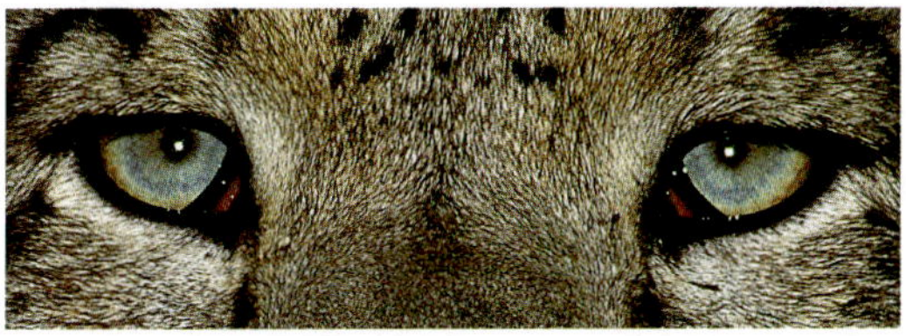

NIGHT SIGHT

The pupils (dark centres) of cats' eyes close to a slit or small circle during the day to keep out the glare. At night they open up to let in as much light as possible. This enables a cat to see well at night as well as during the day.

The Lion and the Saint

St Jerome was a Christian scholar who lived from about AD331 to 420. According to legend, he found an injured lion in the desert with a thorn in its paw. Instead of attacking him, the lion befriended the saint when he removed the thorn. St Jerome is often shown with a lion sitting at his feet.

Did you know? Some Arctic peoples believe that cats represent the spirits of the dead.

CATS EARS

A cat's ears are set high on its head. This gives a keen hunter the best possible chance of picking up sounds. The ears have a rounded shape, which enables sounds to be picked up from many directions. Cats can also rotate their ears to face towards the source of a sound.

The Big Cats

Scientists classify (arrange) the members of the cat family into related groups. The two main groups are small cats (the domestic cat and many wild cats) and the big cats (the tiger, lion, leopard, snow leopard and jaguar). The clouded leopard and the cheetah are each grouped separately, but many people regard them as big cats. Big cats differ from small cats not only because of their size. Big cats can roar, but small cats cannot. Small cats purr. They have a special bone, the hyoid, at the base of their tongues that enables them to breath and purr at the same time. Big cats have elastic cartilage instead and can only purr when they breathe out. The puma is in fact a very large small cat. It is included here because of its size.

LION
The lion is the only social cat and lives in a family group called a pride. Adult male lions, unlike other big cats, have a long, thick mane of hair. Female lions do not have manes.

PUMA
The puma is also called the cougar or mountain lion. Although it is about the same size as a leopard, a puma is a small cat because it can purr. Pumas live in North and South America.

CHEETAH
The tall cheetah is built like a slim athlete and is able to chase prey at great speed. Cheetahs are different from other cats in that they have retractable claws, but no sheaths to cover them. It was thought that cheetahs were related to dogs, but scientists now think that their closest cousins are pumas.

LEOPARD

The leopard is built for bursts of speed and for climbing trees. Heavier than a cheetah, this cat is not as large and bulky as a tiger or a lion. Its spotted coat helps to hide the cat as it hunts in wooded grassland. Black leopards are called panthers. They are the same species, but their spots are hidden.

SNOW LEOPARD

Snow leopards are a different species from true leopards. These rare cats have very thick coats to keep them warm in the high mountains of central Asia. They have very long tails, which help them to balance as they leap from rock to rock in their mountainous surroundings.

JAGUAR

The jaguar is sometimes confused with a leopard, but it is stockier and not as agile. This big cat lives throughout South America in forested habitats where it needs power to climb rather than speed to run.

TIGER

The most powerful and largest of all the big cats is the tiger. A tiger reaches on average a length of over 2m and weighs 230kg. The biggest tigers live in the snowy forests of Siberia in Russia. A few tigers also live in tropical forest reserves and swamps in Asia.

Did you know? Although lions are called the King of the Jungle they do not live there.

Bones and Teeth

The skeleton of a cat gives it its shape and has 230 bones (a human has 206). Its short and round skull is joined to the spine (backbone), which supports the body. Vertebrae (the bones of the spine) protect the spinal cord, which is the main nerve cable in the body. The ribs are joined to the spine, forming a cage that protects a cat's heart and lungs. Cats' teeth are designed for killing and chewing meat. Wild cats have to be very careful not to damage their teeth, because with broken teeth they would quickly die from starvation.

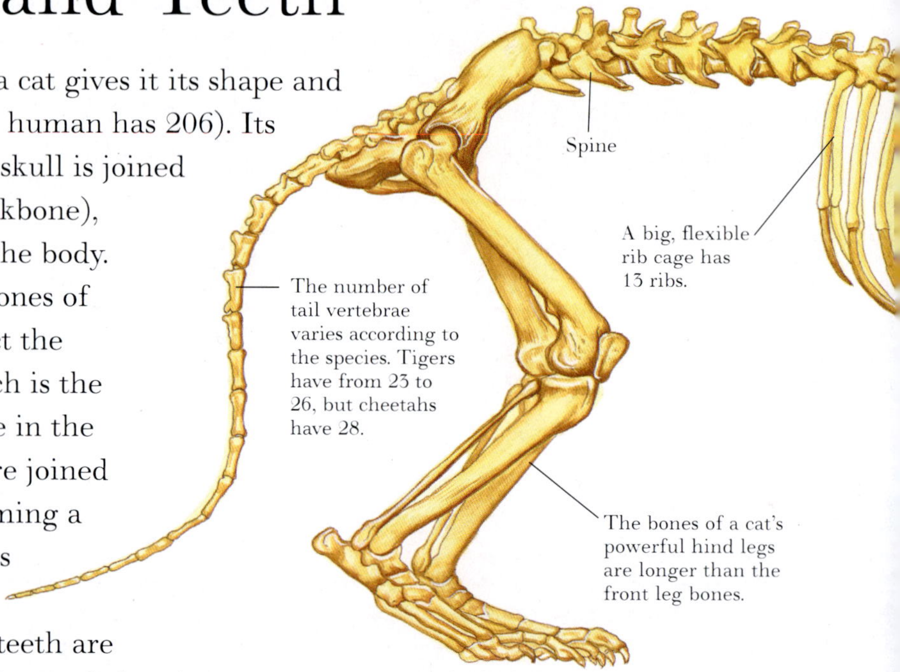

THE FRAME
The powerfully built skeleton of a tiger is similar to all cats' skeletons. Cats have short necks with seven compressed vertebrae. These help to streamline and balance the cat so that it can achieve greater speeds. All cats have slightly different shoulder bones. A cheetah has long shoulder bones to which sprinting muscles are attached. A leopard, however, has short shoulder bones and thicker, tree-climbing muscles.

CANINES AND CARNASSIALS
A tiger reveals its fearsome teeth. Its long, curved canines are adapted to fit between the neck bones of its prey to break the spinal cord. Like all carnivores, cats have strong back teeth, called carnassials. These do most of the cutting by tearing off pieces of meat.

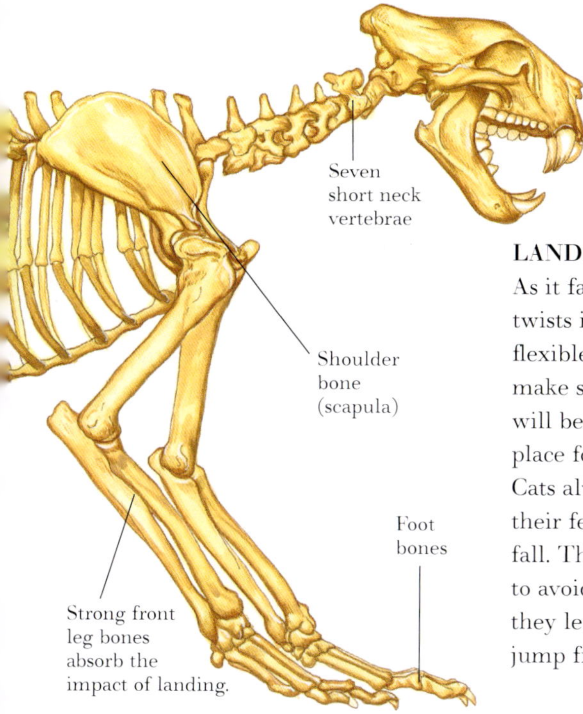

LANDING FEET

As it falls, this cat twists its supple, flexible spine to make sure its feet will be in the right place for landing. Cats always land on their feet when they fall. This helps them to avoid injury as they leap on prey or jump from a tree.

CHEWING ON A BONE

Ravenous lions feast on the carcass of their latest kill. Cats' jaws are hinged so that their jaw bones can move only from side to side, not up and down. Because of this, cats eat on one side of their mouths at a time and often tilt their heads when they eat.

CAT SKULL

Like all cats' skulls, this tiger's skull has a high crown at the back giving lots of space for its strong neck muscles. Big eye sockets allow it to see well to the sides as well as to the front. Its short jaws can open wide to deliver a powerful bite.

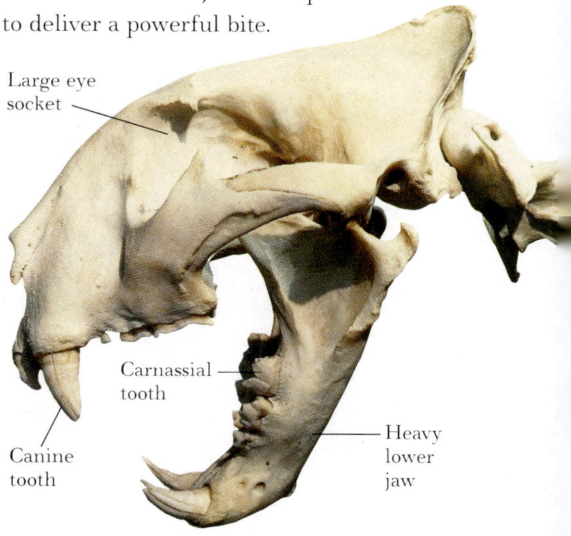

Muscles and Claws

Both inside and out, cats are designed to be skilled hunters and killers. Thick back and shoulder muscles help them to be excellent jumpers and climbers. Sharp, curved claws that grow from all of their digits (toes) are their weapons. One of the digits on a cat's front foot is called the dew claw. This is held off the ground to keep it sharp and ready to hold prey. Cats are warm-blooded, which means that their bodies stay at the same temperature no matter how hot or cold the weather is. The fur on their skin keeps them warm when conditions are cold. When it is hot, cats cool down by sweating through their noses and paw pads.

Heracles and the Nemean Lion
The mythical Greek hero Heracles was the son of the god Zeus and immensely strong. As a young man he committed a terrible crime. Part of his punishment was to kill the Nemean lion. The lion had impenetrable skin and could not be killed with arrows or spears. Heracles chased the lion into a cave and strangled it with his hands. He wore its skin as a shield and its head as a helmet.

KNOCKOUT CLAWS

Cheetahs have well developed dew claws that stick out from the front legs. They use these claws to knock down prey before grabbing its throat or muzzle to strangle it. Other cats use their dew claws to grip while climbing or to hold on to prey. Cats have five claws, including the dew claw, on their front paws. On their back paws, they have only four claws.

Did you know? Cats cannot digest sugar so they prefer not to eat sweet things.

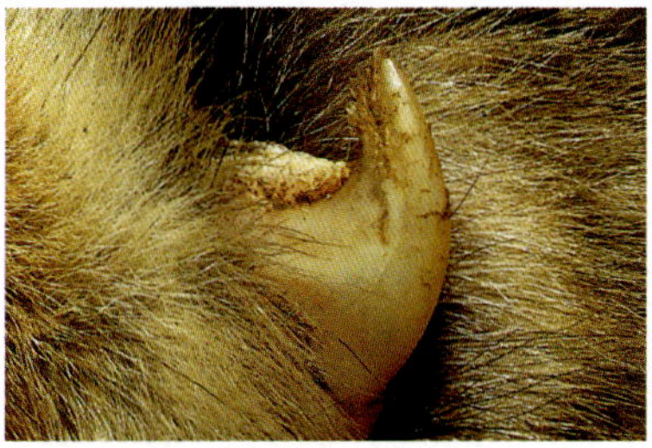

TIGER CLAW

This is the extended claw of a tiger. Cats' claws are made of keratin, just like human fingernails. They need to be kept sharp all the time.

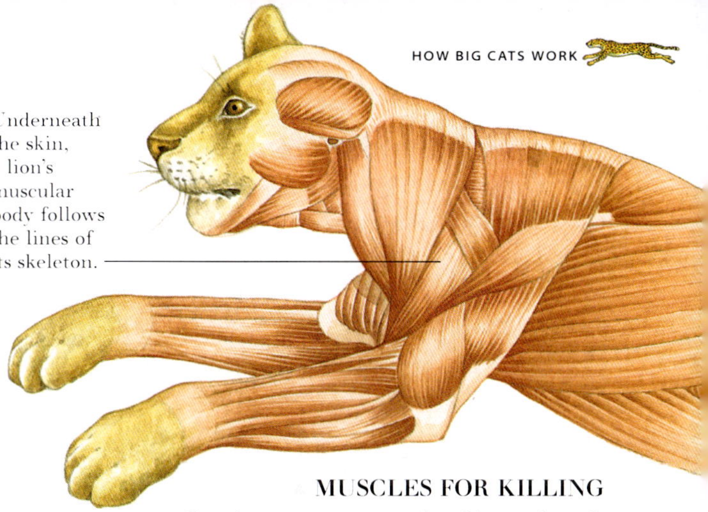

MUSCLES FOR KILLING

Cats have very strong shoulder and neck muscles for attacking prey. The muscles also absorb some of the impact when the cat pounces.

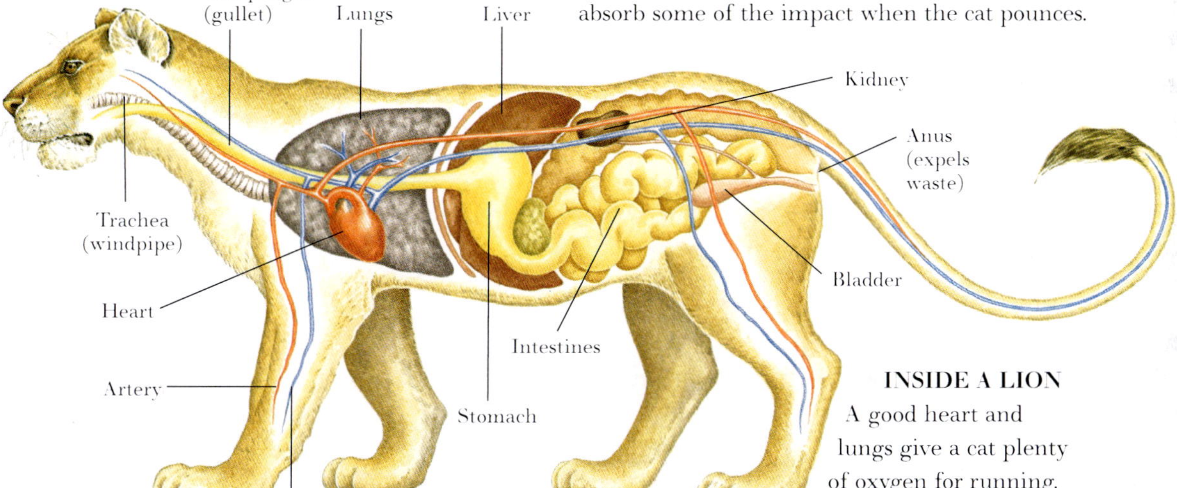

INSIDE A LION

A good heart and lungs give a cat plenty of oxygen for running. Its stomach and intestines (gut) are developed to digest meat easily, but not plants. Its intestines are quite short, so food passes through quickly. This means as soon as it needs more food, a cat is light enough to run and pounce. Once a lion has had a big meal, it does not need to eat again for several days.

CLAW PROTECTION

Cats retract (pull back) their claws into fleshy sheaths to protect them. This stops them getting blunt or damaged. Only cheetahs do not have sheaths.

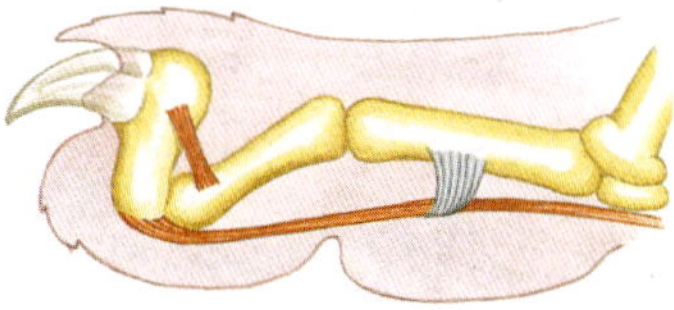

Sheathed claw is protected by a fleshy covering.

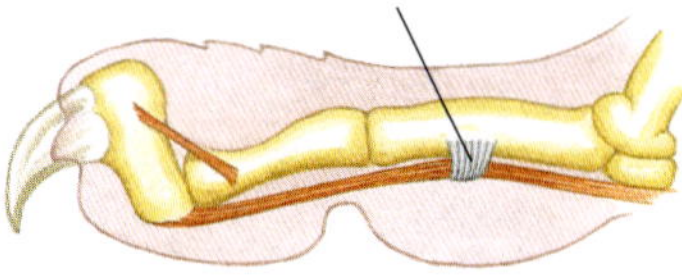

The claw is unsheathed when a muscle tightens.

Sight and Sound

To hunt well and not be seen or heard by prey or enemies, cats use their senses of sight, hearing and touch. Cats' eyesight is excellent. Their eyes are adapted for night vision, but they can also see well in the day. Cats' eyes are big compared to the size of their heads. They have good binocular vision, which allows them to accurately judge how far away objects are. At night, cats see in black and white. They can see colours in the day, but not as well as humans can. Cats have very good hearing, much better than a human's. They can hear small animals rustling through the grass or even moving around in their burrows underground.

CAUGHT IN BRIGHT LIGHT
Cats' eyes are very sensitive to light. During the day in bright light, the pupils of the eyes close right down, letting in only as much light as is needed to see well. A domestic cat's pupils close down to slits, while most big cats' pupils close to tiny circles.

GLOWING EYES
Behind the retinas (light sensitive areas) in this leopard's eyes is a reflecting layer called the tapetum lucidum. This helps to absorb extra light in the dark. When light shines into the eyes at night the reflectors glow.

Did you know? A cat's pupils open wide when it is frightened and close up when angry.

PREY IN SIGHT ►

As it stalks through the long grass a lion must pounce at just the right moment if it is to catch its prey. Binocular vision helps the cat to judge when to strike. Because its eyes are set slightly apart at the front of the head their field of view overlaps. This enables a cat to judge the position of its prey exactly.

▲ ROUND-EYED

This puma's rounded pupils have closed down in daylight. In dim light, the pupils will expand wide to let in as much light as possible.

SHARP EARS ►

Cats' ears are designed for them to hear very well. This Siberian lynx lives in snowy forests where the sound is soft and muffled. It has specially shaped, big ears to catch as much sound as possible.

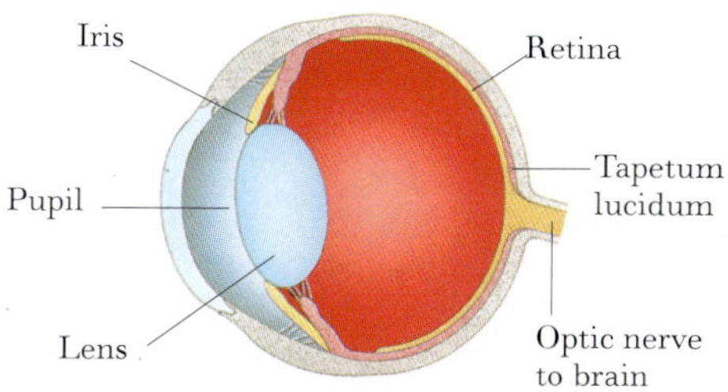

▲ INSIDE THE EYE

The lens focuses light rays to produce a sharp image on the retina. Impulses from the retina are carried to the brain by the optic nerve. Cats have a membrane that can be pulled over the surface of the eye to keep out dirt and dust.

Touching, Tasting, Smelling

Like all animals, cats feel things with nerves in their skin, but they have another important touching tool – whiskers. These long, stiff hairs on the face have very sensitive nerve endings at their roots. Some whiskers are for protection. Anything brushing against the whiskers above a cat's eyes will make it blink. Cats use smell and taste to communicate with each other. A cat's tongue is a useful tool and its nose is very sensitive. Thin, curled bones in the nose carry scents inwards to smell receptors. Unlike most animals, cats have a special place on the roof of their mouth to taste smells, especially the scent of other cats.

▲ **TONGUE TOOL**
A leopard curls the tip of its tongue like a spoon to lap up water. After several laps it will drink the water in one gulp. As well as drinking, the tongue is used for tasting, scraping meat off a carcass and grooming.

◄ **ROUGH TONGUE**
A tiger's bright pink tongue has a very rough surface. Cats' tongues are covered with small spikes called papillae. The papillae point backwards and are used by the cat, together with its teeth, to strip meat off bones. Around the edge and at the back of the tongue are taste buds. Cats cannot taste sweet things, but can actively recognize pure water.

The tiger raises its head and grimaces to taste the air.

Cats twitch their tails from side to side as they concentrate. When angry, the tail lashes to and fro.

WHO PASSED BY?

By tasting the air a tiger uses his Jacobson's organ (the special scent centre on the roof of the mouth) to detect the scent left by another tiger. To get as much of the scent as he can, he wrinkles his nose, curls his lips upwards, bares his teeth and lifts his head. This action is known as flehmen. Males use it especially to locate females ready to mate.

Did you know? Hairballs coughed up by lions are worn as talismans in parts of Africa.

THE CAT'S WHISKERS

This snow leopard's face is surrounded by sensitive whiskers. Cats use their whiskers to judge how far away objects are. The most important whiskers are on the sides of the face. These help a cat to feel its way in the dark, or when it is walking through tall grass.

COAT CARE

The long, rough tongue of a lion makes a very good comb. It removes loose hairs and combs the fur flat and straight. Cats wipe their faces, coats and paws clean. They need to keep well groomed and spend a lot of time looking after their fur. Hair swallowed by grooming is spat out as hairballs.

Spots and Stripes

A cat's fur coat protects its skin and keeps it warm. The coat's colours and patterns help to camouflage (hide) the cat as it hunts prey. Wild cats' coats have two layers – an undercoat of short soft fur and an outercoat of tougher, longer hairs, called guard hairs. Together these two layers insulate the cat from extreme cold or extreme heat. Some guard hairs are sensitive and help a cat to feel its way. Cats have loose skin, making it difficult for an attacker to get a good grip and helping to prevent injury. The colours and patterns of a wild cat's coat depend on where it lives.

TIGER IN THE GRASS
The stripes of a tiger's coat are the perfect camouflage for an animal that needs to prowl around in long grass. The colours and patterns help to make the cat almost invisible as it stalks its prey. These markings are also very effective in a leafy jungle where the dappled light makes stripes of light and shade.

Did you know? Domestic cats have a wider range of colours and markings than wild cats.

KING OF THE HILL
King cheetahs were once thought to be different from other cheetahs. They have longer fur, darker colours and spots on their backs that join up to form stripes. Even so, they are the same species. All cheetahs have distinctive tear stripes running from the corners of their eyes down beside their muzzles.

NON-IDENTICAL TWINS

Many big cats of the same species come in variations of colour, depending on where they live. These two leopard cubs are twins, but one has a much darker, blackish coat. Black leopards are called panthers. (Black jaguars and pumas are also sometimes called panthers.) Some leopards live deep in the shadows of the forest, where a darker colouring would allow them to hide more easily. Panthers are most common in Asia.

SPOT THE DIFFERENCE

Spots, stripes or blotches break up the outline of a cat's body. This helps it to blend in with the shadows made by the leaves of bushes and trees, or the lines of tall grass. In the dappled light of a forest or in the long grass of the savanna, cats are very well hidden indeed.

A leopard's spots are in fact small rosettes.

The tiger has distinctive black stripes.

A jaguar has rosettes with a central spot of colour.

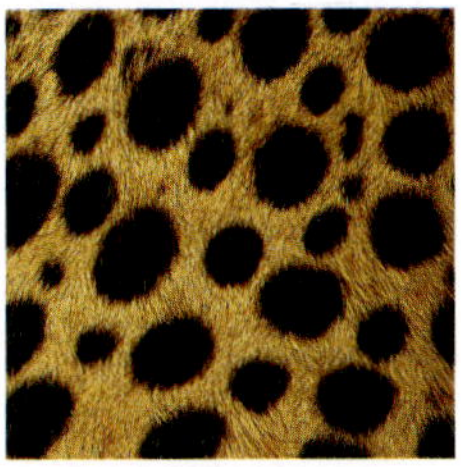

The cheetah has lots of spots and no rosettes.

WHITE FOR SNOW

A snow leopard has a shaggy, off-white coat with darker spots. This colouring helps the snow leopard to stay well hidden in the rocky, mountainous terrain where it lives. It moves about early in the morning or late afternoon, blending with its habitat as it looks for prey.

A snow leopard's pale, thick coat has dark irregular spots and streaks. This helps it hide between the rocks and snow.

On the Move

Cats run and jump easily and gracefully. They have flexible spines and long, strong hind legs. With long, bouncy strides, they can cover a lot of ground very quickly. Big cats are not good long-distance runners, they are sprinters and pouncers. They use their long tails for balance when climbing trees and running fast. All cats can swim very well, but some do not like the water and will only swim to escape danger. Others, such as tigers and jaguars, live near water and often swim to hunt their prey.

▲ **THRILL OF THE CHASE**
A lion chases its prey through the scrub. When lions stalk, run and pounce, they make use of their flexible backs, strong back legs, powerful chests and cushioning pads under their paws. Cats' back legs are especially powerful. They provide the major thrust for running. Cats can outpace their prey over short distances before launching into a final jump.

◄ **TREE-CLIMBING CAT**
Leopards spend a lot of time in trees and are designed for climbing. They have very powerful chests and front legs. Their shoulder blades are positioned to the side to make them better climbers. A leopard can leap 3m without difficulty and, in exceptional circumstances, can leap over 6m.

SOFT PADDING

The thick pads under a lion's paw are like cushions. They allow the lion to move very quietly and also act as shock absorbers for running and jumping. Hidden between the pads and fur are the lion's claws, tucked away safely until they are needed.

As it leaps, a bobcat pinpoints its landing position. The front feet land separately in quick succession.

Did you know? In the 1500s, rich people kept cheetahs as hunting animals like dogs.

GRACE AND AGILITY

A bobcat leaps with great agility off a rock. All cats have flexible backs and short collarbones to help make their bodies stronger for jumping off things. Bobcats are similar to lynxes. Both cats have an extensive coating of fur on their feet to give them extra warmth. The fur also prevents them from slipping on icy rocks.

KEEPING COOL CAT

A Bengal tiger swims gracefully across a river. Many tigers live in warm climates, such as India and Southeast Asia. As well as swimming to get from one place to another, they often look for pools of water to bathe in during the heat of the day. They are one of the few cats that actively enjoy being in or near water. Tigers are excellent swimmers and can easily cross a lake 5km wide.

Focus on the

A cheetah can run at 115km per hour over short distances – a speed equivalent to a fast car. This makes it the world's fastest land animal. The cheetah's body is fine-tuned for speed. It has wide nostrils to breathe in as much oxygen as possible and specially adapted paws for running fast. Most cheetahs today live in east and southern Africa, with a small number living in Asia – in Iran and Pakistan. They live in many different kinds of habitats, from open grassland to thick bush and even in desert-like environments.

1 A pair of cheetahs creep up stealthily on a herd of antelope. Cheetahs hunt their prey by slinking towards the herd, holding their heads low. Cheetahs are not pouncing killers, like other cats. Instead, they pull down their prey after a very fast chase. In order to waste as little energy as possible, cheetahs plan their attack first. They pick out their target before starting the chase.

2 The cheetah begins its chase as the herd of antelope starts to move. It can accelerate from walking pace to around 70km per hour in two seconds. Cheetahs have retractable claws, but unlike other cats they have no protective sheaths. The uncovered claws act like the spikes on the bottom of track shoes, helping the cheetah to grip as it runs. Ridges on their paw pads also help to improve grip.

3 At top speed a cheetah makes full use of its flexible spine and lean, supple body. A cheetah's legs are very long and slender compared to its body. It can cover several metres in a single bound.

Hunting Cheetah

4 As the cheetah closes in on the herd, the antelope spring in all directions. The cheetah changes direction without slowing down. If a cheetah does not catch its prey within about 400m, it has to give up the chase. Cheetahs usually hunt in the morning or late in the afternoon, when it is not too hot. They have short lifespans in the wild, because their speed declines with age.

5 As the cheetah closes in on its prey it may have to make several sharp turns to keep up. The cheetah's long tail gives it excellent balance as it turns. The cheetah knocks its victim off balance with a swipe of its front paw. It uses its big dew claw to pull the victim to the ground.

6 Once the prey animal is down, the cheetah grabs the victim's throat. A sharp bite suffocates the antelope. Cheetahs are not strong enough to kill by biting through the spinal cord in the prey's neck like other cats. The cheetah will hang on to the victim's throat until the antelope is dead.

Communication

All big cats communicate with one another. They tell each other how old they are, whether they are male or female, what mood they are in and where they live. Cats communicate by signals such as smells, scratches and sounds. The smells come from urine and from scent glands. Cats have scent glands on their heads and chins, between their toes and at the base of their tails. Every time they rub against something, they transfer their special smell. Cats make many different sounds. Scientists know that cats speak to each other, but still do not understand much about their language. Cats also communicate using body language. They use their ears to signal their mood and twitch their tails to show if they are excited or agitated.

A MIGHTY ROAR
The lion's roar is the loudest sound cats make. It is loud enough for all the neighbourhood lions to hear. Lions roar after sunset, following a kill and when they have finished eating. Lions make at least nine different sounds. They also grunt to each other as they move around.

HISSING LEOPARD
An angry leopard hisses at an enemy. Cats hiss and spit when they feel threatened, or when they are fighting an enemy. The position of a cat's ears also signals its intentions. When a cat is about to attack, it flattens its ears back against its neck.

EAR SIGNALS

Many wild cats, such as this tiger, have white markings on the back of their ears. They turn their ears to show the markings to an enemy when they are angry.

CAT SPRAY

A king cheetah marks its territory by spraying urine at points along its trails. Scent marks left by a male tell other males to keep away. The scent left by a female will tell a male passing through her range if she is ready to mate.

BABY TALK

Mothers talk to their cubs a lot. The sounds are quiet so that enemies do not hear. The softest and safest sound of all is purring.

MARKS FOR SHOW

Cats like to scratch things to clean their claws and stretch their limbs. At the same time they leave a scented mark for others to both see and smell. When this lioness scratches, she leaves her own personal scent from the glands between her toes on the scratch marks.

Did you know? When they are close together, lions chirrup, meow and yowl to each other.

Hunting Prey

All cats, big and small, are carnivores – they eat meat. Their bodies are not designed to digest plants. Big cats must hunt down and kill their own food. Most big cats, however, are only too happy to eat someone else's meal and steal kills from other animals whenever they can. Cheetahs are an exception, and eat only animals they have killed themselves. To catch and kill their food, big cats must hunt. Some, like cheetahs, patrol their neighbourhoods, looking for prey. Others, such as jaguars, hide in wait and then ambush their victims. Many cats, such as leopards, do both.

King Solomon
Solomon ruled Israel in the 900s BC and was reputed to be a very wise ruler. His throne was carved with lions because of his admiration for these big cats who killed only out of necessity. In law, if a man was said to have fallen into a lions' den, it was not proof of his death.

THE MAIN COURSE
A big lion can kill large, powerful animals like this buffalo. A big cat usually attacks from behind, or from the side. If the prey is too big to grab straight away, the cat will knock it off balance, hold on to it and bite into its neck.

CHOOSING A MEAL
A herd of antelope and zebra graze while keeping watch on a lioness crouched in the grass. She lies as close to the ground as possible, waiting to pounce. Finally, when focused on a victim, she will bring her hind legs back into position and dart forwards.

WARTHOG SPECIAL
Four cheetahs surround an injured warthog. The mother cheetah is teaching her three cubs hunting techniques. The cheetah on the right is trying a left paw side swipe, while another tries using its dew claw. Cheetahs love to eat warthogs but also catch antelope and smaller animals such as hares.

CAT AND MOUSE
A recently killed capybara (a large rodent) makes a tasty meal for a jaguar. Jaguars often catch their food in water, such as fish and turtles. On land they hunt armadillos, deer, opossums, skunks, snakes, squirrels, tortoises and monkeys.

Did you know? Cheetahs will only chase prey if it runs. If it stops, so does the cheetah.

SLOW FOOD
If a lion has not been able to hunt successfully for a while, it will eat small creatures such as this tortoise. Lions usually hunt big animals such as antelope, wildebeest, warthogs, giraffe, buffalo, bush pigs and baboons. They work together in a group to hunt large prey.

Killing Prey

The way a big cat kills its prey depends on the size of the cat and the size of its meal. If the prey is small with a bite-sized neck, it will be killed with a bite through the spinal cord. If the prey has a bite-sized head, the cat will use its powerful jaws to crush the back of its skull. Large prey is killed by biting its throat and suffocating it. Lions often hunt together and use a combined effort to kill large prey. One lion may grab the prey's nose to suffocate it, while other lions attack from behind.

Did you know? Lions try to flip porcupines on to their backs to avoid the sharp spines.

OLD AGE
When big cats get old or injured it is very difficult for them to hunt. They will eventually die from starvation. This lion from the Kalahari Desert in South Africa is old and thin. It has been weakened by hunger.

FAIR GAME
A warthog is a small, delicious meal for a cheetah. The bigger the cat, the bigger its prey. A cheetah is quite a light cat, so to kill an animal the cheetah first knocks it over, then bites the prey's neck to suffocate it.

A DEADLY EMBRACE
A lioness immobilizes a struggling wildebeest by biting its windpipe and suffocating it to death. Lions are very strong animals. A lion weighing 150 to 250kg can kill a buffalo more than twice its weight. Female lions do most of the hunting for their pride.

SECRET STASH

A cheetah carrying off its prey, a young gazelle, to a safe place. Once it has killed, a cheetah will check the area to be sure it is secure before feeding. It drags the carcass to a covered spot in the bushes. Here it can eat its meal hidden from enemies. Cheetahs are often driven off and robbed of their kills by hyenas and jackals or even other big cats.

A SOLID MEAL

These cheetahs will devour as much of this antelope as they can. Big cats lie on the ground and hold their food with their forepaws when they eat. When they have satisfied their hunger, cheetahs cover up or hide the carcass with grass, leaves or whatever is available in order to save it for later.

LIONS' FEAST

A pride of lions gather around their kill, a zebra. They eat quickly before any scavenging hyenas and vultures can steal the meat. Each lion has its place in the pride. Even if they are very hungry, they must wait until it is their turn to eat. Usually the dominant male lion eats first.

Focus on the Lone

Leopards are one of the most widespread of all the big cats, but are also the most secretive. They live in many different habitats throughout Africa and southern Asia, in open, rocky country as well as forests. Not much is known about them because they are nocturnal animals, coming out to hunt at night. They sometimes creep up on prey on the ground, then pounce. At other times they ambush their quarry from a tree.

CAT NAP
Leopards usually sleep all day in a tree, especially when it is very hot. Their spotted coat is excellent camouflage in the patchwork of light and shade in the forest. It is so good that, when they are resting, they are especially hard to see.

LONE LEOPARD
Leopards are loners. They come together only when a female signals to a male that she is ready to mate. After mating they separate again. The mother brings up the cubs until they are able to fend for themselves.

Leopard

BRUTE STRENGTH
A leopard drags its dead victim across the ground. Leopards have strong jaws, chests and front legs so that they can move an animal as big as themselves.

AMBUSH
Leopards like to ambush prey. They climb on to a low branch and wait for an animal to walk underneath. Then they jump down and grab it. The leopard uses its great strength to drag its victim high up into the tree. Prey includes pigs, antelope, monkeys, dogs and many other animals.

TOP MEAL
This leopard has dragged its kill up into a tree. This is to prevent the carcass being stolen. Other big carnivores that live in the same area cannot climb trees as well as leopards. Once the prey is safe, a leopard can finish its meal.

Living Together

Most big cats live alone. They hunt alone and the females bring up their cubs alone. Big cats come together only when they want to mate. They are solitary because of the prey that they hunt. There is usually not enough prey in one area for a large group of big cats to live on. Lions are the exception. They live in family groups called prides. (Male cheetahs also sometimes live in groups with up to four members.) All wild cats have territories (home ranges). These territories are a series of trails that link together a cat's hunting area, its drinking places, its lookout positions and the den where it brings up its young. Females have smaller home ranges than males. Males that have more than one mate have territories that overlap with two or more female home ranges.

BRINGING UP BABY
Female snow leopards bring up their cubs on their own. They have up to five cubs who stay with their mother for at least a year. Although snow leopards are loners, they are not unsociable. They like to live near each other and let other snow leopards cross their territories.

Did you know? Big cats' territories range from a few kilometres to over 1,000 km.

THE LOOKOUT
A puma keeps watch over its territory from a hill. Pumas are solitary and deliberately avoid each other except during courtship and mating. The first male puma to arrive in an area claims it as his territory. He chases out any other male that tries to live there.

A PRIDE OF LIONS
The lions in a pride drink together, hunt together, eat together and play together. A pride is usually made up of related females and their young. Prides usually try not to meet up with other prides. To tell the others to keep out of their territory, the pride leaves scent markings on the edge of their range.

Daniel and the Lions' Den
A story in the Bible tells how Daniel was taken prisoner by Nebuchadnezzer, king of Babylon. When Daniel correctly interpreted the king's dreams he became a favourite of the king. His enemies became jealous of his position and had him thrown into a lions' den, a common punishment for prisoners at the time. But instead of eating Daniel, the lions befriended him. They were tamed by his great faith in God.

FAMILY GROUPS
A cheetah mother sits between her two cubs. The cubs will stay with her until they are about 18 months old. The female then lives a solitary life. Males, however, live in small groups and defend a territory. They only leave their range if there is a drought or if food is very scarce.

WELL GROOMED
Cats that live together groom each other. They do this to be friendly and to keep clean. They also groom to spread their scent on each other, so that they smell the same. This helps them to recognize each other and to identify strangers.

Focus on

Lions are the second largest of the big cats after tigers. They like to live in open spaces, sometimes in woodland, but never in tropical forests. Lions are usually found on the savanna (grassy plains) and on the edges of deserts. Female lions live together in prides (family groups) of up to 12 lionesses and their cubs. The size of the group depends on how much food there is available. Male lions may live together in groups, called coalitions, which look after one or more prides. The coalitions defend their prides, fighting off any other males who want to mate with the females of their pride.

FATHER AND SON
Male lions are the only big cats that look different from the females. They have long shaggy manes to look larger and fiercer and to protect their necks in a fight. A male cub starts to grow a mane at about the age of three. At that age he also leaves the pride to establish his own territory.

THE FAMILY
A pride of lions rests near a waterhole. The biggest prides live in open grasslands where there are large herds of antelope, wildebeest and buffalo. If a foreign male takes over a pride, the new lion kills all the cubs under six months old. This is to ensure all the cubs are his.

a Pride

NURSERY SCHOOL
Young lions play tag to learn how to chase things and to defend their pride. The pride does not usually allow strange lions to join the family group. Young lions need to be prepared in case other lions come to fight with them.

FIRST AT THE TABLE
Male lions usually eat first, even though the females do most of the hunting. He can eat up to 30kg of meat at one time, but he will not need to eat again for several days.

CAT SCRAP
Two lionesses fight each other to decide who will be top cat. There is usually a dominant female in each pride, even when there are males around. This chief female rules the family.

MOTHER AND CUBS
Lionesses give birth to a litter of between one and six cubs. The cubs start learning to hunt when they are about 11 months old, but stay with their mother for over two years. In dry areas, lions live in small prides because less food is available.

Lionesses help to raise the young together. They even suckle each other's cubs.

Finding a Mate

Big cats roam over large areas, so it can be quite difficult for them to find a partner. When they are ready to mate, they use scent markers. These are like advertisements to all the other cats in the district. A female also calls loudly, in the hope that a nearby male will come to her. Often more than one male will follow a female. This almost always leads to fights between the interested males. The winner of the fight then begins to court the female. In a pride of lions, one male establishes his dominance over the group. In this way he avoids having to fight every time a female is ready to mate. Many big cats will mate several times a day for up to a week to make sure that the female is pregnant.

SIGNS OF LOVE

A male lion rubs against a female and smells her all over. He knows that the lioness is ready to mate from her scent. Having fought off rivals, he must now persuade the lioness to mate with him. He courts her by being attentive. Their courtship may last for several days before they mate.

Did you know? Pumas are also called mountain screamers after the female's mating call.

THE HAPPY COUPLE

When the female is ready to mate, she crouches on the ground with her hindquarters slightly raised. The male sits behind and over her and sometimes holds the scruff of her neck between his jaws. Large cats, such as leopards and lions, may mate up to 100 times a day. Smaller cats, such as cheetahs, are more vulnerable to predators and so mate for a shorter time.

KEEPING HIM IN HIS PLACE

After mating the lioness is aggressive and often lashes out at the lion. As soon as the two have mated the male jumps back very quickly. He remains close by her side to stop other males approaching. Once she has calmed down she rolls on her back and they mate again. Each mating lasts only a few seconds.

Did you know? A wild big cat may have up to 5 litters in an average lifespan of 12 years.

LEAN AND HEALTHY

This lioness is only just pregnant. She has not put on much weight and can still hunt efficiently. At the end of her pregnancy (about three to four months long) she will hunt small, easy-to-catch prey. Lionesses in a pride also get to share in the pride's kill.

CAT ATTRACTION

Two courting tigers often make a great deal of noise. They roar, meow, moan and grunt as they mate. Female tigers mate every other year. The male stays close by the female for a few days until he is sure she is pregnant. Then the pair separate and live on their own again.

Giving Birth

The cubs (babies) of a big cat are usually born with spotted fur and closed eyes. They are completely helpless. The mother cat looks after them on her own with no help from the father. She gives birth in a safe place called a den. For the first few days after birth, she stays very close to her cubs so that they can feed on her milk. She keeps them warm and cleans the cubs by licking them all over. The cubs grow quickly. Even before their eyes open they can crawl, and they soon learn to hiss to defend themselves.

SNOW CUB

Snow leopard cubs have white fur with dark spots. They are always born in the spring and open their eyes one week after birth. The cubs begin to follow their mothers around when they are about three months old. By winter, they will be almost grown up.

MOTHERLY LOVE

Tiger cubs are capable killers by the time they are 11 months old. They stay with their mothers, however, until they are two or even three years of age. In the wild, the mother does all she can to protect her young, but often at least half of the litter dies. Predators may kill the cubs, or sometimes they starve to death if the mother cannot catch enough food.

IN DISGUISE

A cheetah cub is covered in long, woolly fur. This makes it look similar to the African honey badger, a very fierce animal, which may help to discourage predators. The mother cheetah does not raise her cubs in a den, but moves them around every few days.

A cheetah cub, unlike the adult, has a mane of fur. This may help to disguise it as it hides in the long grass.

BRINGING UP BABY

Female pumas give birth to up to six kittens (babies). The mother has several pairs of teats for the kittens to suckle from. Each kitten has its own teat and will use no other. They will suckle her milk for at least three months and from about six weeks they will also eat meat.

ON GUARD

Two lionesses guard the entrance to their den. Lions are social cats and share the responsibility of keeping guard. Dens are kept very clean so that there are no smells to attract predators.

MOVING TO SAFETY

If at any time a mother cat thinks her cubs are in danger, she will move them to a new den. She carries the cubs one by one, gently grasping the loose skin at their necks between her teeth.

Focus on

PLAY TIME
A cub plays with its mother's tail. As soon as cubs can see, they begin to play. Play helps to build up muscles, improve co-ordination and develop good reflexes. It is valuable early preparation for learning how to hunt when the cubs are older.

The number of cubs in a big cat's litter depends on the species and where it lives. Most big cats have two or three cubs, but cheetahs have five or more. All cubs are born helpless, but it is not long before their eyes open and they can wobble about, learning to balance on their uncertain legs. Within a few weeks they begin to play with their mothers and each other. There is a lot to find out, but they learn very quickly. By the time they are six months old they will have learned how to keep safe, what food tastes good and how to catch it. They will start to understand the language of smells and sounds. For the next year and a half they will stay close to their mothers, practising their new skills until they are experts.

SAFETY FIRST
For the first two years of their lives, cubs remain close to their mother. She protects them and helps them when they make mistakes. A good mother may rear all her cubs successfully in a good year. A poor mother may lose most or even all of her cubs.

BATH TIME
Cubs must learn to clean themselves, but while they are still young their mother washes them with her tongue. As she licks, she spreads her scent on the cubs so that all of her family has the same smell.

Cute Cubs

FAMILY BLISS
Lion cubs are spotted all over to help hide them from predators. The spots gradually fade as the lions grow older. Adult lions have only very faint spots on their legs and stomachs. Lion cubs are lucky because they have many companions to play with. Cubs of solitary cats have to grow up without much company. Some do not even have any brothers or sisters. Lion cubs learn through play how to get on with their fellows.

MOVING HOME
To move her cubs a lioness carries each one gently in her mouth. Not only do the cubs have loose skin at their necks, but the lioness has a special gap in her mouth behind the canines. This allows her to lift the cub off the ground without biting it.

LION LESSONS
These cubs are working together to kill an injured warthog. One grabs the neck, while the other starts tearing at the hind leg. The mother lioness watches over them. She is the cubs' teacher. They must learn to hunt as soon as possible and this warthog is a small animal for them to begin with. The lioness brought down the warthog so that the cubs could learn to kill it.

Growing Up

Growing cubs have to learn all about life as an adult so that they can look after themselves when they leave their mother. She teaches them as much as she can and the rest they learn through play. Cub games depend on their species, because each type of cat has different things to learn. In play-fighting cheetahs use their paws to knock each other over. This is a technique they will need for hunting when they are older. Cubs need to learn how to judge distances and when to strike to kill prey quickly, without getting injured or killed themselves. Their mother introduces them to prey by bringing an animal back to the den to eat. Mothers and cubs use very high-pitched sounds to communicate. If she senses danger, she growls at them to tell them to hide.

▲ **PRACTICE MAKES PERFECT** These cheetah cubs are learning to kill a Thomson's gazelle. When the cubs are about 12 weeks old, a mother cheetah brings back live injured prey for them to kill. They instinctively know how to do so, but need practice to get it right.

▼ **FOLLOW MY LEADER** Curious cheetah cubs watch an object intently, safe beside their mother. At about six weeks the cubs start to go on hunting trips with her. They are able to keep up by following her white-tipped tail through the tall grass.

THE CLASSROOM

A group of lion cubs relax in the shade on a fallen tree. From here they watch the adults hunt, as if in a big, open-air classroom. Female cubs often stay in the pride, but young males are chased off by the dominant male.

SCRATCH AND SNIFF

Three young lions sniff at the shell of a tortoise. Cubs learn to be cautious when dealing with unfamiliar objects. First the object is tapped with a paw, before being explored further with the nose. Cubs' milk teeth are replaced with permanent canine teeth at about two years old. Not until then can they begin to hunt and kill big animals.

TAIL TOY

A mother leopard's tail is a good thing for her cub to learn to pounce on. She twitches it so the cub can develop good co-ordination and timing. As the cub grows, it will practise on rodents and then bigger animals until it can hunt for itself. Once they leave their mothers, female cubs usually establish a territory close by, while males go farther away.

Enemies of Big Cats

Big cats are perfect killing machines and are feared by all their prey. They do, however, have enemies. Big cats have to watch out for other carnivores taking their food or attacking their cubs. Wolves are a problem for pumas; wild dogs are a threat to tigers; hyenas and jackals prey on the cubs of African big cats. Even prey animals can be a danger to big cats. Buffaloes are very aggressive and can attack and kill a young lion. Humans, however, are the main enemies of wild cats. As people move farther into the wilderness to build homes and farms, they destroy the precious habitats of the big cats. People kill the big cats' prey, leaving them with less to eat. They also hunt big cats for their beautiful and valuable fur coats.

SCAVENGING HYENAS

A spotted hyena finishes off the remains of a giraffe. Hyenas live in Africa and western Asia. They eat whatever they can find. This is often carrion (the remains of dead prey) that animals such as big cats have killed for themselves. They will also kill cubs. Hyenas have strong jaws and bone-crunching teeth and look for food at night.

Did you know? Despite their dog-like appearance, hyenas are more closely related to cats.

PACK POWER

Wolves live in Europe and North America. They live in the mountains as well as on open plains, hunting in packs of up to 20 animals. Wolves usually eat deer, elk, moose, or small animals such as hares. But they never ignore a good meal caught by a puma.

BIG CAT THREAT

Leopards live in the same areas as cheetahs, but they are very hostile towards them. In fact, if they get a chance, leopards prey on cheetahs and their cubs. In turn, leopards have to be very wary of lions. Lions will attack and kill a leopard to protect the pride or their territory. Big cats do not like others because of competition for food in an area.

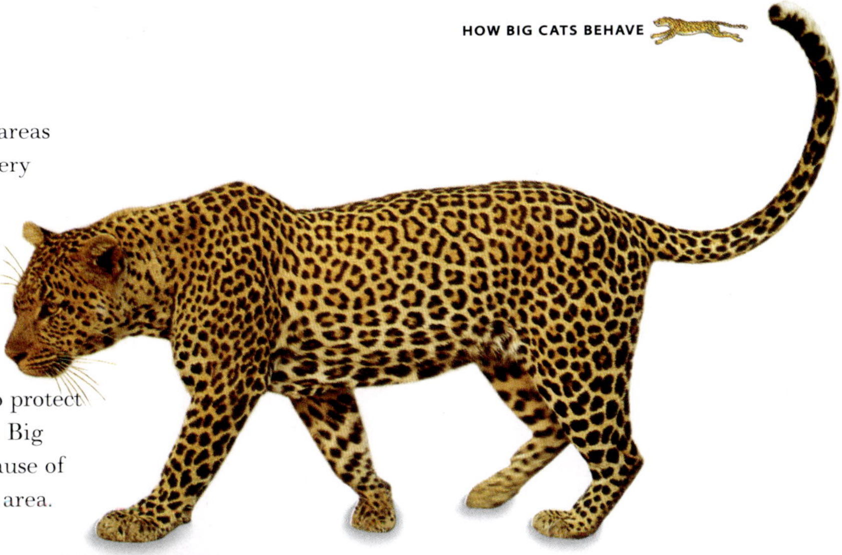

HUMAN TRAPS

Experts examine a tiger trap. Poachers (people who kill animals illegally) often use traps to catch big cats. When the trap snaps shut, the animal is stuck until it dies or until the poacher returns to kill it. These traps cause great pain. A cat may try to chew off its trapped leg to escape.

DOG-LIKE JACKALS

Jackals (a relative of the dog) are half the size of hyenas and live in Africa. They will eat most things and will steal a big cat's kill. If they come across an unprotected den, they quickly kill and eat all the cubs.

INTRUDER PERIL

Sometimes big cats become cannibals and eat their own kind. These cheetahs are eating another cheetah that has invaded their territory. Male lions also eat all the young cubs in a pride when they take over dominance from another male.

Mountain Cats

To live in the mountains, cats need to be hardy and excellent rock climbers. They also have to cope with high altitudes where the air is thin and there is less oxygen to breathe. Big cats that live in the mountains include leopards and the rare snow leopard. Small cats include the puma, mountain cat, bobcat and lynx. Mountain climates are harsh and the weather can change very quickly. To survive, mountain cats need to use their wits and to know where to find shelter. They mate so that their cubs are born in the spring. This is to ensure that they will be almost grown by the time winter closes in.

▲ **MOUNTAIN CAT**
The mountain cat is a secretive, shy creature and seldom seen. It is about 50cm long and has soft, fine fur. It is also known as the Andean mountain cat, since it lives in the high Andes mountains of Chile, Argentina, Peru and Bolivia. This cat is found at altitudes of up to 5,000m above sea level.

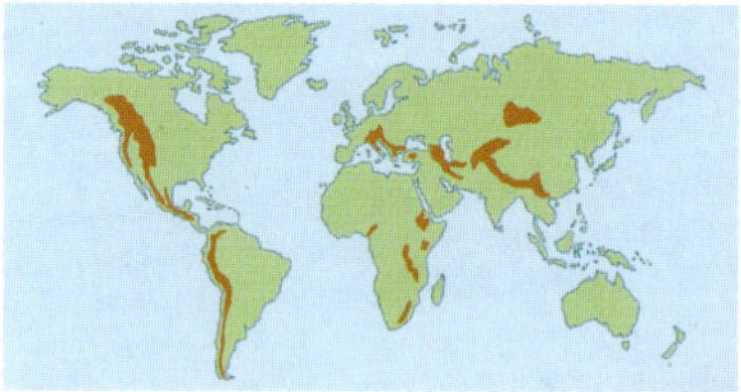

This map shows the world's major mountain ranges. The puma, mountain cat and lynx live in the Americas. Lynx also live in Europe and Asia, while the snow leopard lives in Asia.

◀ **MOUNTAIN LION**
A puma keeps watch over its vast territory. Pumas are also known as mountain lions and cougars. Male pumas can grow to 2m long and weigh 100kg. They are good at jumping and can easily leap 5m on to a high rock or into a tree. Pumas are found over a wide area, from Canada to the very tip of South America. They live along the foothills of mountains, in forests on mountain slopes and all the way up to 4,500m above sea level. Depending on where they live, pumas will eat deer, porcupines, hares, beavers and armadillos.

WINTER LYNX

Lynx live in mountainous regions of Europe, Asia and North America. They have unusually short tails and tufted ears. Lynx are well designed to live in very cold places. In winter they grow an especially long coat, which is light coloured so that they are well camouflaged in the snow. The bobcat of North America looks similar to the lynx.

PUMA CHASE

A snowshoe hare darts this way and that to shake off a puma. To catch the hare, the puma must make full use of its flexible back and its long balancing tail. Pumas hunt by day as well as by night.

LONG-TAILED SNOW LEOPARD

The snow leopard is one of the rarest big cats, found only in the Himalaya and Altai mountains of central Asia. It can live at altitudes of 6,000m, the highest of any wild cat. Snow leopards feed on wild goats, hares and marmots. Their bodies measure just over 1m long, with tails that are almost as long. They wrap their bushy tails around themselves to keep warm when they are sleeping. Snow leopards are particularly agile jumpers and said to be able to leap a gap of 15m. Their long tails help them balance as they jump.

Did you know? A female snow leopard makes her den cosy by lining it with her own fur.

Forest Dwellers

Dense, wet rainforests are home to lots of small creatures, such as insects and spiders. These animals and forest plants provide a feast for birds, snakes, frogs and small mammals, which in turn are a banquet for big cats. Jaguars, tigers, leopards and clouded leopards all live in rainforests. Small cats include ocelots and margays. Although there is plenty of food in a forest, the dense trees make it a difficult place to hunt. There is little space among the trees and prey can escape easily in the thick undergrowth.

This map shows where the world's tropical rainforests are located. They lie in a band either side of the Equator.

CLOUDED LEOPARD
The clouded leopard is a shy and rarely seen Asian big cat. It lives in forests from Nepal to Borneo, spending most of its time in the trees. The Chinese call the clouded leopard the mint leopard because of its unusual leaf-like markings. Male clouded leopards reach about 1m long, with an equally long tail, and weigh about 30kg. They are perfectly built for tree climbing, with a long, bushy tail for balance and flexible ankle joints.

JAGUAR
Although jaguars live in grasslands and semi-deserts, they prefer the thick forests of South and Central America. They are the third largest big cat (after tigers and lions), growing to 2.4 m in length and weighing up to 120kg.

Did you know? The jaguar was the symbol of the Sun for the Maya of Central America.

SUMATRAN TIGER

A tiger walks stealthily into a jungle pool on the island of Sumatra. Tigers are good swimmers and a forest pool is a good place to hunt as well as to cool off from the tropical heat. Tigers often hide the carcasses of their prey in water or the dense undergrowth.

The sleek margay is very striking in appearance, with beautifully marked fur and large eyes.

MARGAY

Margays live in the tropical forests of Central and South America. They are the best of all cat climbers, with broad, soft feet and exceptionally flexible ankles and hind legs. They feed largely on birds and so need to be good at moving around in the tops of trees.

JAVAN LEOPARD

At one time, many leopards lived in the tropical rainforests of Asia. But now, like this Javan leopard, they are endangered. They are threatened by over-hunting and the destruction of their forest habitat.

A TURTLE TREAT

A jaguar catches a river turtle in a pool. Jaguars are such good swimmers that they hunt some of their prey in water. They love to eat fish and turtles. Their jaws are powerful enough to crack open a turtle's shell like a nut. They have been known to kill cayman (a type of crocodile).

BLACK JAGUAR

Forest jaguars are darker than their grassland cousins. Some can be black and are so well camouflaged that they can disappear in the shadows of their forest habitat.

Focus on

SOLITARY SIBERIAN
Siberian tigers live alone in huge territories of over 1,000km. They do not like to fight and try to avoid each other. A tiger will kill another if it invades its territory.

Tigers are the largest of all the big cats and the largest of all the tigers is the Siberian tiger. An adult male can reach up to 2.6m long and weigh as much as 270kg. Siberian tigers live in the snow-covered forests of Siberia, which is part of Russia, and Manchuria in China. They are also sometimes known as Amur tigers. Although there is only one species of tiger, they can differ significantly in their appearance. Siberian tigers have a relatively pale coat with few stripes. Bengal tigers from India, however, have shorter fur and are more strikingly marked. As humans destroy more of their habitats, the number of tigers in the wild is declining dramatically. Today, there are only about 400 Siberian tigers left in the wild.

OPEN WIDE
A Siberian tiger shows its long, sharp canine teeth in a wide yawn. Canines are used to catch and kill prey. Tigers ambush prey and kill it by biting the neck or strangling.

LUNCH TIME
A group of Siberian tigers devour a black ox. Despite being solitary animals, tigers do sometimes share food. The only other time they come together is to mate. Tigers have been known to roar when they have killed a big animal, just as lions will often roar when they have successfully caught their prey.

Siberian Tigers

ICY TONIC
A mother shows her 18-month-old cub how to get water from melted ice. Tigers have up to four cubs in a litter, every other year. They stay with her for at least two years.

COURTING COUPLE
When a female is ready to mate, she sprays, roars and grunts to tell the male. When tigers want to be friendly, they blow sharply through their nostrils and mouths, rub their heads together and gently bite each other's necks.

A PALE ICE QUEEN
Siberian tigers like this female have a lot of white fur. This makes it more difficult for prey and enemies to see them in the snow. They are powerfully and heavily built, with bodies slung close to the ground.

On the Savanna

Savannas are open, flat areas of grassland. Apart from grasses, the main plants of the savanna are small bushes and clumps of trees. Savanna is the ideal habitat for big herds of grazing animals, such as antelope, zebra and buffalo. In Africa, these herds migrate for thousands of kilometres each year in search of fresh grass and water. They are followed by lions, cheetahs and leopards who prey on the herds. The savanna of South America is home to jaguars. Rodents, such as mice, gerbils and marmots also thrive on the savanna and these are a good food source for smaller cats, such as the serval.

LION IN THE GRASS
A lion walks through the long, dry grass of the African savanna. Its sandy colouring perfectly matches its habitat. Lions hunt their quarry using the cover of grass. Often, only the tips of a lion's ears are seen as it slowly stalks its prey.

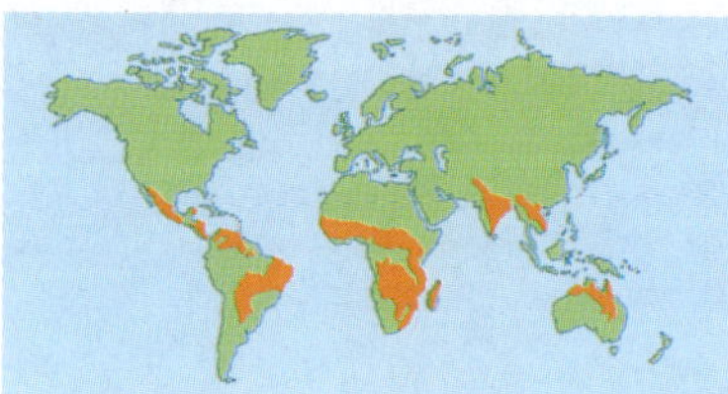

This map shows where the world's savannas (tropical grasslands and dry woodlands) are located. The largest region of savanna lies in Africa.

Did you know? Cats sleep longer than most other animals. Lions sleep for 20 hours a day.

CHEETAH ON THE LOOKOUT
A cheetah stands on the top of a small mound on the Kenyan savanna. Cheetahs are perfectly adapted for life on the plains. The surrounding open, flat terrain lets them make the most of their ability to chase down prey. From its vantage point, a cheetah uses its excellent eyesight to search for prey. It also keeps watch for any other cheetahs who might have invaded its home range.

◄ VIEW FROM A BRANCH
Leopards like to live in areas of grassland where there are trees. Here they can sleep hidden during the heat of the day. They can also enjoy the afternoon breeze and avoid the insect pests that live in the grass below. Leopards also prefer to eat up in a tree, out of the reach of scavengers.

Leo the Lion
People born between July 24 and August 23 are born under the astrological sign of Leo (the lion). They are said to be brave, strong and proud, just like lions.

▲ AT THE WATERHOLE
During the dry season in the African savanna, many grazing animals gather near waterholes to drink. Thomson's gazelle, zebra and giraffe are shown here. Lions congregate around the waterholes, not only to drink, but also to catch prey unawares. Their favourite prey is antelope, buffalo, zebra and warthog, but they also eat giraffe.

SPEEDY SERVAL ►
Servals are small cats that live on the savanna of western and central Africa. They like to live near water where there are bushes to hide in. The servals' long legs enable them to leap over tall grass when they hunt small rodents. They also climb well and hunt birds. With their long legs, servals can run quickly over short distances and so escape from predators.

Desert Cats

Hot deserts are very dry places. Although they are hot during the day, at night they are very cold. Few plants and animals can survive in such a harsh environment, but cats are very adaptable. Cheetahs, lions and leopards live in the Kalahari and Namib deserts of southern Africa. As long as there are animals to eat, the cats can survive. Even the jaguar, a cat that loves water, has been seen in desert areas in Mexico and the southern USA. But they are only visitors in this tough, dry land and soon go back to the wetter places they prefer. The most well-adapted cat to desert life is the sand cat. It lives in the northern Sahara Desert, the Middle East and western Asia.

▲ **DESERT STORM**
Two lions endure a sandstorm in the Kalahari Desert of southern Africa. The desert is a very hostile place to live. There is very little water, not much food and the wind blows up terrible sandstorms. Despite these hardships, big cats like these lions manage to survive.

This map shows where the world's hot deserts and nearby semi-desert areas are located.

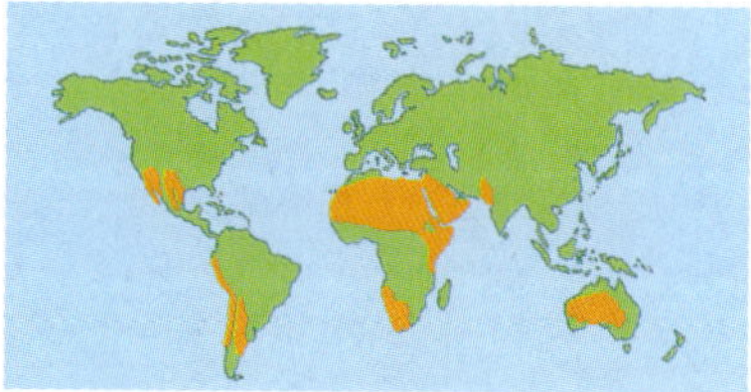

Did you know? Lions follow along dry riverbeds looking for waterholes in the desert.

▼ **A HARSH LIFE**
An old lion drinks from a waterhole in the Kalahari Desert. Even when a big cat lives in a dry place, it still needs to find enough water to drink. This is often a difficult task, requiring the animals to walk long distances. In the desert, prey is usually very spread out, so an old lion has a hard time trying to feed itself adequately.

CHEETAH WALK

A group of cheetahs walk across the wide expanse of the Kalahari Desert. They lead lives of feast and famine. In the rainy season, lush vegetation grows and enormous herds of antelope can graze. The cheetahs have a banquet preying on the grazing herds. But they go very hungry as the land dries up and prey becomes scarce.

SAND CAT

Dense hair on the soles of the sand cat's feet protect it as it walks on hot ground and help it walk on loose sand. All the water the cat needs comes from its food, so it does not need to drink.

ADAPTABLE LEOPARD

A leopard rolls in the desert sand. There are very few trees in the desert, so leopards live among rocky outcrops. Here they can drag their prey to high places to eat in safety. The desert can be a dangerous place. With so little food around, competition can be fierce, especially with hungry lions. Big cats will eat small prey such as insects to keep off starvation.

Egyptian Cat Worship

The Ancient Egyptians kept cats to protect their stores of grain from rats and mice. Cats became so celebrated, they were worshipped as gods. They were sacred to the cat-headed goddess of pleasure, Bast. Many cats were given funerals when they died. Their bodies were preserved, wrapped in bandages and richly painted.

Killer Cats

Humans can sometimes become the prey of big cats. People have been afraid of big cats for thousands of years. From 20,000-year-old cave paintings we know that people lived in contact with big cats and almost certainly feared them. More recently, there have been many reports of big cats killing people. Lions and tigers become bold when they are hungry and there is little other food around. First they prey on livestock such as cattle. When the cattle are taken away, the big cats might kill people. Leopards, who do not have a natural fear of humans may have their killer instinct triggered by an injury.

LION BAITING
The Romans used lions (and bears) for gladiator fights in their amphitheatres (outdoor arenas). When the Romans wanted to kill prisoners, they would feed them to hungry lions. The lions had to be starving and made angry by their handlers, otherwise they would not kill the prisoners. Most captive lions will not kill people.

WRESTLING A TIGER
Tigers are considered the most dangerous of all the big cats. This picture, called *A Timely Rescue*, shows a rather heroic view of killing a tiger. Once a tiger has become used to the taste of human flesh, it will strike at any time. Tigers have killed thousands of people over the centuries. During the early 1900s, tigers killed 800 to 900 people a year in India.

Did you know? In the early 1900s, one Indian leopard killed 125 people in eight years.

TIPPU'S TIGER

This mechanical toy of a tiger attacking a British soldier was made in 1790. It is called Tippu's Tiger and was made for the Sultan of Mysore. The Sultan resisted the British takeover of India and his toy makes growling and screaming noises.

HUNTER WITH A HEART

Jim Corbett was a famous hunter who lived in India in the early 1900s. Unlike most hunters of his day, he did not kill big cats for sport. He shot tigers and leopards that had been eating people.

Jaguar Knights

In the 1500s in Central America, Aztec warriors were divided into military orders. Some of the most prestigious were the jaguar knights who ranked just below the emperor. They wore entire wild cats' pelts, with the still-attached heads worn as helmets. They thought by wearing the pelt they would take on the cat's strength and quiet stealth.

EDUCATION FOR CONSERVATION

Nearly 100 tigers and 50 leopards live in the Corbett National Park in India. The Park runs programmes to teach children all about the big cats and their habitat. The more we know about big cats, the better able we are to respect them.

Cats in Danger

LION HUNT
Egyptian rulers hunted lions from horse-drawn chariots. Hieroglyphics (picture writing) tell us of Pharoah Amenophis III (1405–1367BC) who killed over 100 lions in the ten years of his rule. Some experts now think that the Egyptians may have bred lions specially to hunt them.

The earliest record that we have of people using wild-cat pelts (skins) is from 6,500BC. It comes from the archeological site of Çatal Hüyük in Turkey where there is evidence that dancers wore leopard skins. Much more recently, in the 1800s and 1900s, many wealthy people wanted to hunt big game for the thrill of the chase. Big cat skins were used to decorate the hunters' houses, and their heads hung as trophies on the walls. Today, no one is allowed to hunt the endangered big cats anymore.

TIGER-HUNTING PRINCE
This old painting on cotton shows an Indian prince hunting a tiger from the back of his elephant. Tiger hunting was a very popular pastime for many centuries in India until it was declared illegal in the 1970s.

GREAT WHITE HUNTER

A hunting party proudly displays their tiger trophy. This photograph was taken in the 1860s. When India was under British rule, tiger hunting was considered to be a great sport by the British. Uncontrolled, ruthless hunting was a major cause of the tiger's dramatic fall in numbers.

RITUAL ROBES

The Zulu chief Mangosothu Buthelezi wears wild cat skins on special occasions, like many African leaders and tribal healers. They are a sign of his rank and high status.

Did you know? Until the 1800s, black panthers lived near Los Angeles, California.

CAT'S SKIN

A leopard is skinned, having been shot in the Okavango Delta, Botswana. Some game reserves raise money for conservation by charging huge sums to hunt. This only happens when numbers of a certain species are too large for the reserve.

SECOND SKIN

Some people continue to think it looks good to wear a coat made from the pelts of a wild cat. Many more, however, think that the fur looks much better on the cat. Designers now use fake fur and skins dyed to look like pelts, instead.

Big Cat Relatives

Fossil remains show that the ancestors of today's cats first roamed the Earth 65 million years ago. They were small animals called miacids. Scientists think that the miacids may be the forerunners of all carnivorous mammals on Earth today. About 35 million years ago the first cat-like carnivores appeared. Many of these early cats were large and dangerous creatures that have now become extinct (died out). The best-known are the sabre-toothed cats, which, unlike modern cats, used their canines for stabbing rather than biting. Fossil records of cats similar to modern cats date back about 10 million years. Today, there are 37 species of wild cat, big and small, and over 300 breeds of domestic cat.

ANCIENT ANCESTOR
Miacids are thought to have been the first true carnivorous mammals. They first lived about 65 million years ago, around the same time that the dinosaurs died out. They are the first animals to have carnassial teeth – the strong back teeth used for shearing meat.

Did you know? The bones of sabre-tooth cats were found under Trafalgar Square, London.

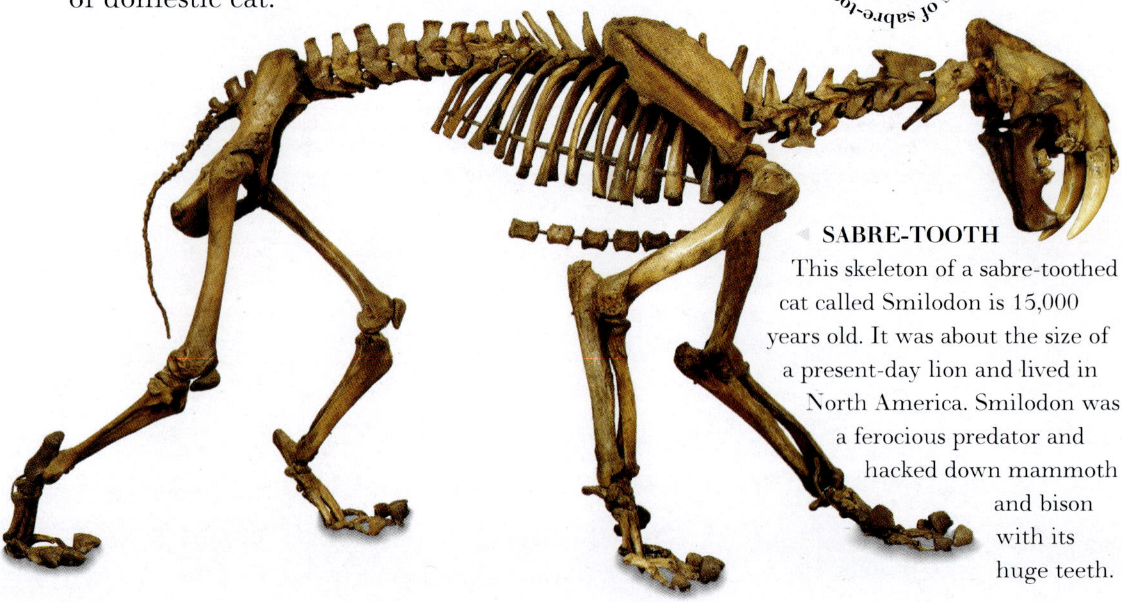

SABRE-TOOTH
This skeleton of a sabre-toothed cat called Smilodon is 15,000 years old. It was about the size of a present-day lion and lived in North America. Smilodon was a ferocious predator and hacked down mammoth and bison with its huge teeth.

DOMESTIC CATS

Small cats are grouped under the genus *Felis.* Domestic or pet cats (*Felis catus*) are like miniature tamed versions of their wild relatives the big cats. They hunt, groom and leave scent marks just like wild cats. When a house cat rubs itself against a human, it is showing affection, but also putting its smell on that person.

Classification Chart

Kingdom	**Animalia** (all animals)
Phylum	**Chordata** (animals with backbones)
Class	**Mammalia** (animals with hair on their bodies that feed their young with milk)
Order	**Carnivora** (mammals that eat meat)
Family	**Felidae** (all cats)
Genus	***Panthera*** (big cats)
Species	***leo*** (lion)

CAT NAMES

Scientists classify (group) every cat within the animal kingdom and give it a Latin name. This chart shows how the lion is classified.

SERVAL

The serval (*Felis serval*) is a spotted, fleet-footed wild cat. It lives near water in southern and central Africa. Although it has long legs like a cheetah, it belongs to a different group and purrs like a domestic cat.

CARACAL

The caracal's (*Felis caracal*) most startling feature is the long tufts on its ears. These may help it locate prey. The caracal is related to both the leopard and the lynx. It lives in grasslands, open woodland and scrub in Africa and parts of Asia.

BOBCAT

The bobcat (*Felis rufus*) is related to the lynx (*Felis lynx*). Both have short tails and live in many sorts of habitats, but prefer the rocks and shrubby plants of mountain slopes. The bobcat is only found in northern North America where it is able to survive the harsh winter conditions.

Protecting Cats

All big cats are in danger of extinction. They are not only hunted for their skins, but for their teeth, bones and other body parts, which are used as traditional medicines in many countries. The Convention for International Trade in Endangered Species (CITES) lists all big cats under Appendix 1, which strictly controls their import and export. For cats particularly at risk, such as the tiger, all trade is banned. There are now many protected areas throughout the world where big cats can live without human interference. These areas are often not big enough, however, so the cats leave in search of food. They attack livestock and sometimes the local farmers.

IN ANCIENT TIMES
A Roman mosaic showing a horseman hunting a leopard. Two thousand years ago, big cats were much more widespread. Until the 1900s, cheetahs lived throughout Africa, central India and the Middle East. Hunting big cats was not a problem when there were many big cats and not so many people, but now the situation is desperate.

GIR NATIONAL PARK
The last remaining Asian lions live in the Gir National Park in western India. There are less than 300 lions living in the park. The Asian lion is slightly different to the African lion. It has a smaller mane and a fold of skin running between its front and back legs.

SERENGETI LION PROJECT
This lion has been drugged so that it can be fitted with a radio collar, checked and then released. In the Serengeti National Park in Tanzania, scientists use methods like this to study lion behaviour.

TIGER DETERRENT

Villagers in the Sundarbans mangrove forests in India wear masks on the backs of their heads. Tigers attack from behind, but will not usually strike if they see a face. The largest remaining tiger population in India is in the Sundarbans. Here 50 to 60 people die each year from attacks by tigers. There is obviously not enough food for the tigers, so conservationists are trying to improve the situation. Another deterrent is to set up dummies that look and smell like humans, but give out an electric shock if attacked. There are also electrified fences in some areas, and pigs are bred and released as tiger food.

RADIO TRACKING

Biologists attach a radio collar to a tigress in the Chitwan National Park, Nepal. To save wild cats we need to understand their habits and their needs. For this reason, many scientists and conservationists are studying them. It is a very difficult task since cats are secretive and often nocturnal animals. One way of gathering information is to put a radio collar on a big cat and then follow its movements. By doing this, the animal can be tracked at long range.

The Great Sphinx

In Egypt, an enormous statue with a human head and the body of a lion guards the Great Pyramids at Giza. This statue is the Great Sphinx. A story written in 1419BC on the Sphinx tells of a prince called Thutmose IV who fell asleep between the paws of the statue. He dreamt that the Sun god told him to take away the sand covering part of the Sphinx and he would become a king. When he awoke, Thutmose did as he had been instructed, and the dream came true.

GLOSSARY

altitude
The height of a place given in metres (or feet) above sea level.

ambush
When an animal hides, waiting for prey to walk past, then pounces on it in a surprise attack.

artery
A blood vessel that carries blood away from the heart.

binocular vision
The ability to see things with both eyes at the same time. This enables cats to judge distances and depth accurately.

bladder
Where waste urine is stored in the body before being expelled.

breed
An animal that belongs to one species, but has definite characteristics, such as colouring, body shape and coat markings.

camouflage
Colours or patterns that allow an animal to blend in with its surroundings to avoid detection.

canines
The sharp, pointed teeth that cats use for killing and holding prey.

carnassials
The strong, shearing teeth at the back of a cat's mouth.

carnivore
An animal that feeds mainly on the flesh of other animals.

classification
Arranging animals according to their similarities and differences in order to study them and suggest how they may be related.

dew claw
The digit (toe) on a cat's front foot that is held off the ground to keep it sharp. It is used to knock down and hold on to prey.

digestion
The process by which food is broken down so that it can be absorbed by the body.

domestic cat
A species of cat whose wild ancestors were tamed by people and bred in captivity.

dominance
A system between social big cats, such as lions, in which one or a few animals rule the group and have first choice over the other, more junior members.

Equator
An imaginary line running around the centre of the Earth, separating north from south.

evolve
When an animal or plant species develops and changes over a long time. As conditions alter, living things change to become better adapted (suited) to surviving. Those that cannot adapt may become extinct (die out completely).

extinct
When a whole species or larger group of animals or plants has disappeared, dying out completely.

fossil
The remains of a once-living plant or animal that have turned to stone over thousands or millions of years.

gland
An organ in the body that produces chemicals for a particular use.

habitat
A type of living place that has certain kinds of animals and plants living there, such as tropical rainforest or semi-desert.

incisors
The front teeth, which a cat uses for cutting up meat.

insulation
A covering, like a cat's fur coat, that does not allow too much heat in or out of the body.

intestine
Part of the gut of an animal specialized to digest food.

jungle
The dense undergrowth found in a rainforest.

kidney
An organ of the body that filters blood to remove waste products, called urine.

liver
An organ that processes food from the digestive system (gut). One of the liver's main tasks is to remove any poisons from the blood.

lungs
An organ of the body that takes in oxygen from the air.

mammal
A warm-blooded animal that has an internal skeleton, a fur-covered body and mammary glands that produce milk to feed its babies.

mangrove
Trees that grow in muddy swamps near the sea.

mating
The pairing up of a male and female big cat to produce young. During mating, the fertilization (joining together) of a male sperm and a female egg takes place, which starts a new life.

membrane
A thin layer of skin that separates one area from another.

milk teeth
A young animal's first teeth, which are replaced by permanent teeth.

nerves
Fibres that carry electrical impulses to and from the brain.

oesophagus
Part of the gut of an animal, usually long and tube-shaped. It transports swallowed food from the mouth to the stomach.

predator
An animal that catches and kills other animals (prey) for its food.

prey
An animal that is hunted for food by another animal (predator).

rainforest
A tropical forest where it is hot and wet all year round.

rodent
An animal with chisel-shaped incisors (front teeth) used for gnawing. Rats, mice, beavers and porcupines are all types of rodent.

savanna
Open grasslands with a few scattered trees in tropical regions.

scavenger
An animal that feeds mainly on the remains left behind from another animal's meal.

social animal
An animal that lives in a group, usually with others of its own kind. Social animals co-operate with other group members.

species
All living things are grouped into species. Animals of the same species are similar to each other and can breed with each other. They produce young that in turn can breed together. All lions belong to one species, tigers to another and leopards to another. Each species is given a unique name, usually in Latin. The species name for lions is *Panthera leo*, for tigers it is *Panthera tigris* and for leopards it is *Panthera pardus*.

stalk
To follow prey quietly and carefully so that a predator is within striking distance.

sweat glands
Small organs beneath an animal's skin that produce sweat. Sweat helps to keep the body cool.

taste buds
Tiny bumps on a animal's tongue, which have nerve endings that pick up taste signals.

territory
An area in which an animal or group of animals live. Cats use scent to mark the borders of their territories so that other cats know to keep out.

trachea
The windpipe running from the nose and mouth to transport air to the lungs.

tropics
Warm, wet regions of the Earth that lie near the Equator.

vein
A blood vessel that carries blood back towards the heart.

warm-blooded
An animal that is able to maintain its body temperature at the same level all the time.

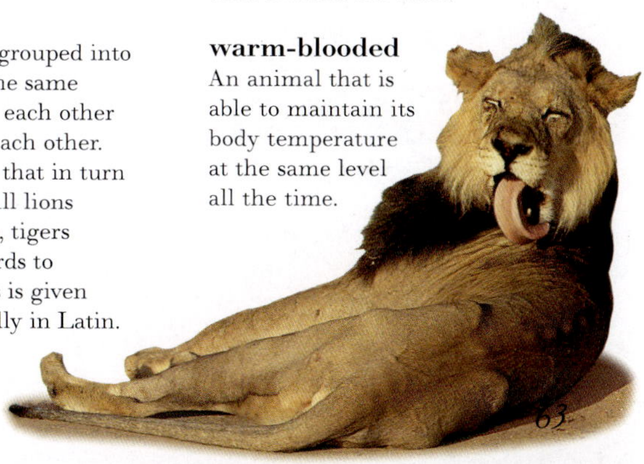

INDEX

A
Andean mountain cat 44
Aztecs 55

B
balance 4–5, 7, 18, 21
big cats 6–7
bobcat 19, 44–5, 59
body language 4, 15, 22–3
bones 8–9

C
camouflage 16–17, 28, 39, 45, 50
cannibalism 43
caracal 59
carnivores 4, 24–5, 58
cheetah 6, 8, 10–11, 20–1, 24–7, 37, 40, 43, 50–3
  king 16, 23
  social behaviour 30–1
  spots 17
CITES convention 60
claws 4, 10–11
  retractable 4, 6, 11, 20
climbing 8, 10, 18, 29, 44–6
clouded leopard 6, 46
communication 4, 14–15, 22–3, 48
conservation 55–7, 60–1
cougar *see* puma
courtship 23, 34–5, 49
cubs 23, 30–3, 36–41, 49

D
Daniel in the lions' den 31
dens 36–7
desert dwellers 52–3
dew claws 10, 21
digestive system 11, 24
domestic cat 4, 6, 58–9
drinking 14, 30, 49, 51–2

E
ears 5, 13, 22–3
Egypt, ancient 53, 56, 61
endangered cats 55–7, 60–1
eyes 4–5, 9, 12–13

F
fighting 22, 33–4, 48
flehmen 15
food 4, 8, 11, 14, 24–7, 37
forest dwellers 46–7
fur 4–5, 16–17, 19, 39, 42
  killing for 43, 56–7, 60

G
grooming 5, 15, 31, 38
guard hairs 16

H
hairballs 15
hearing 4–5, 12–13, 22
Heracles 10
hissing 22, 36
humans and cats 42–3, 54–7, 60–1
hunting 4–5, 10–13, 18–21, 24–7, 45
  at night 28, 61
  learning 38–41
hyoid bone 6

J
Jacobson's organ 15
jaguar 7, 17–18, 24–5, 46–7, 50, 52, 55
  black 17, 47
jaws 9
Jerome, St 5
jumping 18–19

K
keratin 11
killing prey 21, 24, 26–7, 39–40, 48
kittens 37

L
Leo (astrological sign) 51
leopard 6–8, 12, 14, 18, 22, 24, 28–9, 34, 41, 43–4, 46–7, 50–4, 57, 60
  black panther 17
  Javan 47
life expectancy 21, 26
lion 4–6, 9–11, 13, 15, 18–19, 22–7, 31–5, 37–9, 41–3, 50–2, 54, 56, 60
  coalitions 32
  prides 6, 27, 30–3, 41, 43
litters 36–9
lynx 13, 44–5

M
mane 6, 32, 37
margay 46–7
mating 15, 23, 28, 30, 32, 34–5, 49
miacids 58
mountain cat 44
mountain dwellers 44–5
mountain lion *see* puma
movement 4, 18–19
muscles 4, 8, 10–11, 38

N
Nemean lion 10
nose 10, 14, 20

O
ocelot 46

P
panther 7, 17, 47
papillae 14
paws 4–5, 9–11, 18–20
play 33, 38–40
predators 36–7, 39, 42–3
pregnancy 35
prey 10–13, 20–1, 24–7, 29, 39–40, 54–5
prides 6, 27, 30–3, 41, 43
puma 6, 13, 30, 37, 44–5
  black 17
pupils 5, 12–13
purring 6, 23

R
roaring 22, 48
Rome, ancient 54, 60
running 18–21

S
sabre-toothed cats 58
sand cat 52–3
savanna dwellers 50–1
scent glands 22–3
scent marking 22–3, 31, 38
scratching 22–3, 41
senses 12–15
serval 50–1, 59
sight 4–5, 9, 12–13, 50
  binocular vision 12–13
  colour vision 12
skeleton 8–9
small cats 6, 59
smell, sense of 14–15, 22, 31, 34, 38
Smilodon 58
snow leopard 6–7, 15, 17, 30, 36, 44–5
social behaviour 6, 25–7, 30–3, 48
Solomon, King 24
Sphinx 61
spine 8–9, 18, 20

T
tail 4–5, 7, 15, 18, 21–2, 45
tapetum lucidum 12–13
taste, sense of 14–15
tear stripes 16
teeth 4, 8–9, 39
  canine 4, 8–9, 41, 48, 58
  carnassials 8–9, 58
  milk 41
territorial behaviour 30–2 41, 48
tiger 4, 6–8, 11, 14–19, 23, 35–6, 46–9, 54–7, 61
  Bengal 19, 48
  Siberian 4, 48–9
  Sumatran 47
Tippu's Tiger 55
tongue 14–15, 38
touch, sense of 12, 14–15

W
whiskers 4, 14–15

NATURE WATCH

SHARKS

MICHAEL BRIGHT

Consultant: Ian K. Fergusson CBiol MIBiol

The Shark Trust

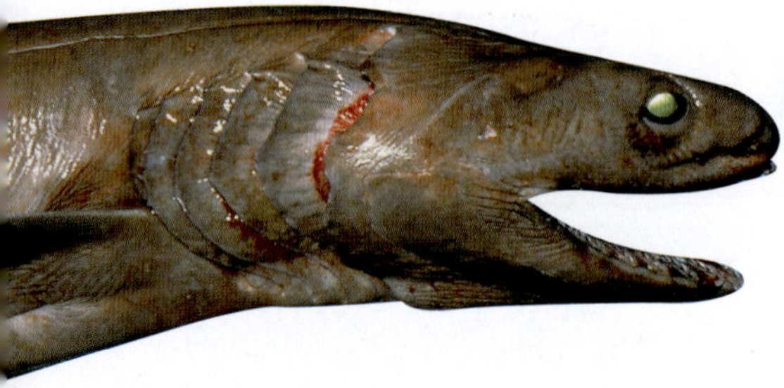

C O N

Published by Anness Publishing Ltd,
Blaby Road, Wigston, Leicestershire LE18 4SE
Email: info@anness.com
Web: www.annesspublishing.com

Anness Publishing has a new picture agency outlet for images for publishing, promotions or advertising. Please visit our website www.practicalpictures.com for more information.

Publisher: Joanna Lorenz
Managing Editor, Children's Books: Gilly Cameron Cooper
Senior Editor: Nicole Pearson
Editors: Peter Harrison; Sarah Eason
Designer: Ann Samuel
Picture Researcher: Gwen Campbell
Illustrators: Stuart Carter, Vanessa Card, David Webb
Production Controller: Yolande Denny
Editorial Reader: Joy Wotton

ETHICAL TRADING POLICY
Because of our ongoing ecological investment programme, you, as our customer, can have the pleasure and reassurance of knowing that a tree is being cultivated on your behalf to naturally replace the materials used to make the book you are holding. For further information about this scheme, go to www.annesspublishing.com/trees

PICTURE CREDITS
b=bottom, t=top, c= centre, l= left, r= right
BBC Natural History Unit: A James, 25top; D Hall, 2br, 41cl J Rotman, 9br, 11tl, 17tl; 17tr; 21tl; 49cl; 57tl; 64t; M Dohrn, 44tl; T Krull, 44br; Bridgeman Art Library: 32br; 45tl; British Museum: 5br; Bruce Coleman/Pacific Stock: 51br; FLPA: DP Wilson, 25br; 33tr; 51cr; LS Sorisio, 29bl; Gallo Images: G Cliff, 35tr; J Rotman, 9bl; 10bl; Heather Angel: 55cr; Innerspace Visions: A Nachoum, 8bl; B Cranston, 5tl; , 17br; , 18bl; , 19bl; 28tl; 29br; 48tl; 57bl; B Cropp, 21cl; B Rasner, 15tr; D Fleetham, 12tr; 40tl; 50br; 58tl; 63b; D Flettmann, 1; D Perrine, 3br; 12bl; 13bl; 14cl; 15tl; 15bl; 16bl; 17cl; 17bl; , 27c; 31bl; 37tl; 38br; 39bl; 40br; 45tr; 49br; 50tl; 54t; 61c; 61bl; 61br; D Shen, 3tr; 6tl; 62b; F Schulke, 57br; H Hall, 12cr; 32tl; 32bl; I Rutzky, 49bl; J Campbell, 59b; J Jaskolski, 55cl; J Knowlton, 30bl; J Morrissey, 35bl; , 42br; J Rotman, 31br; , 43bl; 49cr; 59tl; 64b; JC Carrie, 31cl; JD Watt, 4bl; 26tr; , 39br; M Conlin, 31tl; 33br; 43cr; 53cl; 55tr; M Snyderman, 10br; 19tl; 19cl; 20br; M Strickland, 61tl; MP O'Neill, 34bl; MS Nolan, 21bl; 22tl; N Wu, 34tl; P Humann, 28bl; R Cranston, 11bl; R Ellis, 49tl; R Herrmann, 18tr; R Kuiter, 2tl; , 6br; 42tl; R Troll, 48b; 63t; S Drogin, 20tl; S Gonor, 43tl; 58bl; SH Gruber, 60bl; T Haight, 25bl; W Schubert, 14tl; Mary Evans Picture Library: 53cr; National Geographic: 60tl; N Calogianis, 39tl; NHPA: 7cr; 7bl; 24tl; 24bl;
K Schafer, 22bl; N Wu, 23all; 27bl; , 41tl; 51cl; 59cl; Oxford Scientific Films: D Fleetham, 15br, 38cl; 51tr; 57cr; G Soury, 21br; 26bl; H Hall, 5tr; 39cr; 46all; 47tl K Gowlett, 11tr; M Deeble&V Stone, 36, 37bl; M Gibson, 56bl; N Wu, 6bl; 62c; R Herrmann, 20bl; 53tl; R Kuiter, 2cr; 42cl; W Wu, 47tr; 55b; Papilio Photographic: 16tl; 53bl; Planet Earth:
D Perrine, 36br; 37br; G Bell, 7l; 10tl; G Douwma, 22br; J Seagrim, 33tl; J&A Moran, 59cr; K Amsler, 54b; K Lucas, 47cl; 52tl; 62tl; M Snyderman, 5bl; 9tr; 29tl; 47bc; N Wu, 7tr; 11br; P Atkinson, 13tr; Scott Mycock & Rachel Cavanagh: 45cr; Seitre Bios: 8br; South American Pictures: R Francis, 45bl; Universal Pictures: 56tr; Zefa: M Jozon, 27tl

INTRODUCING SHARKS
4 What is a Shark?
6 Shapes and Sizes

HOW SHARKS WORK
8 Light and Strong
10 Tough Teeth
12 Never Stop
14 Wings in Water
16 Brain and Senses
18 Focus on the Blue Shark Hunt

SHARK BEHAVIOUR
20 Feeding
22 Focus on Tiger Sharks
24 Filter Feeders

SHARK SOCIAL LIFE
26 Keep in Line
28 Focus on Hammerheads

T E N T S

REPRODUCTION

30 Courtship and Mating

32 Inside and Outside

34 Into the World

36 Focus on Lemon Shark Birth

WHERE SHARKS LIVE

38 Look in any Ocean

40 Upwardly Mobile

42 The Ocean Depths

44 Freshwater Sharks

46 Life on the Seabed

RELATIVES AND ANCESTORS

48 Prehistoric Seas

50 Shark Relatives

52 The Eight Families

SHARK FACTS

54 Friends and Enemies

56 Sharks and People

58 Focus on the Great White Shark

60 Conservation

62 Glossary

64 Index

What is a Shark?

There are about 400 different kinds of shark in the world. Some are as big as whales, others as small as a cigar. Whatever their size, they all eat meat. Some sharks eat tiny plants and animals called plankton. Others hunt down fish, squid, and even seals. Many sharks will also feed off the remains of another's meal or eat animal carcasses. They live at all depths, in every ocean, from tropical waters to cold polar seas. Some sharks can survive in the fresh water of rivers and lakes. Like other fish, sharks take oxygen from the water as it passes over their gills. Although some sharks like to live alone, others survive as part of a group.

▼ CLASSIC SHARK
This blue shark (*Prionace glauca*) is how most people imagine sharks. However, there are many different families of sharks in the seas and oceans and with a variety of body shapes.

◄ WHITE DEATH
The awesome great white shark (*Carcharodon carcharias*) is the largest hunting fish in the sea. It has an exaggerated reputation as a killer, partly because a killer shark appears in the film, *Jaws*. In reality great white sharks do often eat large prey, but they attack people only occasionally, in cases of mistaken identity.

► SHARK SCHOOL

Some sharks live alone, others live in schools (groups). Every day schools of female scalloped hammerhead sharks (*Sphyra lewini*) like these gather off the Mexican coast. At night, the sharks separate and hunt alone.

▲ GENTLE GIANT

Although the basking shark (*Cetorhinus maximus*) is the second largest fish after the whale shark, it is not a hunter. It funnels water through its huge mouth, using gill rakers (giant combs) to filter out the tiny plankton that it eats.

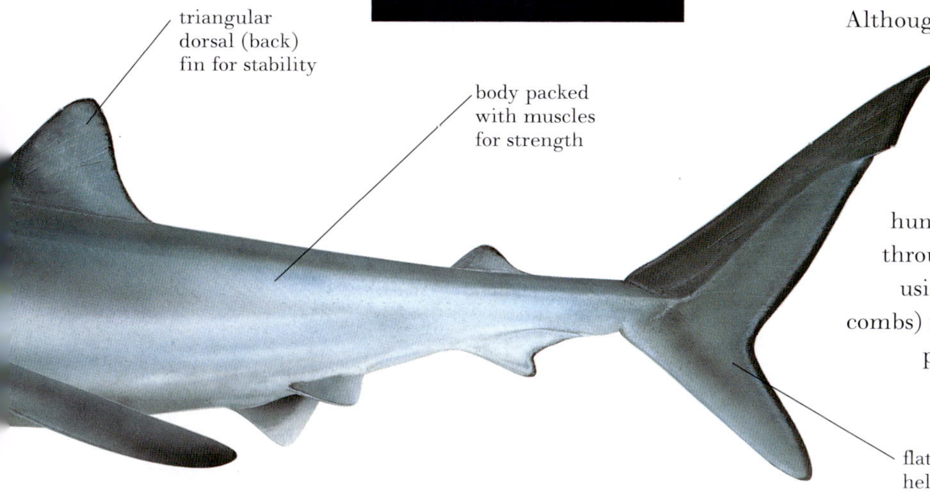

▲ REEF HUNTERS

Whitetip reef sharks (*Triaenodon obesus*) are one of the smaller species (kinds) of shark. They rarely grow over 2m long and hunt along tropical coral reefs at night.

Maya Origins

This monkey head from Central America has been decorated with shark teeth. It is thought that the word shark comes from the Maya people of Central America. It may be based on the Maya word xoc, *(fury). The Maya symbol (word picture) for* xoc *is a shark-like creature.*

Light and Strong

Although most fish are bony, the skeleton of a shark is made up almost entirely of cartilage, which is also found in the human nose. It is lighter and more elastic than bone, and it is this that makes the shark skeleton very flexible. This cartilage structure is strong enough to support a shark's huge muscles, and bendy enough to allow it to move with ease. Because sharks' skeleton cartilage and soft body parts decay (rot away) so quickly after they die, it is unusual to find complete shark fossils (preserved bodies) in ancient rocks. Only the hard teeth and spines are fossilized.

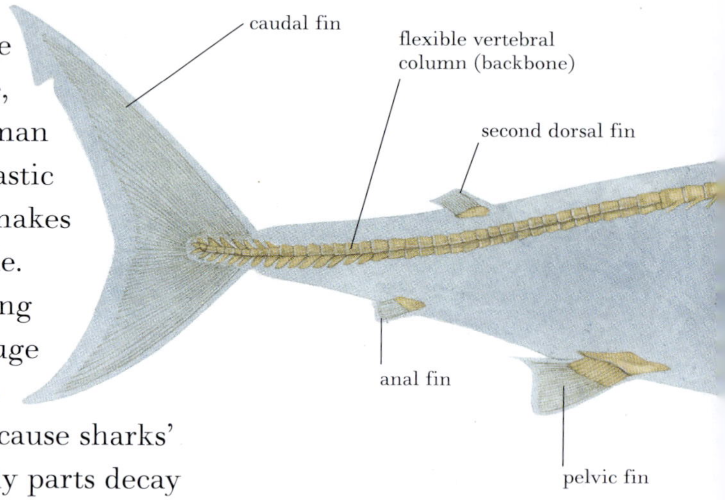

▼ PROTRUDING JAWS
A shark's jaws are attached to the skull by flexible ligaments. These allow some sharks to thrust their jaws forward when taking a bite.

▲ DARK TRIANGLE
Like the keel of a yacht, a shark's stiff dorsal fin helps it to balance in the water and stops it from slipping sideways. Most sharks have two dorsal fins, one at the front and one at the back, but some have just one.

▶ AEROPLANE FINS
A shark's pectoral fins, one on each side, act like the wings of an aeroplane. As water passes over them, the fins give lift.

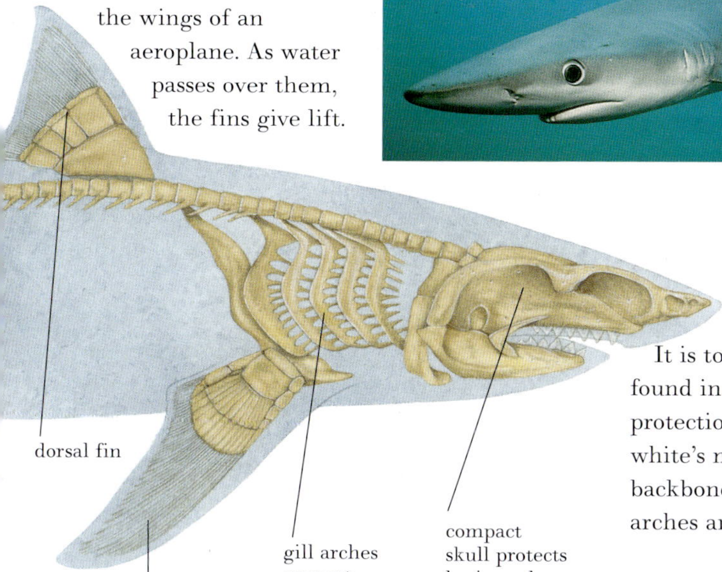

◀ SHARK SKELETON
The skeleton of a great white shark. It is tough, flexible and typical of that found in most sharks, providing support and protection for the entire body. The great white's muscles are attached to a long backbone, the gills are supported by gill arches and a box-like skull protects the brain.

▲ CARTILAGE SOUP
These fins have been cut from sharks and are drying in the sun. The cartilage in a shark's fin helps to make it stiff. When boiled in water it makes a gluey substance that is used in the Far East to make shark's fin soup.

▶ HARD NOSE
The shark pictured above is an adult basking shark. At birth, this shark has a strange, hooked nose, like that of an elephant. When the basking shark starts to grow, the cartilage in its snout gradually straightens.

Tough Teeth

A shark species can be identified by the shape of its teeth alone. Each species has its own distinctive shape, designed for the type of food it eats. Some have sharp, spiky teeth that can hold on to slippery fish and squid. Others have broad, grinding teeth that can crack open shellfish. The teeth of some species of shark change as they get older and hunt different prey. Although sharks lose their teeth regularly, the teeth are always replaced. Behind each row of teeth lie more rows. If a front tooth is dislodged, an extra tooth simply moves forward to take its place!

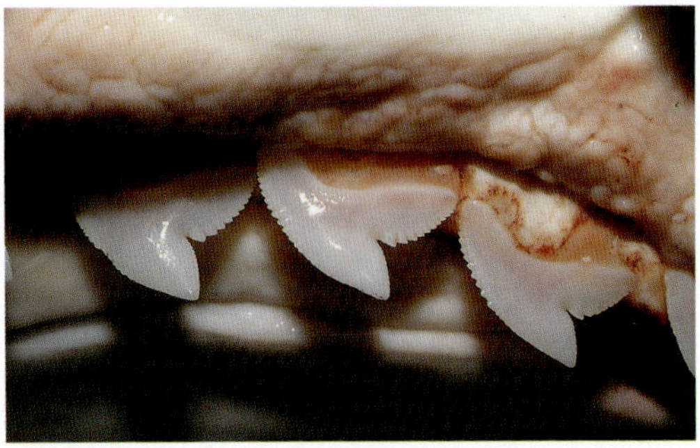

▲ SHARK SAW MASSACRE
The teeth of a tiger shark are shaped like the letter L. They can saw through skin, muscle and bone, and can even crack open the hard shell of a sea turtle. A tiger shark eats its prey by biting hard and shaking its head from side to side, slicing into its food like a chain saw.

▼ AWESOME JAWS
When it is about to grab its prey, a sandtiger shark opens and extends its awesome jaws. The rows of spiky teeth inside are perfect for grabbing and holding slippery fish and squid. Once caught, the prey is swallowed whole.

sandtiger shark
(Eugomphodus taurus)

▲ NEEDLE POINT
This 2m-long leopard shark (*Triakis semifasciata*) has rows of small, needle-sharp, teeth. Although it is thought to be harmless, in 1955 a leopard shark sunk its tiny teeth into a skin diver in Trinidad Bay, California. This was an unprovoked attack, and the diver escaped.

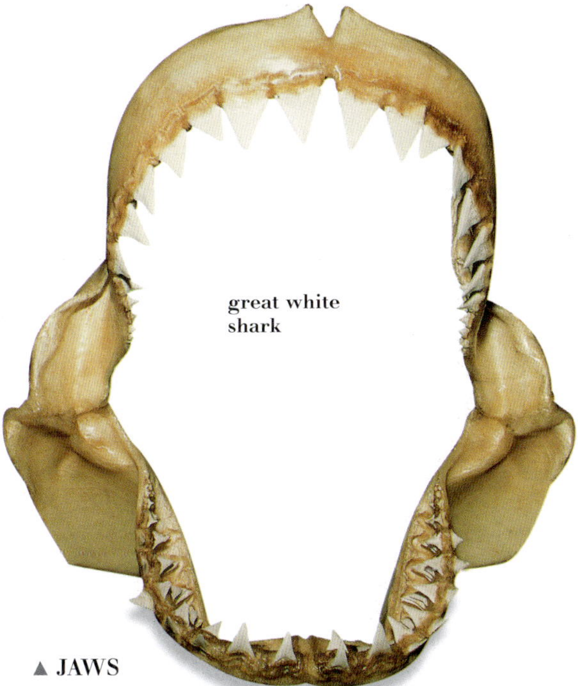

great white shark

▲ **JAWS**

The awesome jaws of a great white shark are filled with two types of teeth. The upper jaw is lined with large, triangular teeth that can slice through flesh. The lower jaw contains long, pointed teeth that are used to hold and slice prey.

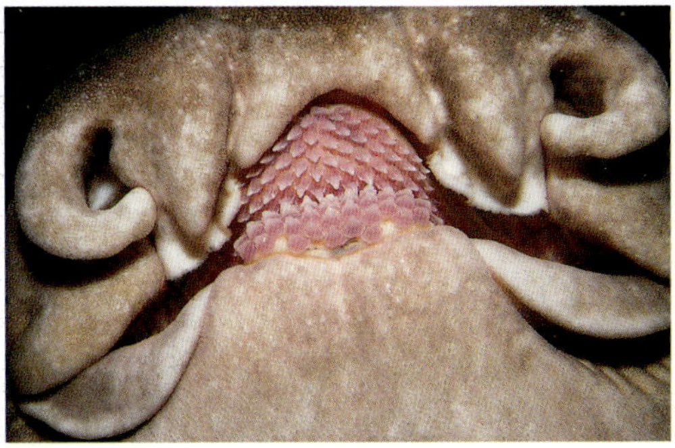

▲ **DUAL SETS OF TEETH**

The Port Jackson shark (*Heterodontus portusjacksoni*) has small, sharp teeth for catching small fish and broad, crushing teeth that can crack open shellfish.

▼ **BIG TEETH**

The cookie-cutter is a small shark, reaching only 50cm long. However, for its body size, it has the biggest teeth of any shark known. It uses them to cut round chunks out of its prey, which includes dolphins, whales and large fish.

Shark Man

Ceremonial carvings such as this one were used in ritual dances performed in the South Pacific Solomon Islands. From one dance master to another, these traditional dances were handed down through many generations. They told of myths in which sharks turned into men, and men turned back into sharks again.

cookie-cutter shark *(Isistius brasiliensis)*

Shark Bodies

Sharks are incredible machines, packed with muscle. Some sharks, such as the great white and mako, can even keep their muscles, gut, brain, and eyes warmer than the temperature of the seawater around them. They do this with special blood vessels, which work like a radiator to collect the heat in the blood and send it back into the body. These make muscles more efficient, allowing the sharks to swim faster. They also help these sharks to hunt in seas of different temperatures. Sharks have a huge, oil-filled liver that helps to keep them afloat. However, like an aeroplane, they must also move forwards in order to stay up. Open ocean sharks must swim all the time, not only to stop them from sinking, but also to breathe. Some sharks can take a rest on the seabed by pumping water over their gills to breathe.

▲ **GILL BREATHERS**
Like this sixgill shark (*Hexanchus griseus*), most sharks breathe by taking oxygen-rich water into their mouths. The oxygen passes into the blood and the water exits the gill slits.

▲ **OCEAN RACER**
The shortfin mako shark (*Isurus oxyrinchus*) is the fastest shark in the sea. Using special, warm muscles, it can travel at speeds of 35–50 kph and catch fast-swimming swordfish.

◄ **SUSPENDED ANIMATION**
The sandtiger shark (*Carcharias taurus*) and a few others can hold air in their stomachs. The air acts like a life jacket, helping the shark to hover in the water. Sandtiger sharks stay afloat without moving, lurking among rocks and caves.

► KEEP MOVING

Like many hunting sharks, the grey reef shark (*Carcharinhus amblyrynchos*) cannot breathe unless it moves forwards. The forward motion passes oxygen into its gills. If it stops moving, the shark will drown.

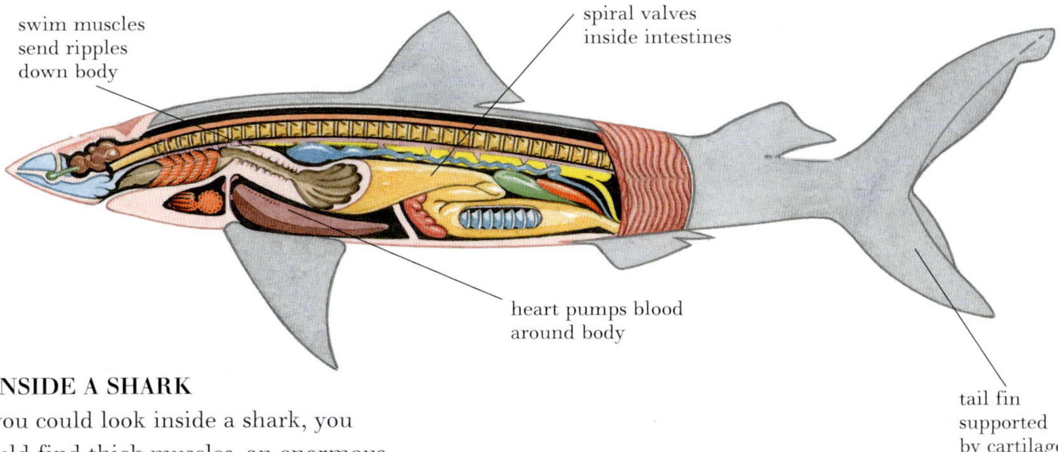

▲ INSIDE A SHARK

If you could look inside a shark, you would find thick muscles, an enormous liver, an intestine with a special, spiral valve, and a complicated system of blood vessels that supply the shark's gills.

Did you know? Mako sharks have been seen to leap 6m clear of the water.

◄ ABLE TO REST

The nurse shark (*Ginglymoostoma cirratum*) pumps water over its gills by lifting the floor of its mouth. This allows it to rest on the seabed, yet still breathe. Whitetip reef sharks, lemon sharks, catsharks and nursehounds also do this.

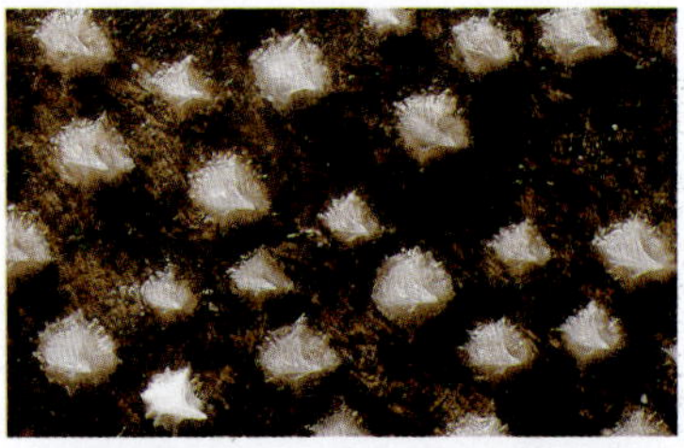

skin teeth of Greenland shark

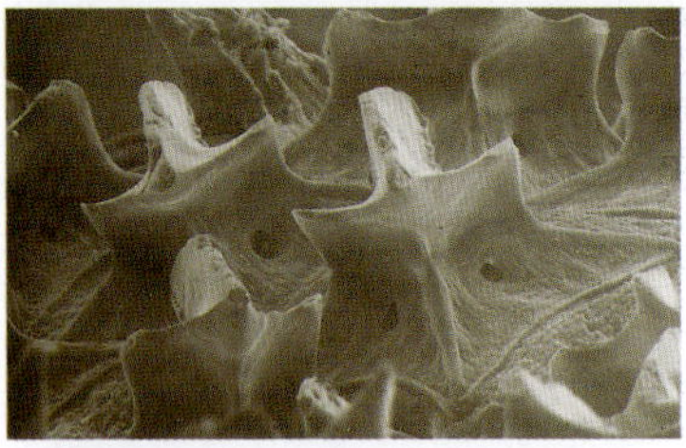

skin teeth of spiny dogfish

skin teeth of dusky shark

Wings in Water

A shark has two pairs of fins (pectoral and pelvic) that work like an aeroplane's wings, lifting the shark as it moves forward. Its dorsal fins and anal fin stop it from rolling sideways, like the tail fin of an aircraft. A shark moves forward in an S shape by rippling a series of waves down its entire body. These waves increase in size as they reach the shark's tail, helping it to propel the body forwards. The shape of the tail can vary from shark to shark, depending on the area of water it inhabits. Sharks that live at the bottom of the sea, such as the nurse shark, tend to have large, flat tails. Sharks that swim in open oceans, such as the tiger shark, usually have slimmer, more curved tails. Both types have a larger upper part to their tail. Sharks that stalk and dash to catch their prey, such as the great white and mako shark, have crescent-shaped tails with top and bottom parts the same size.

▲ SKIN TEETH
A shark's skin is covered with tiny skin teeth called dermal denticles. These teeth help to speed the shark through the water by controlling the flow of water over its body.

► STREAMLINED SHARK
The upper part of the grey reef shark's tail is slightly larger than the lower. Because of this, the tail's downward movement is so powerful that it balances the lift from the pectoral wings. Scientists believe that this helps the shark to move evenly through the water.

▲ **SEABED SWIMMERS**
Hammerhead sharks have unusually small pectoral fins, which allow them to swim and feed close to the seabed. The wings of the hammer-shaped head give the shark extra lift in the water and allow it to turn very tightly. Hammerhead sharks are very adaptable and skilful hunters.

▲ **OCEAN TRAVELLER**
The blue shark (*Prionace glauca*) has long pectoral fins that help it to sail through the sea like a glider plane, making long journeys easy. It swims to the surface, then glides effortlessly to the depths before swimming back to the surface again.

▲ **DEEP SEA GIANT**
The megamouth shark (*Megachasma pelagios*) was only discovered in 1976. It lives in deep water and swims very slowly. Megamouth does not chase anything. It eats deep-sea shrimps filtered through its gills.

▲ **OCEAN CRUISER**
The oceanic whitetip shark (*Carcharhinus longimanus*) swims in the open oceans and is often present at the scene of sea disasters. It is a very distinctive shark, and can be easily recognized by its dorsal and pectoral fins, which are shaped like rounded garden spades.

Brain and Senses

A shark's brain is small for its size, but its senses are highly developed. Sharks see well, and see in colour. They can also recognize shapes. Just as amazing are a different range of senses that allow sharks to pick up sounds and vibrations from miles around. They can detect changes in the ocean currents, recognize smells and follow the trail of an odour right back to its source. Some species have shiny plates at the backs of their eyes that collect light to help them see as they dive to deep, dark water. They also have membranes of dark colour that they draw across the shiny plates to avoid being dazzled by the light when they return to the surface. Sharks even have special nerves in their noses that can detect minute electrical fields, such as those produced by the muscles of their prey.

▲ **ELECTRICAL SENSE**
Like all sharks, sandtiger sharks have tiny pits in their snouts, known as the ampullae of Lorenzini. Inside these pits are special nerves. These help the shark to find food by detecting minute electrical fields in the muscles of its prey.

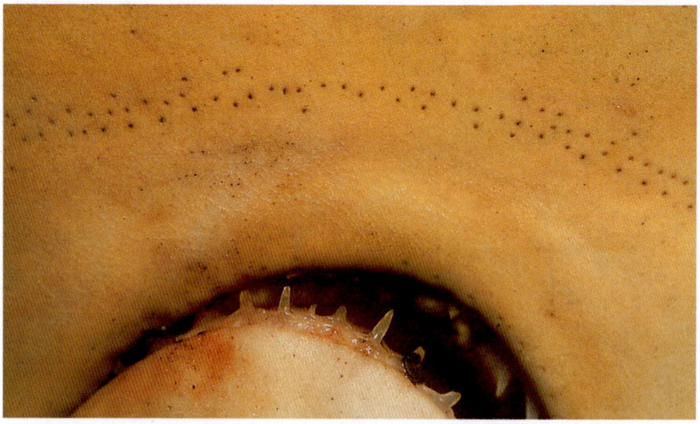

◄ **PREY DETECTOR**
In a hammerhead shark the special pits that can sense electrical fields in its prey are spread across the hammer of the shark's head, helping it to scan for prey across a wide area. The hammerhead searches for food by sweeping its head from side to side, rather as if using a metal detector. It can find any prey buried in the sand below.

scalloped hammerhead shark
(Sphyrna lewini)

◄ SIGHT, SMELL AND SOUND

The nostrils of the hammerhead shark are positioned wide apart on its head. This gives the shark 'stereo smelling' with which it can more easily track odours to their source. But, because its eyes are at the ends of its hammer, it must turn its head from side to side in order to see forwards.

▲ EYE PROTECTION

When a shark bites, its eyes can easily be injured by the victim's teeth, spines or claws. To prevent this, sharks such as this tiger shark have a special membrane (sheath) that slides down across the eye during the attack.

eye of blacktip reef shark
(*Carcharhinus melanopterus*)

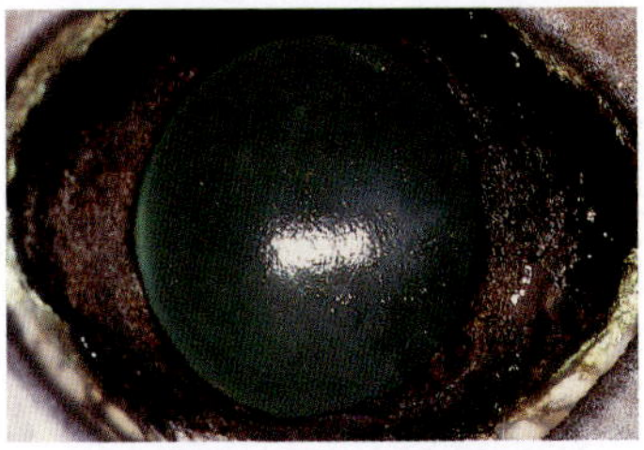

eye of bluntnose sixgill shark
(*Hexanchus griseus*)

◄ DEEP AND SHALLOW

The blacktip reef shark has a small eye with a narrow, vertical slit. This type of eye is often found in shallow-water sharks. Sharks that swim in deeper waters, such as the sixgill shark, tend to have large, round pupils.

Did you know? Sharks find their way through mazes as fast as rabbits.

Shark Callers

On the islands of the south-west Pacific, sharks are the islanders' gods. To test their manhood, young shark callers attract sharks by shaking a coconut rattle under the water. Sensing the vibrations, a shark will swim close to the canoe. It is then wrestled into the boat, and its meat divided among the villagers as a gift from the gods.

Focus on the Blue

The blue shark is an open ocean hunter. Continually looking for food, it can pick up the sounds and vibrations of a struggling fish from over a kilometre away. From half a kilometre, it can smell blood and other body fluids in the water. As it gets closer, the shark can sense changes in the water that help it locate moving prey. Finally, vision takes over. First, only movements are seen, but then the prey itself. As the blue shark closes in for the kill, it pulls down its eye protectors and swims blind. Its electrical sensors then lead it to its prey.

1 When hunting, the blue shark uses all its senses to search constantly the ocean ahead for prey. It will also watch the behaviour of other blue sharks in the water, sometimes joining them to hunt in packs.

2 Smells, sounds, vibrations, and water movements attract the blue shark. The movements made by a school of jack mackerel will initially lead the shark to them. It then uses its sight to find an easy target.

Shark Hunt

3 Sharp eyesight, quick reactions and an ability to speed through the water all help the blue shark to chase its chosen target. In an attempt to escape, this group of mackerel fish will dart all over the place, then crowd close together to confuse the pursuing shark.

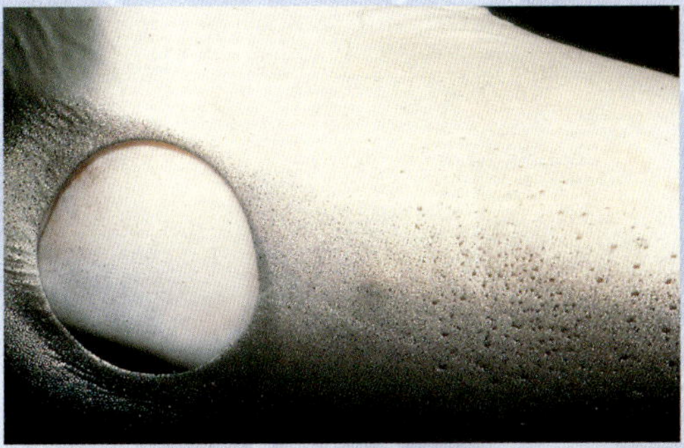

4 As the blue shark closes in to grab its target, a protective membrane covers each eye. At this stage the shark is swimming blind and relies on the electrical sensors in its snout to guide it to its prey. These home in on the electrical field made by the fish's muscles, leading the shark for the last few centimetres.

5 As the shark bites, it extends its jaws and impales its prey on the teeth of the lower jaw. Next, the upper jaw teeth come into action, clamping down on the fish. The shark then removes its lower jaw teeth from the prey, and pushes them forward to pull the fish back into its mouth. The prey is then swallowed by the shark.

Feeding

Sharks catch a variety of foods, eating whatever they can find in their area. Most sharks eat bony fish and squid, but they can be cannibalistic (eat each other). They often feed on other smaller sharks, sometimes even on their own species. Some sharks prefer particular kinds of food. Hammerheads like sting rays, while tiger sharks will crunch a sea turtle's shell for the meat inside. Shortfin mako sharks hunt bluefish and swordfish. Great white sharks eat fish, but as they get older will also hunt seals and sea lions. Sharks will scavenge (feed on dead animals) whenever they can. The bodies of dead whales are food for many sharks that swim in open waters, including tigers, blues and oceanic whitetips.

▲ **FEEDING FRENZY**
Large quantities of food will excite grey reef sharks, sending them into a feeding frenzy. When divers hand out food, the sharks will circle with interest, until one darts forward for the first bite. Other sharks quickly follow, grabbing at the food until they seem out of control.

▲ **FOREVER EATING**
A large shoal of mating squid will send blue sharks into a frenzy. The sharks feed until full, then empty their stomachs to start again!

▲ **FISH BALL**
A group of sharks will often herd shoals of fish into a tight ball. The sharks will then pick off fish from the outside of the ball, one by one.

▶ BITE A BROTHER

Sharks do not look after their relatives! Big sharks will often eat smaller sharks, and sharks that swim side-by-side in the same school will often take a nip out of each other. The remains caught on a fishing line of this blacktip shark show that it has been eaten by a large bull shark.

Did you know? The great white shark sometimes eats crabs and lobsters.

◀ OCEAN DUSTBIN

Tiger sharks are well known as the sharks that will eat not just living things such as fish, other sharks, or dead animals floating in the sea. Tiger sharks have been known to eat coal, rubber tyres and clothes. They are found all over the world and grow to a length of 5.5m. Not surprisingly, tiger sharks have been known to try eating humans.

▼ BITE-SIZE CHUNKS

The cookie-cutter shark feeds by cutting chunks out of whales and dolphins, such as this spinner dolphin. The shark uses its mouth like a clamp, attaching itself to its victim. It then bites with its razor-sharp teeth and swivels to twist off a circle of flesh.

spinner dolphin
(Stenella longirostris)

▲ OPPORTUNISTS

Sharks will often follow fishing boats, looking for a free meal. This silvertip shark is eating pieces of tuna fish that have been thrown overboard.

Focus on

1 Huge groups of albatrosses nest on the ground close to the shore of Hawaiian islands, including the island of Laysan. The birds in each group breed, nest, and hatch their babies at the same time. When it is time, the young birds all take their first flight within days of each other.

Sharks can be found wherever there is food in or near the sea. Tiger sharks are rarely seen around some Hawaiian islands in the Pacific Ocean until the islands' young seabirds start to fly. Then the sharks arrive. Any birds that fall into the sea are quickly eaten. The waters are too shallow for the sharks to attack from behind and below as most sharks do. Instead, the sharks leap clear of the surface then drag the birds underwater to drown and eat them. Sharks arrive for their island feast at the same time each year. How they remember to do so is yet to be explained.

2 When ready to fly, a baby bird practises by flapping its wings in the face of the islands' fierce winds. Eventually, the baby must make its first real flight over the ocean. When it does so, the tiger sharks are waiting in the water below.

3 Tiger sharks patrol the clear, shallow waters close to the albatross nests. Their dark shapes can be seen clearly against the sandy sea floor. Every now and again, a tiger shark's triangular dorsal fin and the tip of its tail can be seen breaking the water's surface.

Tiger Sharks

4 Any baby bird that dips into the sea is prey for the waiting tiger shark. At first, the shark tries its usual attack, from below and behind. However, in the shallow waters the shark cannot make a full attack. Rather than hitting its prey at force, the shark just pushes the bird away on the wave made by its snout.

5 After failing to catch a meal, the shark soon realizes its mistake and tries another approach. Its next style of attack is to shoot across the surface of the water, slamming into its target with its mouth wide open. This technique seems to be more successful, and the shark usually catches the bird.

6 The shark then attempts to drag the bird below the surface, to drown it. If a bird is pushed ahead on the shark's bow wave, it will bravely peck at its attacker's broad snout and, sometimes, may even escape. Some birds also manage to wriggle free as the shark grapples with them underwater.

7 Many albatross babies do not manage to escape a shark attack. They are grabbed by the sharks and drowned. Inside the tiger shark's jaws are rows of sharp teeth that can slice into a bird's body like a saw. Sometimes the tiger shark tears off the bird's wings and leaves them aside to eat the body whole.

Filter Feeders

▲ **FOOD CHAIN**
Eggs and sperm released on the same night by the corals of the Ningaloo Reef in Australia are eaten by the larvae (young) of crabs and lobsters. The larvae are eaten by fish and krill. The fish and krill in turn become food for the hungry whale sharks that swim off the coral reefs.

Some of the biggest fish in the sea eat some of the smallest living things there. Giant species, like the whale shark and basking shark, use their gill rakers to comb plankton (tiny animals and plants) from the water. In the same way as hunting sharks, they use their sharp senses to track down huge areas of food. Whale sharks are often seen near coral reefs, where, at certain times of the year, large amounts of animal plankton can be found. Basking sharks often swim in the area between ocean currents, feeding on plankton that gathers on the boundary. Whale and basking sharks swim in the upper layers of the sea. The giant megamouth shark lives deeper down, sieving out the shrimps that live in the middle layers of the ocean.

◄ **WHALE OF A FEAST**
Exactly two weeks after the coral has spawned at Ningaloo, the whale sharks appear. By this time, each creature, from the smallest larvae to the reef's fish and krill, will have fed upon the coral's rich food. Each night, the whale sharks swim with their huge mouths wide open, scooping up food which they sieve from the sea's surface.

▲ BIG GULP
The basking shark swims slowly through the sea, funnelling food-filled water into its great mouth. In one hour, it can filter 7,000 litres of seawater! When enough food has been trapped, the shark closes its mouth and swallows with one gulp.

▲ COLD WATER SKIMMERS
The basking shark has gill slits that almost encircle its gigantic body. These are used to filter food, such as shellfish larvae and fish eggs, from the water. The shark passes water through its gill chamber, where enormous gill rakers comb the food from the water.

▲ BIG MOUTH
Patrolling the middle waters of the deep, the megamouth shark scoops up tiny shrimps as they cross its path. Since this shark's discovery in 1976, a further 13 examples have been discovered and some of these have been examined by scientists.

▲ PLANKTON
Plankton is made up of tiny plants and animals that float together in huge clouds on and just below the sea's surface. Both animal and plant plankton are eaten by the basking shark.

Keep in Line

No shark is alone for long. Sooner or later, one shark will come across another, including those of its own kind. In order to reduce the risk of fights and injury, sharks talk to each other, not with sound, but with body language. Sharks have a clear pecking order. The bigger the shark, the more important it is. Not surprisingly, small sharks tend to keep out of the way of larger ones. Many species use a sign that tells others to keep their distance. They arch their back, point their pectoral fins down and swim stiffly. If this doesn't work, the offending shark will be put in its place with a swift bite to the sides or head. Bite marks along its gill slits can be a sign that a shark has stepped out of line and been told firmly to watch out.

▲ **GREAT WHITE CHUMS**
Great white sharks were once thought to travel alone, but it is now known that some journey in pairs or small groups. Some sharks that have been identified by scientists will appear repeatedly at favourite sites, such as California's Farallon Islands, 42km off the coast of San Francisco. There they lie in wait for seals.

Did you know? Some great whites return to the same spot every year.

◄ **BED FELLOWS**
Sharks, like these whitetip reef sharks, will snooze alongside each other on the seabed. They search for a safe place to rest below overhanging rocks and coral, where, as fights rarely break out, they seem to tolerate each other. The sharks remain here until dusk, when they separate to hunt.

◄ **PECKING ORDER**
This grey reef shark has swum too close to another, larger shark and has been bitten on its gill slits as a punishment. The marks on its skin show that its attacker raked the teeth of its lower jaw across the sensitive skin of the grey reef's gill slits. A shark's injuries heal rapidly, so this unfortunate victim will recover quickly from its wounds.

► **REEF SHARK GANGS**
Sharks have their own personal space. As they patrol the edge of a reef, schools of blacktip reef sharks will tell others that they are too close by moving their jaw or opening their mouth. During feeding, order sometimes breaks down and a shark might be injured in the frenzy.

◄ **SHARK SCHOOL**
Every day schools of scalloped hammerhead sharks gather close to underwater mountains in the Pacific Ocean. They do not feed, even though they come across shoals of fish that would normally be food. Instead, they swim repeatedly up and down, as though taking a rest.

Focus on

BAD-TEMPERED SHARKS
The larger a female hammerhead becomes, the less likely she is to get on with her neighbours! Older and larger hammerhead sharks like more space than smaller, younger sharks. In hammerhead schools, the relationship between sharks seems to be controlled by constant displays of threat and small fights.

By day scalloped hammerhead sharks swim in large schools around underwater volcanoes in the Pacific, the Gulf of California off Mexico, and off the Coco and Galapagos islands. This species of shark cannot stop swimming or it will drown. Schools are a safe resting place for them. Even sharks have enemies, such as other sharks and killer whales, so there is safety in numbers. In schools, scalloped hammerheads can also find a mate. At night, they separate to hunt. They swim to favourite feeding sites, using their electric sensors to follow magnetic highways made by lava on the seabed.

FEMALES ONLY
The sharks in this huge school of hammerheads are mainly females. Larger sharks swim in the centre, and smaller sharks on the outside. Large sharks dominate the group, choosing the best positions in which to swim. Not only is the middle safer, but it is also the place where male sharks look for a mate.

Hammerheads

READY FOR A SCRUB
At some gathering sites, such as Cocos Island in the eastern Pacific, sharks drop out of the school and swoop down to cleaning stations close to the reef. From the reef, butterfly fish dart out to eat the dead skin and irritating parasites that cling to the outside of the shark's body.

BODY LANGUAGE
Larger sharks within a school perform strange movements and dances to keep smaller sharks in their place. At the end of the movement, a large shark may nip a smaller one on the back of the head!

STRANGE HEAD
The scalloped hammerhead is so named because of the groove at the front of its head, which gives it a scalloped (scooped out) appearance. The black tips on the underside of its pectoral fins are another way of identifying this particular shark.

Courtship and Mating

Male sharks find female sharks by their smell. The female gives off odours that drift in the currents of the sea, attracting every male shark within smelling distance. The males follow her closely, until one grabs hold of a pectoral fin with his mouth and hangs on tightly in preparation for mating. Fortunately, the female has thickened skin on her pectoral fins, which prevents her from being hurt. The male has a pair of claspers (sex organs) one of which he places inside the female's sexual opening. During mating, the male shark shakes occasionally, to make sure that the female accepts his presence. Once he is sure of this, the male will complete his mating with her.

▲ ENGAGEMENT IN THE SHALLOWS
A group of male whitetip sharks will be stimulated by the sexual smell of a female ready to mate. Following her in the shallow waters of a coral reef, the males compete for the female. Eventually, one will win possession by grabbing hold of one of her pectoral fins.

◄ HANGING AROUND
Sandtiger sharks hover in the water at special meeting places, waiting for members of the opposite sex. At these sites, lots of shark teeth are found on the sea floor. It is believed that they fall out during the rough and tumble of courtship and mating.

◄ **THREE IS A CROWD**
Male sharks are usually smaller than females of the same age. Here two males have each seized a female's pectoral fin, but only one male will be able to mate with her.

Did you know? Only one in ten of nurse shark matings are successful.

► **NURSE SHARK NUPTIALS**
Nurse sharks (*Gingly-mostoma cirratum*) travel to traditional mating sites. The male nurse shark grips the female's pectoral fin and arches his body alongside hers. He will then insert his right or left clasper, depending on which pectoral fin he has seized.

▲ **MALE SEX ORGANS**
The claspers (sex organs) of male sharks are pelvic fins that have been adapted for mating. Similar to the penises of mammals, they are used to transfer sperm from the male into the female.

▲ **MATING SCARS**
The courtship and mating of sharks can be a rough affair. Female sharks, like this tiger shark, can be scarred with bite marks made by her mate. However, females have thicker skin than males, which prevents further damage.

▲ **EGG WITH A TWIST**
The horn shark (*Heterodontus francisci*) egg case has a spiral-shaped ridge. The mother shark uses her mouth like a screw-driver to twist the case round into the gaps in rocks.

Inside and Outside

Sharks bring their young into the world in two ways. Most sharks grow their eggs inside the mother's body, and give birth to breathing young called pups. Others lay eggs in which the pup grows outside of the mother's body. Catsharks and nursehound sharks grow their young in cases called mermaid's purses, which they lay outside their bodies. These can sometimes be found washed up on beaches after a storm. Each mating season, catsharks lay up to 20 mermaid's purses. Inside each is one pup. When the case has been laid in the sea, the mother shark does not guard or look after it in any way. Instead, she relies on the tough, leathery case to protect the pup inside.

▲ **TIME TO LEAVE**
When it is ready to leave its egg, the baby horn shark uses special scales on its snout and pectoral fins to cut its way out of the tough egg case. On its dorsal fins are tough spines that protect it from the moment it emerges.

Mermaid's Purses
The mermaid is a mythical undersea creature with a woman's body and a fish's tail. In legends, the mermaid lured men to their death with beautiful songs. Catshark and skate cases that are washed up on beaches look like pouches, and are often called mermaid's purses.

50 days 100 days 150 days 200 days

▲ IN THE SAC

In the earliest stages of development, the catshark pup is tiny. It is attached to a huge, yellow yolk sac from which it takes its food. Inside the egg case, the growing catshark pup makes swimming movements, which keeps the egg fluids and supply of oxygen fresh. After nine months, the pup emerges, with diagonal stripes that eventually turn into spots as it grows.

► SWELL SHARK

The length of time it takes the swell shark (*Cephaloscyllium ventriosum*) pup to grow depends on the temperature of the sea water around it. If warm, it can take just seven months. If cold, it might take ten months. As it emerges, it uses special skin teeth to tear its capsule open.

Into the World

Most sharks give birth to fully formed, breathing pups. However, pups grow in many different ways. Baby nurse and whale sharks start their life in small capsules. The pups then hatch from the capsules inside their mother's body, where they continue to grow before being born. Other shark pups, like blue sharks, also grow inside their mothers, in a womb. A sandtiger shark might have just two pups, but a blue shark can grow up to 135 at one time. The length of time it takes a pup to grow also varies. Nurse sharks take just five months, but frilled sharks take two years. Some pups feed on unfertilized eggs inside the womb. Baby sandtiger sharks go one better. They eat each other.

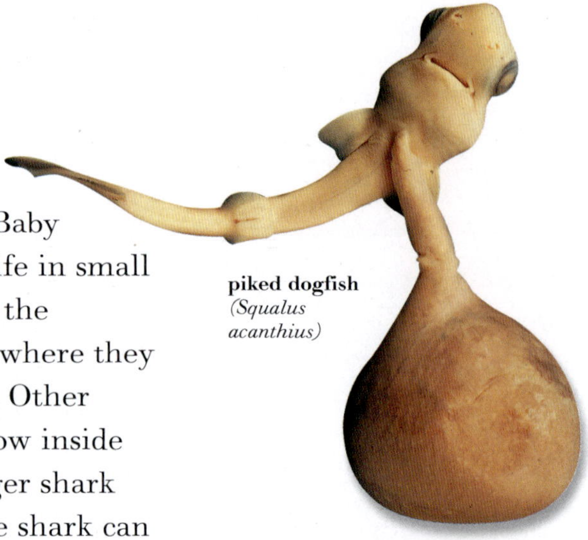

piked dogfish
(Squalus acanthius)

▲ **DOGFISH YOLK SAC**
Up to 12 piked dogfish pups can grow inside one mother. At first, all the pups are enclosed in one capsule that breaks after six months. Each pup then feeds off its own yolk sac until it is born three months later.

► **RESTING**
When birth is near, a pregnant whitetip reef shark will rest in a protective area. This pregnant shark is resting on rocks near Cocos Island in the Pacific. Inside her womb, she may develop up to five pups. Each will be fed by a placenta attached to the womb wall. The pups are born after five months.

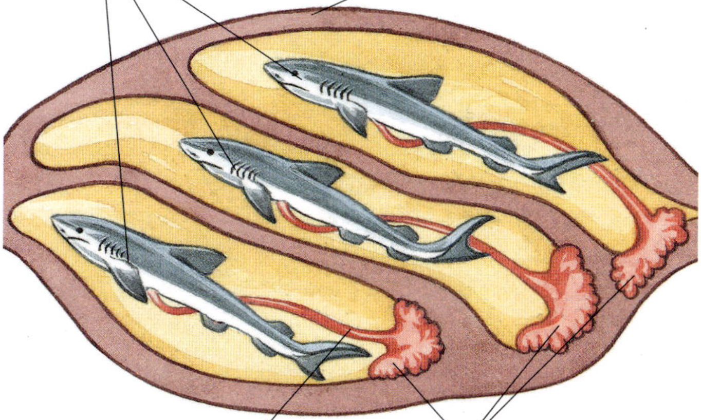

▲ WOMB MATES

In some species of shark, such as the whaler shark, embryos develop inside the pregnant female as they do in mammals. At first, each embryo has its own supply of food in a yolk sac, but when this is used up its sac turns into a placenta that attaches itself to the wall of the womb. Nutrients and oxygen pass directly from the mother across the placenta and along the umbilical cord to its own body. Waste products go the other way.

▲ BIG BELLIES

Some baby sharks will eat unfertilized eggs in their mothers' wombs. They eat so much so fast their stomachs become swollen. These baby shortfin mako sharks, caught off the coast of South Africa, have gorged on unfertilized eggs, filling their tiny stomachs.

◀ STRIPED SHARK

Safe inside its mother, this baby tiger shark will have fed on both the nutrients from its yolk sac, and on a fluid produced by the wall of the womb. The pattern of blotches on the newborn baby's skin will form stripes, which will gradually fade as it gets older.

Focus on Lemon

A year after mating, pregnant lemon sharks arrive at Bimini Island in the Atlantic Ocean. Here, they give birth to their pups in the island's shallow lagoon where males do not enter. An adult male is quite likely to eat a smaller shark, even one of its own kind. In many species of shark, pregnant female sharks leave the males and swim to safer nursery areas to give birth. Some scientists even believe that females lose their appetite at pupping time, to avoid eating their own young. After birth, however, the lemon shark pups live on their own.

1 By pumping sea water over her gills, a pregnant lemon shark (*Negaprion brevirostris*) can breathe and rest on the seabed at Bimini Island. She gives birth on the sandy lagoon floor to the pups that have developed inside her for a year.

2 Baby lemon sharks are born tail first. There might be 5–17 pups in a mother's litter (family). Each pup is about 60cm long. After her litter is born, a female lemon shark will not be able to mate again straight away. Instead, she will rest for a year.

Shark Birth

3 A female lemon shark can give birth to her pups as she swims slowly through the shallows. The pups are still attached to the umbilical cord when born, but a sharp tug soon releases them. The small remora fish that follow the shark everywhere will feast on the afterbirth.

4 After birth, a baby lemon shark makes for the safety of the mangroves at the edge of the lagoon. Here, it spends the first few years of its life in a strip of mangrove 40m wide and 400m long feeding on small fish, worms and shellfish, taking care to avoid sharks larger than itself. Its home overlaps the territory of other young sharks.

5 To avoid being eaten, young lemon sharks gather with others of the same size. Each group patrols its own section of the lagoon at Bimini. This young lemon shark is about one year old. When it is seven or eight, it will leave the safety of the lagoon and head for the open reefs outside.

Look in any Ocean

Sharks live throughout the world's oceans and seas, and at all depths. Some sharks, like bull sharks, even swim in rivers and lakes. Whale, reef and nurse sharks are all tropical species that prefer warm waters. Temperate-water sharks, such as the mako, horn and basking sharks, live in water that is 10–20°C. Cold-water sharks often live in very deep water. The Portuguese shark, frilled shark, and goblin shark are all cold water sharks. A few species will swim in extremely cold waters, such as the Greenland shark which braves the icy water around the Arctic Circle.

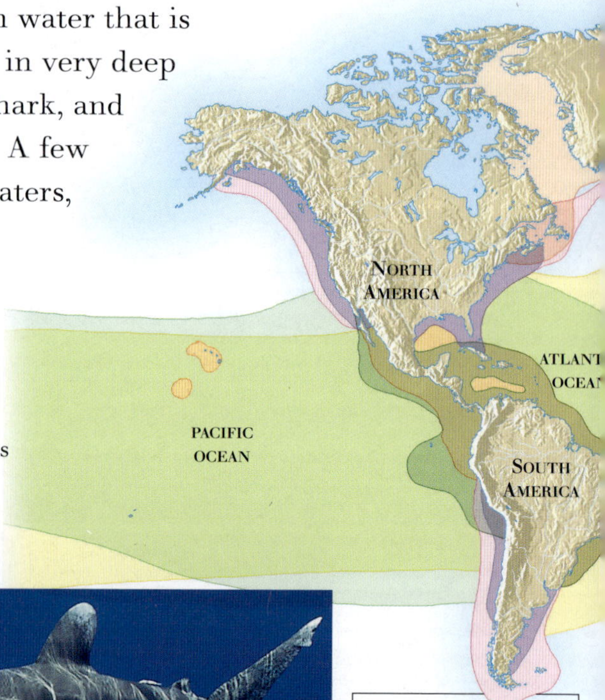

► SWIMMING POOLS
This map shows the main parts of the world's seas in which different kinds of sharks live. The key beneath the map shows which sharks live where.

► OCEAN WANDERER
The oceanic whitetip shark swims the world's deep, open oceans, in tropical and sub-tropical waters. It is also one of the first sharks to appear at shipwrecks.

◄ ISLAND LIVING
The Galapagos shark (*Carcharhinus galapagensis*) swims in the waters of the Galapagos islands, on the Equator. It also swims around tropical islands in the Pacific, Atlantic and Indian oceans.

◄ UNDER THE ICE

The Greenland shark (*Somniosus microcephalus*) is the only shark known to survive under polar ice in the North Atlantic seas. It has a luminous parasite attached to each eye that attracts prey to the area around its mouth.

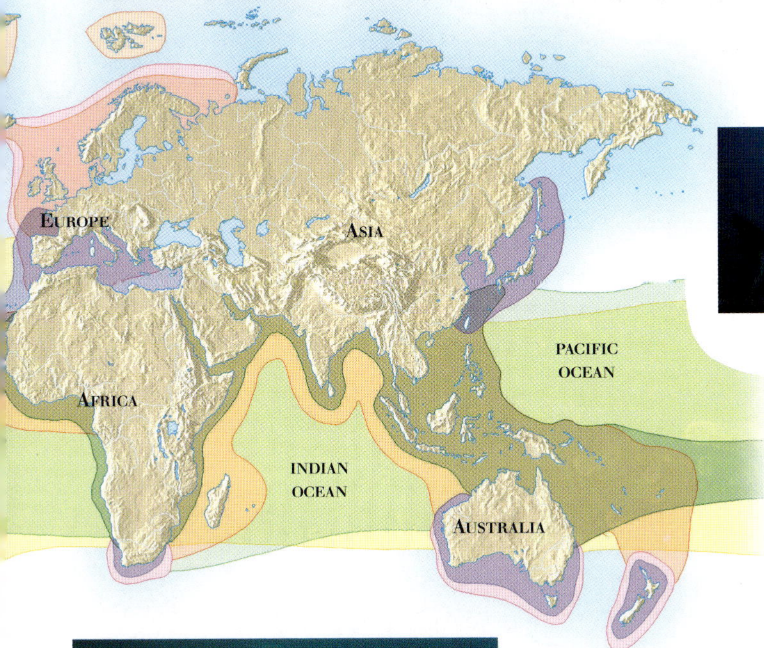

▲ TEMPERATE PREDATOR

The great white shark lives in temperate, sub-tropical and tropical seas, including the Mediterranean. It usually swims in coastal waters rather than the open sea.

▲ TIGER OF THE SEAS

The tiger shark swims in mainly tropical and warm temperate waters, both in open ocean, and close to shore. Tiger sharks have been seen off Morocco and the Canary Islands.

◄ REEF SHARK

The blacktip reef shark patrols reefs in the Indian and Pacific oceans. It also lives in the Mediterranean and Red Sea, and as far west as the seas off Tunisia, in North Africa.

Upwardly Mobile

Did you know? A blue shark once travelled a total distance of 7,146 km.

sandbar shark
(Carcharinus plumbeus)

Not all sharks travel far afield. Some prefer to stay close to home, swimming only in one small area. Others have a daily routine, spending the day in deep waters, but moving closer to shore to feed at night. A few deep sea sharks make a different daily journey, spending the day in the deep, and rising to the surface to feed at night. Some sharks travel vast distances, crossing oceans. This has only recently been discovered with the tagging of sharks. Rather than killing sharks when they catch them, scientists and fishermen now give them a special tag. Each tag has its own number, which identifies the shark. So, when the shark is caught again, scientists can see how much it has grown, and also how far it has travelled.

▲ INTO THE GULF
Atlantic sandbar sharks can travel over 3,000km, from the Atlantic coastline of the USA to the coast of Mexico. These amazing sharks grow incredibly slowly, only about 3cm a year. They reach adulthood very late in life, when they are 30 years old.

◄ GIVEN A NAME
This tiger shark has been tagged (marked) by scientists and is being released back into the sea. Tagging has shown that tiger sharks travel great distances across oceans. Previously people had believed that they stayed in one place.

◄ **OCEAN MIGRATOR**
Female blue sharks in the North Atlantic go on a very long migration. They circle the Atlantic Ocean, mating off North America and then giving birth near Spain and Portugal at the end of their journey.

► **EPIC JOURNEY**
Female blue sharks travel from North America to Europe where they give birth to their pups. Then they turn back towards the USA. They travel at about 40 km per day. A shark swimming fast might cover the round trip of 15,000km in 15 months.

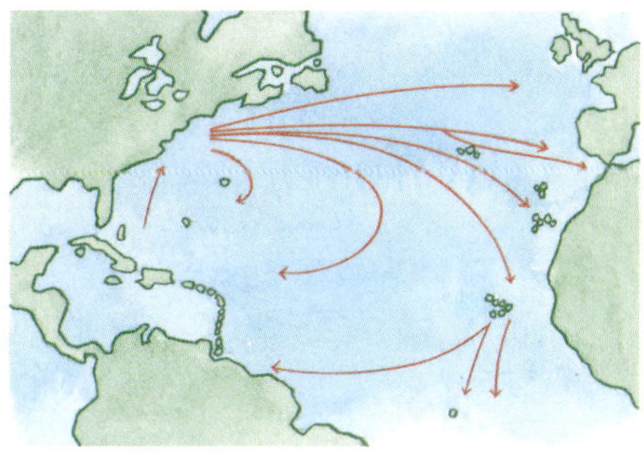

▼ **FOLLOW THE TEMPERATURE**
Shortfin mako sharks travel into the North Atlantic, but rarely swim the whole way across. They like to swim in an exact temperature of 17–22°C. They follow thermal water corridors through the ocean to winter in the Sargasso Sea.

shortfin mako shark
(Isurus oxyrinchus)

► **DOUBLE BACK**
Migrating mako sharks travel to the middle of the Atlantic Ocean and then turn back towards the USA. The sharks do not go further because from the middle of the ocean to Europe the water is not the temperature they prefer to swim in.

The Ocean Depths

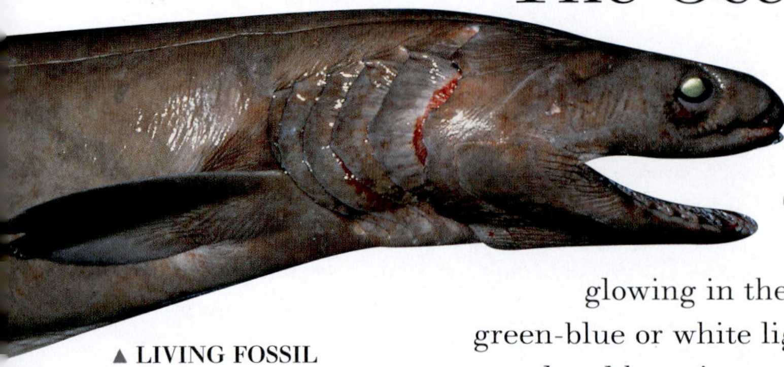

frilled shark
(Chlamydoselachus anguineus)

Many sharks are rarely seen because they live in the darkness of the deep. Catsharks and dogfish live in these gloomy waters, glowing in the dark with a luminous green-blue or white light. Some of these species travel and hunt in packs, following their prey to the surface at night, returning into the depths by day. Most of the world's smallest sharks live here. Pygmy and dwarf sharks no bigger than a cigar travel up and down the ocean for several kilometres each day. On the deep sea floor are such enormous sharks as the sixgill, sevengill and sleeper sharks. These eat the remains of food that sinks down from the sea's surface. Many deep sea sharks look primitive, but strangest of all are the frilled and horned goblin sharks. These look like the fossilized sharks that swam the seas 150 million years ago.

▲ LIVING FOSSIL
The frilled shark is the only shark shaped like an eel. It has six feathery gill slits, 300 tiny, three-pointed teeth and a large pair of eyes. Instead of a backbone, it has a firm, but flexible, rod of cartilage. These features tell us that the frilled shark resembles sharks that lived in the oceans millions of years ago.

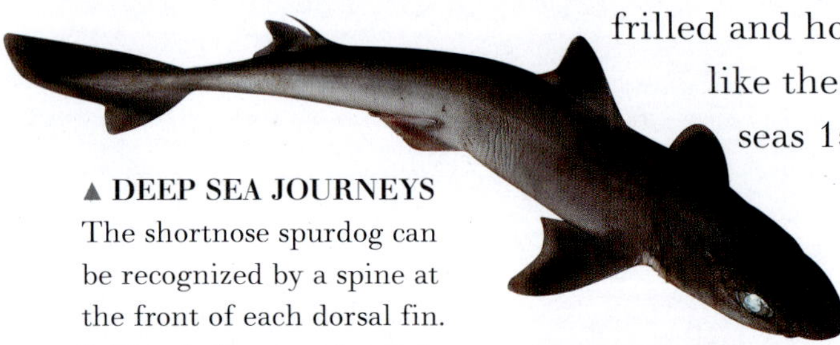

shortnose spurdog
(Squalus megalops)

▲ DEEP SEA JOURNEYS
The shortnose spurdog can be recognized by a spine at the front of each dorsal fin. It lives in large packs made up of thousands of sharks. It swims at depths of 800m in the northern waters of the Atlantic and Pacific oceans. Seasonally, the packs make a daily migration, from north to south and from coastal to deeper waters.

▼ DEEPEST OF THE DEEP
The Portuguese dogfish holds the record for living in the deepest waters. One was caught 2,718m below the sea's surface. At this depth, the water temperature is no higher than a chilly 5–6°C.

Portuguese dogfish
(Centroscymnus coelolepis)

◀ SIXGILL SLITS

Most modern sharks have five gill slits, but primitive sharks, like bluntnose sixgill sharks (*Hexanchus griseus*), have more. These sharks are found at huge depths around the world. They have evolved (developed) slowly, and still have the features of sharks that lived millions of years ago.

▼ SEVENGILL SLITS

Broadnose sevengill sharks have seven gill slits. They have primitive, sharp teeth that look like tiny combs. They use these to slice up ratfish, small sharks and mackerel. Because some of their prey live near the surface, sevengill sharks travel to the sea's surface to hunt at night.

Broadnose sevengill shark
(Notorynchus cepedianus)

Did you know? Many deep-sea sharks have light-organs on their bodies.

velvet belly
(Etmopterus spinax)

◀ SLIMY COAT

The velvet belly is 66cm long. It lives in the Atlantic and Mediterranean, at depths of 70–2,000m. The velvet belly is covered with luminous slime, and the underside of its body has special organs that give out light. It eats deep-sea fish and shrimps.

Freshwater Sharks

Although most sharks live in the salt water of the sea some, such as the bonnethead and sandbar sharks, swim to the mouths of rivers to give birth. There are a few sharks that swim all the way up rivers, and some swim into freshwater lakes. Atlantic sharpnose and spadenose sharks, and Ganges and Borneo river sharks swim in fresh water. Bull sharks are the species most often seen in fresh water. How this species' body copes with fresh water is not known. A fish that usually swims in salt water needs to find a way of coping with water that is not salt. On entering a river or lake, a fish used to salt water would be expected to absorb water and blow up like a balloon, but bull sharks do not. Somehow, they have found a way to keep the levels of salt in their blood low, so reducing water absorption in fresh water.

▲ **HOLY RIVER**
Shark attacks on pilgrims bathing in the holy Ganges River in India were once blamed on the Ganges river shark. Instead, they were probably made by the bull shark, which feeds on cremated bodies thrown into the river.

◄ **RIVER JOURNEYS**
Bull sharks have been seen in the Amazon, Congo and Mississippi rivers, and in other tropical rivers and lakes around the world. They gather at the river mouth, where edible rubbish is found during floods. Bull sharks are also sometimes called Zambezi sharks because they make regular journeys up Africa's Zambezi river.

Vishnu

In Hindu religion, the god Vishnu is believed to be the protector of the world. In legendary tales, Vishnu saved the sacred religious texts of India when the whole world was covered in water. This Indian picture shows him emerging from the mouth of a monstrous fish, probably a shark.

▲ KEYS SWIMMER

This bonnethead shark (*Sphyma tiburo*) lives near river mouths in the Florida Keys, in Florida Bay in the southeast of the USA.

► BORN-AGAIN SHARK

The Borneo river shark (*Glyphis*) was believed extinct until one was caught in 1997 by a fisherman in Sabah in South-east Asia. Until then, the only known specimen was 100 years old and displayed in an Austrian museum.

Borneo river shark
(Glyphis)

◄ LAKE NICARAGUA

Although they do not live all the time in Lake Nicaragua in Central America, bull sharks (*Carcharhinus leucas*) are also called Nicaragua sharks because they travel between the lake and the Caribbean Sea. It is thought that some female bull sharks swim to the lake in order to give birth to their pups.

Life on the Seabed

People once thought that all sharks died unless they kept swimming. This is not true. Many sharks that live close to the sea floor do so without moving for long periods of time. Wobbegong and angel sharks have flattened bodies that help them to lie close to the sea floor. Their skin colour also blends in with their background, hiding them from their prey as they lie in wait. These sharks take in water through a special spiracle (hole) behind their eye to stop their gills becoming clogged with sand. Some sharks that live on the sea floor, such as catsharks and carpet sharks, are not flattened. Whatever their shape, most are camouflaged with spots, stripes or a mottled pattern on their back.

▲ **OCEAN CAMOUFLAGE**
Sharks that live near the surface can also camouflage themselves. From above, the great white's dark back blends in with the ocean depths.

▼ **SPOTTED ZEBRA**
The adult zebra shark (*Stegostoma fasciatum*) has spots instead of zebra stripes. It has stripes on its skin as a pup. These break up into spots as the shark grows. It lives in the Indian and Pacific Oceans.

▲ **CORAL COPYCAT**
This tasselled wobbegong (*Eucrossorhinus dasypogon*) is invisible to its prey. It copies the colour of rock and coral, and has a fringe of tassels hanging down below its mouth that look like seaweed.

▲ **AMBUSH EXPERT**
The Pacific angel shark (*Squatina californica*) buries itself in the sand, and watches for prey. When a fish comes close, the shark rises up and engulfs the fish in its huge mouth. It then sinks back to the seabed, swallowing its food whole.

► **SHELLFISH EATER**
The leopard shark lives in the shallow waters of the Pacific, along the west coast of the USA. It swims slowly, searching the sea floor for the molluscs that it eats.

◄ **JAGGED SNOUT**
The common sawshark (*Pristiophorus cirratus*) has a long snout with tooth-like barbs along each side. Two sensitive barbels (bristles) hang beneath its snout. It mows through sand and seaweed on the seabed, catching its prey by slashing about with its barbed snout.

Prehistoric Seas

▲ **CIRCULAR SAW**
This strange spiral of fossilized teeth comes from the jaws of a whorl-tooth shark. This creature lived during the golden age of the sharks. Unlike modern species, this shark did not lose its teeth. Instead, they were rotated along its spiral system of teeth, then stored in a special chamber underneath the lower jaw.

The first recognizable sharks lived in the sea 400 million years ago. These early sharks developed from primitive jawless fish. The golden age of the sharks did not take place for another 100 million years. Then, sharks of every shape and size filled the sea and ruled the oceans. They grew into both incredible hunters and giant filter feeders.

About 150 million years ago, all sharks of the golden age began to die out. They were replaced with new breeds of sharks, the ancestors of today's sharks. When the sea-going relatives of the dinosaurs became extinct, sea mammals (warm-blooded animals) began to evolve, and with them eventually arrived a gigantic predator – megalodon. This shark grew to 18m and is thought to be the ancestor of the great white shark. It fed on whales and dolphins, but eventually outgrew its food supply and it died out, too.

▼ **SCISSOR TOOTH**
This is an artist's impression of the ancient whorl-tooth shark. This shark had a single row of sharp teeth in its lower jaw, and a scattering of broad, crushing teeth in the upper jaw.

Did you know? Fossil sharks' teeth detected poison.

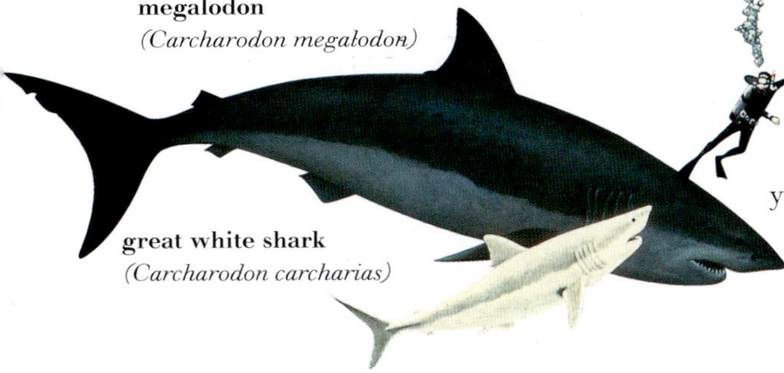

megalodon
(Carcharodon megalodon)

great white shark
(Carcharodon carcharias)

◀ MONSTER SHARK
Today's great white shark might seem a monster, but megalodon was three times bigger. It first appeared about 18 million years ago, disappearing two million years ago. It was the terror of the oceans.

◀ GIANT TOOTH
Megalodon's teeth (left) were huge. They make the teeth of a great white shark (right) seem tiny in comparison. The triangular shape and jagged edges of both teeth are similar, but a giant mako shark that lived earlier than megalodon is believed to be the direct ancestor of the great white.

▲ RECONSTRUCTED JAWS
Fossilized teeth and some pieces of backbone are the only megalodon remains ever found. However, by using the great white as a model, scientists have been able to reconstruct its huge jaws. Over 2m wide, megalodon's jaws could have eaten several people in one big gulp.

▶ FOSSIL TEETH
These fossilized teeth belong to an ancestor of the salmon and porbeagle sharks. It lived about 40 million years ago.

▶ BIZARRE SHARK
This artist's impression of the spine-brush shark is based on fossils found in Glasgow. *Stethacanthus* is its scientific name. The shark has a strange spiny brush where its dorsal fin should be. No one knows what it is for.

spine-brush shark
(Stethacanthus)

Shark Relatives

▲ **STING IN THE TAIL**
One of 30 known species of sting ray, the southern sting ray (*Dasyatis americana*) moves through the water by rippling its broad pectoral fins. It uses its tail to dish out a barbed sting to any attacker.

▼ **UNDERSEA RATS**
Although they are relatives, ratfish look quite different from sharks. In fact, in some ways they resemble rats. They have a long, thin tail, smooth skin and rodent-like teeth. Male ratfish also have an extra, hooked clasper on their forehead, and use two pairs of claspers around their pelvic fins to grip the female.

Sharks have some close relatives. Skates and rays are especially similar to their shark cousins. Both have features that are found in sharks, including a cartilage skeleton, electrical sensors and skin teeth. In fact, skates and rays look like flattened sharks. They also come in all shapes and sizes, ranging from long guitarfish to giant manta rays. Like whale sharks, manta rays also filter plankton from the surface of the sea, but in a more unusual way. Turning somersaults in the water, the rays guide the plankton into their mouths with flattened horns at either side of their head. Another close relative of the shark is the ratfish. Looking like a cross between a shark and a bony fish, the ratfish is probably the long-lost descendant of mollusc-eating sharks that lived 300 million years ago.

ratfish
(Hydrolagus colliei)

► SHARK OR RAY?

The long, flat shovelnose guitarfish looks like a cross between a shark and a ray. Although it uses its tail to swim, it is more closely related to rays. It swims in the coastal waters of the eastern Pacific. An adult usually grows to 1.5m long.

shovelnose guitarfish
(Rhinobatus productus)

◄ SAW FISH

Saw fish (Pristidae) belong to the ray family. Unlike sawsharks, the pectoral fins of the saw fish grow forwards on its body, and are joined to the side of its head. Its gill openings are found on the underside of its head. It has a broad saw, which is lined with skin teeth that have been especially adapted for hunting.

electric ray
(Torpedo torpedo)

◄ ELECTRIC SHOCK

Torpedo, or electric, rays have special blocks of muscle in their wings that can produce electric shocks of up to 220 volts. They use this hidden weapon to knock out their prey. The shock can stun a human.

Did you know? Manta rays may grow to 7m across and weigh 1,400 kg.

► FLYING RAYS

Spotted eagle rays (*Aetobatus narinaria*) have broad pectoral fins which they flap like birds. The fins of some species span nearly 3m. Spotted eagle rays feed on shellfish and oysters, which they crush with their teeth.

The Eight Families

Sharks fall into eight main orders (groups) divided according to different features. The most primitive order, including frilled and sevengill sharks, have more than five gill slits. Dogfish sharks have long, round bodies, and include luminous (glow in the dark) sharks that live in very deep water. The seven or more species of sawshark have a saw-like snout. Angel sharks look like rays and lie hidden on the seabed. Bullhead sharks have spines on both of their dorsal fins, and carpet sharks, like the wobbegong sharks, have short snouts and bristles on their snouts. Mackerel sharks, with their special, warm muscles, are awesome hunters. These sharks include the great white and mako. The ground sharks include the widest range of all, from catsharks to bull sharks, hammerheads, blue sharks and oceanic whitetips.

▲ REEF WALKER
Two pairs of muscular pectoral fins allow the epaulette shark (*Hemi-scyllium ocellatum*) to walk over its tropical reef home. It feeds on the seabed of shallow waters around the Australian reefs.

▼ TYPES OF SHARKS
Modern sharks are divided into eight large family groups. These groups are divided into over 30 smaller families, and nearly 400 species. This number will probably rise as more species of shark are discovered.

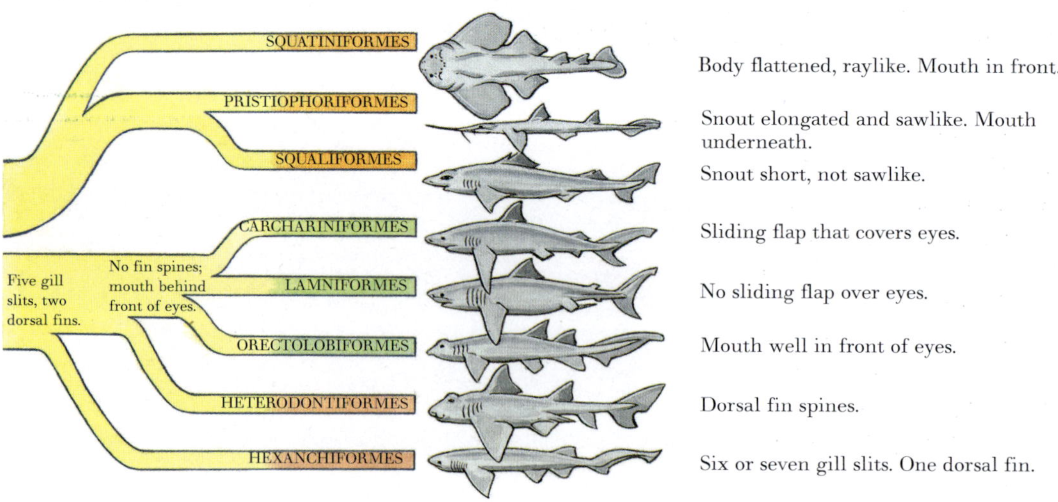

◄ **PRIMITIVE SHARK**

The broadnose sevengill shark (*Notorynchus cepedianus*) is one of five species of primitive sharks. Each has six or seven gill slits. All swim in deep waters.

▲ **GROUND SHARK**

The swell shark (*Cephaloscyllium ventriosum*) is a ground shark. It blows up like a balloon by swallowing water and storing it in its stomach. When it is threatened, this amazing shark wedges itself firmly inside the cracks between rocks. It can be very difficult to remove.

Did you know? Pacific island children ride on nurse sharks' backs.

Food for the sharks

The Carib peoples buried the bodies of their dead relatives by ceremonially putting them into Lake Nicaragua in Central America. Many of the bodies were then eaten by bull sharks in the lake. One thief made a fortune by catching the sharks, slitting them open and removing jewels that had decorated the bodies of the dead. Until he was caught, that is ......

► **REQUIEM SHARK**

The sandbar, or brown, shark (*Carcharhinus plumbeus*) is a requiem shark, meaning 'ceremony for the dead'. All members of this family are active hunters. They rule tropical seas, hunting fish, squid and sea turtles. They are probably the most modern group of shark.

Friends and Enemies

No sharks spend all their time alone. They attract all kinds of hangers-on, including pilot fish, bait fish and a wide range of parasites, both inside their bodies and out! Sandtiger sharks are often seen surrounded by a cloud of baby bait fish. Too small for the shark to eat, the bait fish crowd around it for protection. Basking sharks are sometimes covered with sea lampreys, which clamp on to the shark's skin with their sucker-like mouths. To get rid of these pests, the giant sharks leap clear of the water and crash back down to dislodge them. Smaller parasites live inside each shark. These have adapted so well to life with sharks that they can only survive in one species, some in only one part of the shark's digestive system.

▲ STRIPED PILOTS
Tiny pilot fish often ride the bow wave in front of a shark's snout. Young golden trevally fish swim with whale sharks. When they are older and lose their stripes, they leave the shark and return to their reef homes.

▼ HITCHHIKER
A remora fish stays with a shark for most of its life. Its dorsal fin is designed like a sucker, which the fish uses to attach itself to the shark's belly. The fish then feeds on scraps from the shark's meal.

▸ SCRAPING CURE
Some fish use the shark's rough, sandpaper-like skin to remove their own parasites. Rainbow runner fish will rub against a whitetip reef shark's side. Behind one shark might be an entire group of fish lining up for a scrub.

◂ BARBER SHOP
Many sharks visit what scientists call cleaning stations. Here, small fish and shrimps remove dead skin and parasites from the shark's body, even entering the gills and mouth. This hammerhead shark is gliding past a cleaning station where several king angel fish have darted out to clean it.

▾ UNWELCOME FRIENDS
Strings of parasitic copepod eggs trail behind the dorsal fin of this shortfin mako shark. These parasites will have little effect on the shark's life, but if large numbers of parasites grow inside the shark, it can die.

▸ BLOOD SUCKER
A marine leech feeds by attaching itself to any part of the shark's skin and sucking its blood. Other parasites feed only on certain areas of the shark's body, like the gills, mouth and nasal openings.

Did you know? Pilot fish ride sharks' bow waves like dolphins on ships'.

mako shark
(Isurus oxyrinchus)

Sharks and People

Sharks are feared because they attack people. However, only a few such attacks take place each year. People are more likely to be killed on the way to the beach than be killed by a shark in the water. Fortunately, attitudes are changing. Today, people have a healthy respect for sharks, rather than a fear of them. As we come to understand sharks, instead of killing them, people want to learn more about them. Diving with sharks, even such known threats as the great white shark or bull shark, is more accepted. People study sharks either from the safety of a cage or, increasingly, in the open sea without any protection. Such is our fascination with sharks that aquariums are being built all over the world. Here, more people will be able to learn about sharks at first hand, and not even get wet!

Jaws

The book and film Jaws *featured an enormous great white shark that terrorized a seaside town. The film drew great crowds and its story terrified people all over the world. It also harmed the reputation of sharks, encouraging people to see them as monsters, rather than the extraordinarily successful animals that they are.*

◄ FEEDING TIME
At tourist resorts in the tropics, divers can watch sharks being fed by hand. This activity is not always approved of. Sharks come to rely on this free handout, and may become aggressive if it stops. Inexperienced divers may also not know how to behave with sharks, resulting in accidents, although these are rare.

◄ **ANTI-SHARK MEASURES**
Anti-shark nets protect many popular South African and Australian beaches. Unfortunately, these nets not only catch sharks, like this tiger shark, but also other sea life, including dolphins and turtles. Less destructive ways of reducing people's fear of attack have yet to be invented.

► **SHARK POD**
Although a similar system is not yet available to bathers, one anti-shark invention seems to work for divers and, possibly, surfers, too. A shark pod can produce an electric field that interferes with the electrical sensors of a shark, encouraging the animal to keep its distance.

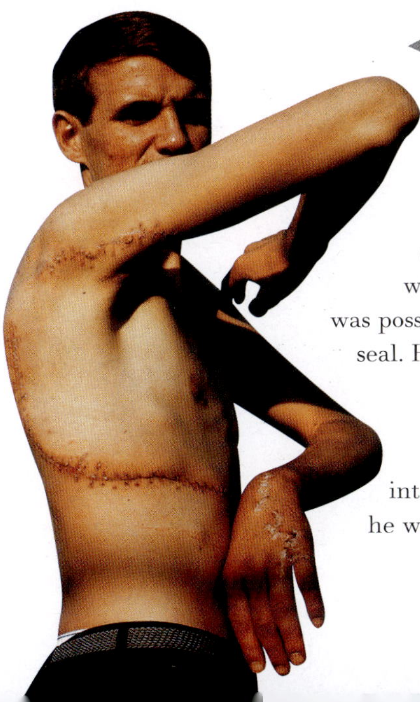

◄ **SHARK ATTACK**
Occasionally, sharks do attack. While diving in Australian waters, Rodney Fox was attacked by a great white shark. Rodney was possibly mistaken for a seal. He is probably alive because he did not have enough blubber on him to interest the shark and he was able to get away.

▲ **SHARK WARNING**
On many beaches, shark warning signs are used to tell people that sharks might be present. During the day, danger of attack is low, but it increases at night, when the sharks move inshore to feed.

Focus on the Great

The great white shark grows to over 6m long and is the largest hunting fish in the sea. Its powerful jaws can bite a full-grown elephant seal in half. Many people believe that the great white will attack people readily. This is not true. In the whole world, only about ten people a year are bitten by great whites. Great whites are aggressive, powerful fish and will attack people when they mistake them for their natural prey, such as seals. If they realize their mistake, there is a chance that a person can survive – that is, if the blood loss from the first bite can be stopped.

INTELLIGENT SHARK
The great white shows signs of 'intelligent' planning. It stakes out places off the Farallon Islands to the west of San Francisco, USA, where young elephant seals swim. In this way it avoids the large, possibly aggressive, adult bulls that could do it damage.

BODY PERFECT
The great white has the torpedo shape typical of a hunting shark. Its crescent-shaped tail, with its equal upper and lower parts, helps the shark to speed through the water. Although it is called the great white, it is not white all over, but grey on top and white underneath.

White Shark

GIANT SHARK ENCOUNTER
A great white shark dwarfs any diver. To a diver in a cage, it can sometimes seem that a shark is trying to attack. In reality, the shark's electrical sensors are probably confused. The diver's metal cage produces an electrical field in seawater – the shark is then likely to react to the cage as if it were prey.

TERRIBLE JAWS
As a great white rises to take bait, its black eyes roll back into their protective sockets. Its jaws thrust forward, filled with rows of triangular teeth ready to take a bite. This incredible action takes place in little more than a second.

SHARP TEETH
The powerful, arrow-shaped teeth in the upper jaw of a great white have a serrated (jagged) edge. These teeth can slice through flesh, blubber and even bone. The shark saws through the tissue of its prey by shaking its head from side to side.

GAME FISH
To fishermen who hunt great whites for sport, the large breeding female sharks are the most attractive. The killing of these sharks has brought them near extinction in some places.

Conservation

Flying Tigers
A Japanese god of storms is known as the shark man. In ancient Japanese legend, the shark man was terrifying. Encouraged by this fear of the shark, Allied airmen fighting the Japanese during World War II painted a tiger shark on their aircraft as a talisman. The pilots of these planes soon became known as "the flying tigers."

Sharks take a long time to grow to adulthood. They have very few offspring and may breed only every other year. Added to these factors, the hunting and killing of sharks can quickly reduce their numbers. This happened at Achill Island, on the Irish coast, where large numbers of basking sharks quickly disappeared, killed for their oil. Off the coasts of South Australia and South Africa, the great white shark was hunted as a trophy for many years. Numbers of great whites were so reduced that the hunting of them has since been banned internationally. A few countries control the fishing of sharks, to try to conserve (protect) them. However, in other countries, sharks are still hunted for shark fin soup, unusual medicines, and souvenirs. They are also sold to supermarkets as shark steak. Sharks, it seems, have more to fear from people than people have to fear from sharks.

◄ **WASTED SHARKS**
Each summer, many sharks are killed in fishing tournaments off the east coast of the USA. Sports fisherman are now learning to tag sharks, returning them to the sea alive instead of killing them. By tagging sharks, our understanding of shark biology is increased.

▶ **CRUEL TRADE**
Caught by fishermen, this whitetip reef shark has had its valuable fins removed. The shark was then thrown back into the sea, still alive. Without its fins, a shark is unable to move and, therefore, feed. It will quickly starve to death. This awful process, called finning, has been banned by some countries.

◀ **TRAVELLING INTO TROUBLE**
This tiger shark is being tagged to track its movements. Shark tagging programmes like this show that many sharks travelling great distances are being netted by several fisheries along their routes. Unless shark fishing is controlled internationally, far-travelling sharks will probably disappear from the sea altogether.

◀ **CARTILAGE PILLS**
Cashing in on the discovery that cartilage can stop the development of cancer, shark cartilage pills have been sold as dietary supplements. However, to prevent cancer, the cartilage has to be applied directly to tumours. Pills taken orally are unlikely to have any effect.

▲ **MULTIMILLION DOLLAR SOUP**
Shark fin soup is made from the dried fins of a shark. It has been prepared by chefs in oriental countries for over 2,000 years. The soup was once served to show favour to an honoured guest, and was also thought to be a health-giving food. Today, it is sold in cans and can be bought in supermarkets.

GLOSSARY

afterbirth
Any birth membranes and other tissues discarded or discharged into the sea when a baby shark is born.

bioluminescence
The production of light by living organisms.

camouflage
The way in which an animal blends in with its background or resembles another animal or object.

cannibalism
Animals eating others of their own kind.

cartilage
The strong but flexible material from which the skeletons of sharks and rays are made, rather than the bone that is found in most other animals with backbones.

catshark
The common name given to a group of sharks that are known in the British Isles as dogfish.

clasper
The male sexual organs in sharks, consisting of two modified pelvic fins.

dorsal fin
The tall triangular fin on a shark's back. Some sharks have two dorsal fins, the front fin larger than the back one.

egg case
The leathery covering that protects a shark embryo developing outside its mother's body.

electrical field
A zone of electricity surrounding an object, such as a muscle or nerve cell, that generates electricity.

embryo
The early stage of an animal before birth.

feeding frenzy
The name for what happens when sharks go berserk, slashing and biting anything that moves, when there is blood in the water or when they are presented with large quantities of food.

filter-feeder
Animals that sieve water through giant combs called gill rakers, for very small particles of food.

fish ball
The ball that schools of fish make when attacked.

gestation
The period of time between conception and the birth of an animal.

gill arch
The part of the skeleton that supports the gills.

gills
The organ used by aquatic animals, such as sharks, for breathing.

gill slit
The vertical openings on either side of the shark, just behind the head, from which the water taken in through the mouth and passed over the gills leaves its body.

gut
The long tube in which food is digested and absorbed, running a winding path through an animal's body.

jawless fish
Primitive fish with sucker-like mouths rather than true jaws. They had their origins 500 million years ago and living descendants include lampreys and hagfish.

krill
Small shrimp-like creatures that swim in huge shoals. Krill form part of the diet of filter feeders such as whale sharks.

ligament
A band of white, fibrous tissue that cannot stretch. It connects bones in a joint and strengthens them.

light organs
Special structures in a fish's skin that produce 'cold' light. They work either by mixing particular chemicals together or with the help of luminous bacteria that do it for them.

megalodon
A gigantic shark ancestor that first appeared 18 million years ago. It is thought to be the ancestor of the great white shark.

nutrients
Chemicals in food that, when digested, build blood, bone and tissue. This tissue maintains growth and strength in the body.

parasite
An organism that lives on or in another living thing (its host), using the host as a source of food and shelter.

pectoral fins
The pair of large wing-like fins on either side of a shark's body.

pelvic fins
The pair of small fins on the underside of a shark's body behind the pectoral fins.

placenta
A disc-shaped organ that is attached to the lining of the womb during pregnancy. It is through this that the embryo receives oxygen and nutrients.

plankton
Tiny aquatic organisms that drift with the water movements in the sea or in lakes.

primitive
Keeping physical characteristics that may have origins millions of years ago.

pup
A young shark, particularly when it has just been born.

pupil
The opening, which can be round or slit-like, through which light passes to the eye of an animal.

receptor
A cell or part of a cell that is designed to respond to a particular stimulus such as light, heat or smell.

remora
A streamlined fish that attaches itself to a shark's body with a sucker, and accompanies its larger companion everywhere.

scavenger
An animal that feeds on animals that have died naturally or were the prey of other predators.

sensory system
The collection of organs and cells by which an animal is able to receive messages from its surroundings.

species
A group of individuals that can breed successfully together. When naming animals scientifically, this is the basic unit of classification.

spiracle
A modified gill slit positioned behind the eye in sharks and rays.

spiral valve
A complicated folding of the tissues in the intestine of sharks that aids efficient digestion of nutrients.

thermal corridor
A layer of water at a particular temperature.

threat display
The aggressive behaviour shown by some species of sharks when confronted by other sharks or sea creatures.

tooth whorl
A spiral arrangement of the teeth in some species of extinct sharks.

vertical migration
The daily movement that sharks make downwards into the deep sea by day and upwards to the surface waters at night.

yolk sac
An outgrowth of the embryo's gut containing food that sustains the shark embryo before it is born. As the yolk is used up the sac is withdrawn into the embryo's body.

INDEX

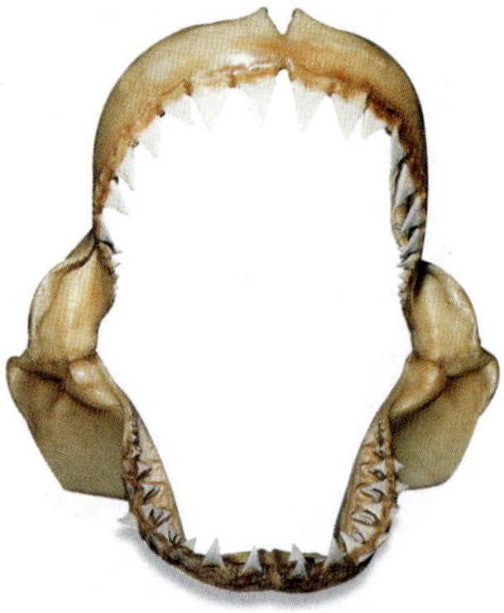

A
ampullae of Lorenzini 16
angel shark 6, 46–7, 52
Atlantic sandbar shark 40
  sharpnose shark 44
  spadenose shark 44

B
basking shark 5, 9, 24–5, 38, 54, 60
blacktip reef shark 17, 21, 27, 39
blue shark 6, 15, 18–19, 20, 34, 40, 41, 52
bluntnose sixgill shark 17
body language 26–7, 29
body temperature 12
bonnethead shark 6, 44–5
Borneo river shark 44–5
breathing 4, 12-13, 36
brown shark *see* sandbar shark
bullhead shark 7, 52
bull shark 6, 21, 38, 44–5, 52–3

C
cannibalism 20, 21, 36
carpet shark 7, 46, 52
cartilage 8–9, 13, 42, 50, 61
catshark 13, 32–3, 42, 46, 52
cleaning stations 55
Colombian lanternshark 6
communication 26–7, 29
cookie-cutter shark 11, 21
coral reefs 5, 7, 24–5, 26–7, 30, 52

D
dogfish 14, 34, 42, 52
dorsal fins 5, 8, 14–15, 22
dwarf shark 42

E
eggs and egg cases 32–3, 34–5
electrical sensors 16–17, 18–19, 28, 50, 59
eyes 16–17, 18–19

F
fighting 26, 27, 28
filter feeders 5, 7, 9, 15, 24–5, 48, 50
fins 4–5, 8–9, 14-15, 31, 61
fossils 8, 42, 48-9
frilled shark 6, 34, 38, 42, 52

G
gills 4, 9, 12–13, 15, 25, 27, 36, 43, 46
great white shark 4, 8–9, 11, 12, 14, 20, 21, 26, 39, 46, 48–9, 52, 57, 58–9
Greenland shark 14, 38–9
grey reef shark 13, 14, 20, 27

H
hammerhead shark 5, 6, 15, 16–17, 20, 27, 28–9, 55
horn shark 6, 32, 38
hunting 6, 13, 14–15, 18–19, 20

J
jaws 8, 11, 13, 59
Jaws (film) 4, 56

L
lakes 4, 38, 44–5
lemon shark 13, 36–7
leopard shark 10, 47

M
mackerel sharks 52
mako shark 12, 13, 14, 20, 35, 38, 41, 52, 55
megalodon shark 48–9
megamouth shark 6, 15, 24–5
muscles 5, 9, 12–13

N
nervous system 16–17
nurse shark 13, 14, 31, 34, 38, 53

O
oceanic whitetip shark 15, 38, 52

P
pectoral fins 4, 9, 14–15, 30–1
pelvic fins 8, 15, 31
piked dogfish 34
plankton 4, 5, 9, 24–5, 50
pups 32–7
pygmy shark 6, 42

R
reef shark 5, 13, 14, 17, 20, 21, 26, 27, 34, 38, 39

S
sandbar shark 40, 44, 53
sandtiger shark 10, 12, 16, 30, 34, 54
sawshark 47, 51, 52
scalloped hammerhead shark 5, 17, 27, 28–9
schools 5, 27, 28–9
senses 16–17, 18–19, 24
sevengill shark 42, 43, 52–3
shortfin mako shark 12, 20, 35, 41, 55
sixgill shark 12, 17, 42, 43
skeleton 8–9, 50
smell 16–17, 18, 30
spiny dogfish 14
spotted eagle ray 51
swimming 12–13, 14–15

T
tail 5, 8, 13, 14
tasselled wobbegong shark 47
tawny nurse shark 13
teeth 5, 8, 10–11, 19, 43, 48–9, 59
tiger shark 7, 10, 14, 17, 20–1, 22–3, 31, 35, 39, 40, 57, 61

V
velvet belly shark 43
Vishnu 45
vision 16-17, 18-19

W
whaler shark 35
whale shark 6, 7, 24, 34, 38, 54
whitetip reef shark 5, 13, 26, 34, 55, 61
whitetip shark 5, 13, 15, 20, 26, 30, 34, 38, 52, 55, 61
whorl-tooth shark 28
winghead shark 6
wobbegong shark 7, 46–7, 52

Z
zebra bullhead shark 7
zebra shark 46

NATURE WATCH

BEARS and PANDAS

MICHAEL BRIGHT

Consultant: Dr. Douglas Richardson,
Zoological Director, Rome Zoo

CON

Published by Anness Publishing Ltd,
Blaby Road, Wigston, Leicestershire LE18 4SE
Email: info@anness.com
Web: www.annesspublishing.com

Anness Publishing has a new picture agency outlet for images for publishing, promotions or advertising. Please visit our website www.practicalpictures.com for more information.

Publisher: Joanna Lorenz
Managing Editor, Children's Books: Gilly Cameron Cooper
Senior Editor: Nicole Pearson
Editor: Charlotte Hurdman
Designer: Caroline Reeves
Picture Researcher: Cathy Stastny
Photographer: John Freeman
Illustrators: David Webb, Vanessa Card, Julian Baker
Production Controller: Ann Childers
Editorial Reader: Diane Ashmore

Special thanks to Nick Lindsay, Curator of the Whipsnade Wildlife Park, for his help on this book.

ETHICAL TRADING POLICY
Because of our ongoing ecological investment programme, you, as our customer, can have the pleasure and reassurance of knowing that a tree is being cultivated on your behalf to naturally replace the materials used to make the book you are holding. For further information about this scheme, go to www.annesspublishing.com/trees

PICTURE CREDITS
b = bottom, t = top, c = centre, l = left, r = right
ADRIAN LISTER: 48tr. B & C ALEXANDER: 33tr, 38tl, 44br, 45bl, 57b; Bob Wilson: 41bl. BBC NATURAL HISTORY UNIT: Doug Allan: 56tl; Gerry Ellis: 23bl; Jeff Foott: 27tl, 33br, 39tl; Martha Holmes: 28tl; Hans Christoph Kappel: 27br; Tom Lazar: 25tr; Mary Ann McDonald: 52bl; Thomas D. Mangelsen: 19tr, 36b; Klaus Nigge: 13bl; Pete Oxford: 60tr; Lynn M. Stone: 11cr; David Welling: 37br. BRIDGEMAN ART LIBRARY: Oxford University Museum of Natural History: 11b. BRITISH MUSEUM: 14bl. BRUCE COLEMAN: Erwin & Peggy Bauer: 4bl, 20br, 47tl, 58tl; Thomas Buchholz: 60bl; Sophy & Michael Day: 37tr, 54tr; Stephen Kraseman: 45br; Joe McDonald: 32tr; Hans Reinhard: 8br, 9tr, 53cl; Leonard Lee Rue: 36tr; John Shaw: 7br, 9br, 50bl; Joseph Van Wormer: 47cr. E T ARCHIVE: 20tr; Archaeological Museum, Sofia: 61tl. FLPA: Rolf Bender: 35bl; P. Davey: 53tr; Gerard Lacy: 12bl, 46tl; Mark Newman: 6br, 12tr, 15bl, 15tl, 24bl, 25br, 34tr, 40tr, 40b, 51br, 55tr; Fritz Polking: 8tl, 51tl; Cordier/Sunset: 19tl; T. Leeson/Sunset: 47bl. HEATHER ANGEL: 5c, 17tl, 23tl, 23tr, 34bl, 42tr, 54bl, 55cl. IMAGE BANK: Paul McCormick: 61tr; Joseph Van Os: 10tr, 18bl. L B & W: 4tr. MARY EVANS: 12br, 16tr, 38bl; J M Gleeson in The Outing Magazine 1906, vol 49, p289: 33tl; Whittaker's Star Atlas, Plate 6: 44tr. NHPA: B & C Alexander: 29bl, 29cr; Daniel Heuclin: 58bl; T. Kitchin & V. Hurst: 6c, 59bl; Stephen Kraseman: 29tl, 53br, 57cl, 59tl; Rob Planck: 59br; B. Hanumantha Rao: 43br; Andy Rouse: 35cr, 51tr, 55br; Kevin Schafer: 26bl; John Shaw: 33bl, 52tr. OXFORD SCIENTIFIC FILMS: Michael Austerman: 7bl; Liz Bomford: 11cl; John Brown: 42bl; John Chellman: 50tl; Judd Cooney: 24br; Daniel J. Cox: 5bl, 15tr, 17br, 17cl, 18tr, 21bl, 24tr, 25bl, 30tr, 34br, 35t, 37tl, 46br, 50br; Richard Foster: 53tl; David C. Fritts: 13br, 26tr, 27bl; Dan Guravich: 16bl, 44bl; Djuro Huber: 31bl; Frank Huber: 42br; Keith & Liz Laidler: 22tl; Zig Leszczynski: 6tl, 15br; Stephen Mills: 31tl; Ben Osborne: 59tr; Stan Osolinski: 41tl; Stouffer Productions: 30bl; Norbert Rosing: 31tr, 37cl, 41c, 61br; David Shale: 38br; Vivek Sinha: 21br; Keren Su: 14tr, 22br; Konrad Wothe: 57cr. PAPILIO: 51bl. PLANET EARTH: Jim Brandenburg: 17tr; Brian Kenny: 4br; Ken Lucas: 11tr, 14br; Louise Murray: 5tl; Nikita Ovsyanikov: 21tr, 31br; Tom Walker: 7tl, 19br, 21tl, 26tr. ROGER TIDMAN: 32b, 39bl, 39c. TONY STONE IMAGES: Tim Davis: 25tl; David Hiser: 56br. WWF FOR NATURE 1986: 55tl.

INTRODUCING BEARS

4 The Bear Facts
6 Size of a Bear
8 Giant Pandas

HOW BEARS WORK

10 Bones and Teeth
12 Strong Muscles
14 Warm Fur
16 Life on the Ice
18 Using Brain and Senses

BEAR BEHAVIOUR

20 Finding Food
22 Focus on Bamboo Bears
24 Climbing Trees
26 Gone Fishing
28 Focus on Seal Hunting
30 Winter Shutdown
32 A Solitary Life

TENTS

LIFE CYCLES

34 Meeting a Mate
36 Nursery Dens
38 Focus on Snow Homes
40 Raising Cubs

WHERE BEARS LIVE

42 Where in the World are Bears?
44 Finding the Way
46 Focus on Super Bears

FACTS ABOUT BEARS

48 Ancient Bears
50 Family Groups
52 Relatives and Impostors
54 Pandas on the Edge
56 Focus on Churchill
58 Bears and People
60 Bears in Danger
62 Glossary
64 Index

The Bear Facts

Bears may look large, cuddly and appealing, but in reality they are enormously powerful animals. Bears are mammals with bodies covered in thick fur. They are heavily built with short tails and large claws. All bears are carnivores (meat-eaters), but most enjoy a very mixed diet with the occasional snack of meat. The exception is the polar bear, which feasts on the blubber (fat) of seals. There are eight living species of bear: the brown bear, American black bear, Asiatic black bear, polar bear, sun bear, sloth bear, spectacled bear and the giant panda. They live in both cold and tropical regions of the world. Bears are loners and their nature is unpredictable, which makes them potentially dangerous to people.

WINNIE-THE-POOH
The lovable teddy bear Winnie-the-Pooh was created by A.A. Milne. Like real bears he loved honey. Teddy bears became popular as toys in the early 1900s. The President of the U.S.A. Theodore Roosevelt refused to shoot a bear cub on a hunting trip. Toy bears went on sale soon after known as "teddy's bears".

◄ **BEAR FACE**
The brown bear shares the huge dog-like head and face of all bears. Bears have prominent noses, but relatively small eyes and ears. This is because they mostly rely on their sense of smell to help them find food.

▲ **BIG FLAT FEET**
A polar bear's feet are broad, flat and furry. The five long, curved claws cannot be retracted (pulled back). One swipe could kill a seal instantly.

◀ POINTS OF A BEAR

The brown bear is called the grizzly bear in North America. Fully-grown brown bears weigh nearly half a tonne. They fear no other animals apart from humans. They can chase prey at high speed, but they rarely bother as they feed mainly on plants.

▲ GIANT PANDA

China's giant panda, with its distinctive black and white coat, is a very unusual bear. Unlike most other bears, which will eat anything, pandas feed almost exclusively on the bamboo plant.

◀ ARCTIC NOMAD

Most bears lead a solitary life. The polar bear wanders alone across the Arctic sea ice. Usually it will not tolerate other bears. The exceptions are bears that congregate at rubbish dumps, or mothers accompanied by their cubs as shown here.

Size of a Bear

The two largest bears in the world are the powerful polar bear and the brown bears of Kodiak Island, Alaska. Kodiak bears grow up to 2.8m long and weigh, on average, up to 443kg, while polar bears have a maximum length of 3m and weigh as much as 650kg. Brown bears in Europe and Asia are generally smaller than grizzlies (American brown bears). The largest is the Kamchatka brown bear of eastern Russia. A full size adult grizzly weighs as much as a bison and even the smallest is bigger than a wolf. The smallest bear is the sun bear at 1.4m long and weighing 65kg. In between are American and Asiatic black bears (both 1.7m long and up to 120kg), the spectacled bear (2.1m long, up to 200kg) and the sloth bear (1.9m long, up to 115kg).

Sloth bear
(Melursus ursinus)

▲ **SLOTH BEAR**
Long curved claws, a mobile snout and long fur are the distinguishing features of the sloth bear. It lives in India and Sri Lanka and feeds mainly on insects called termites and fruit.

American black bear
(Ursus americanus)

▲ **AMERICAN BLACK BEAR**
There are ten times as many black bears as brown bears living in the forests of North America. Black bears resemble small brown bears except that they lack a shoulder hump.

▼ **SPECTACLED BEAR**
The South American spectacled bear gets its name from the distinctive markings on its face. It is the only bear found in South America and is a good climber.

Spectacled bear
(Tremarctos ornatus)

Polar bear
(Ursus maritimus)

◄ ICE GIANT

The male polar bear is a giant among bears. It is bigger than most brown bears, but less robustly built, with a longer head and neck. Female polar bears are much smaller, weighing less than half a fully-grown adult male. Polar bears live in the frozen wastes of the Arctic. They can swim in the icy sea, protected by insulating fur and layers of thick fat.

Brown bear
(Ursus arctos)

► BIG BROWN BEARS

The brown bear is the most widely found bear, living in Europe, Asia and North America. Its size varies in different parts of the world. This is due to diet and climate rather than any genetic differences. For example, large Kodiak bears catch a lot of protein-rich salmon.

Did you know? The heaviest known Kodiak brown bears weigh up to 750kg.

▼ THE SUN BEAR

The sun bear is the smallest of all the bears. It lives in the thick forests of South-east Asia. Because it lives in a hot tropical climate, the sun bear also has a short coat. Its feet have naked soles and long, curved claws, which can grip well when climbing trees.

Sun bear
(Helarctos malayanus)

Giant panda
(Ailuropoda melanoleuca)

Giant Pandas

Scientists have argued for many years about whether or not the giant panda is a bear. In 1869, the first Western naturalist to see a giant panda identified it as a bear. But a year later, scientists examined a panda skeleton and decided it was more like a raccoon. This was because the giant panda's skeleton shared some features with the red panda, an earlier discovery that had been grouped with the raccoon family. The giant panda certainly looks like a bear, but it does not behave like one. It does not hibernate, although it lives in places with very cold winters. It rarely roars like a bear, but tends to bleat. Recent genetic studies and comparisons with other animals, however, indicate that the panda's nearest relatives are bears and that the giant panda is indeed a bear.

▲ UP A TREE

Giant pandas sometimes climb trees to avoid enemies. They also scrape trees with their claws. This is a sign that says KEEP OUT to other pandas. Pandas are about 1.7m long and weigh up to 125kg. They are the only surviving members of the earliest group of bears to evolve. Fossils of giant pandas have only ever been discovered in Thailand and China.

► RED PANDA

The red panda is a member of the raccoon family and lives in the high bamboo forests of southern Asia. It has several similarities to the giant panda, including skull shape, tooth structure and a false thumb to hold bamboo.

Red panda
(Ailurus fulgens)

▸ BAMBOO BEAR

The giant panda of western China spends 10 to 12 hours a day eating bamboo. Its massive head contains large chewing muscles needed to break up the tough bamboo. It has a simple stomach and short gut, however, which are features of a meat-eating carnivore and make digesting bamboo hard work. The giant panda's large shoulders and reduced hindquarters give it a curious, shambling walk.

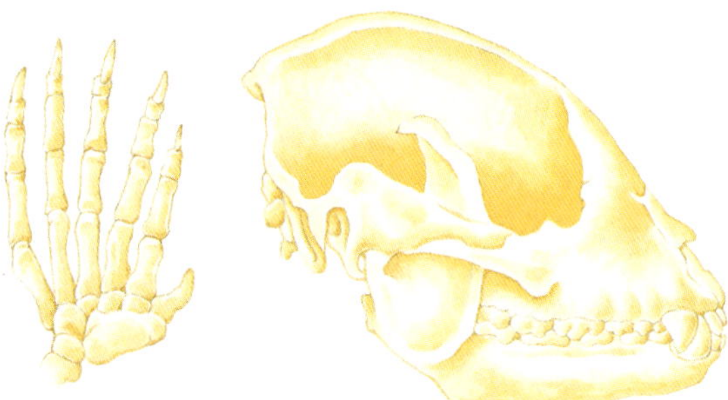

Giant panda
(Ailuropoda melanoleuca)

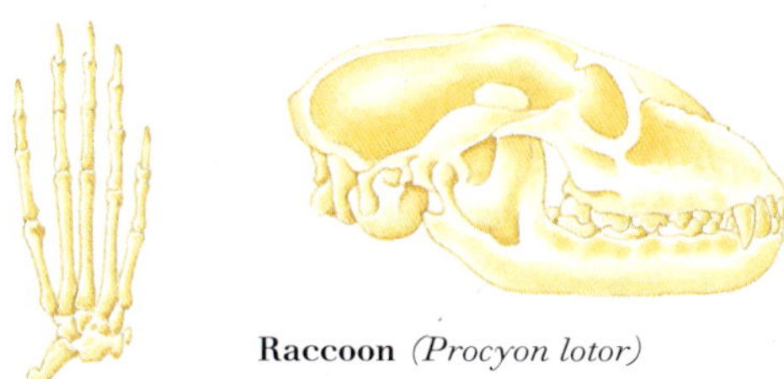

Raccoon *(Procyon lotor)*

▴ FALSE THUMB

The giant panda has a false thumb on its forepaw. This is a modified wrist bone used to hold narrow bamboo shoots. A giant panda's massive skull is quite distinct from the smaller, more slender skull of the raccoon. The raccoon (*Procyon lotor*) does not have a false thumb, unlike its relative the red panda.

▾ AMERICAN BANDIT

The inquisitive raccoon is a close relative of bears. It lives in North America and uses its front paws to capture small aquatic prey, such as freshwater crayfish. It also scavenges through the remains of human waste.

Raccoon
(Procyon lotor)

Bones and Teeth

Bears have a large and massive skull, a solidly built skeleton, relatively short and stocky limbs, a small tail and short feet. Each foot has five equal-sized digits (toes), with strong, curved claws for digging and tearing. The claws cannot be retracted (pulled back) so are constantly worn down. In most bears' jaws, the carnassial teeth (large meat-shearing teeth) common to all carnivores are reduced or even missing. Instead, bears have broad flat molars for crushing plant food. Only the polar bear has flesh-slicing carnassials to deal with its animal prey. The sloth bear is also unusual because it lacks the inner pair of upper incisors (front teeth). This helps it suck up insects from their nests.

▲ A MIGHTY ROAR
As this bear roars, it bares its large canine teeth. However, on average only 20 per cent of a brown bear's diet is made up of animal flesh. Instead, bears rely on their large molars to crush their vegetable food.

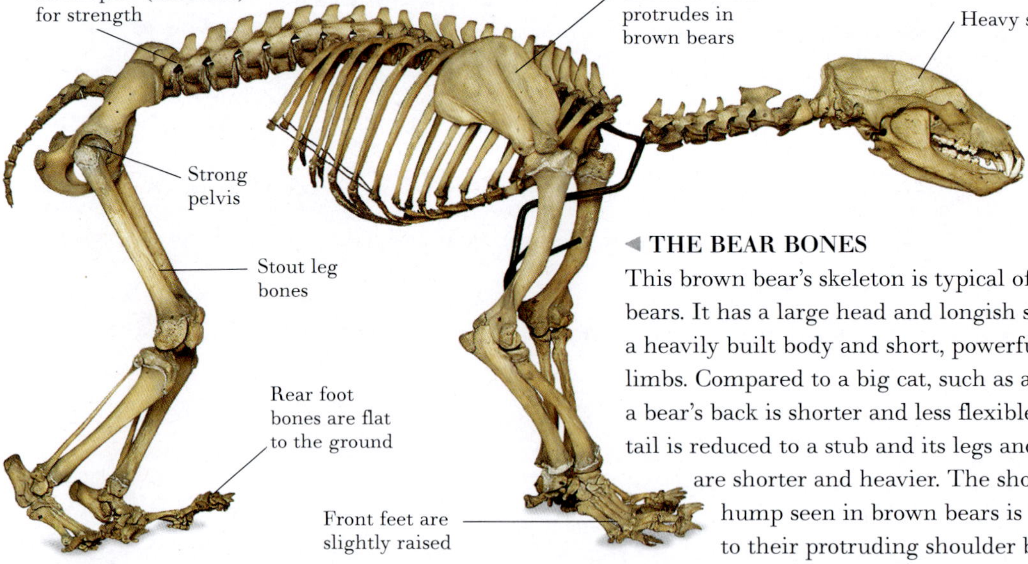

◄ THE BEAR BONES
This brown bear's skeleton is typical of most bears. It has a large head and longish snout, a heavily built body and short, powerful limbs. Compared to a big cat, such as a lion, a bear's back is shorter and less flexible. Its tail is reduced to a stub and its legs and feet are shorter and heavier. The shoulder hump seen in brown bears is due to their protruding shoulder bones.

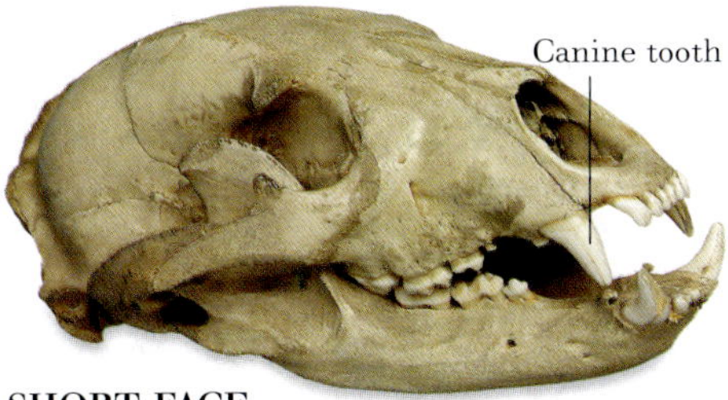

▲ **SHORT FACE**

This brown bear's skull is shorter and more robust than that of the polar bear. North American brown bears tend to have larger skulls than bears in other parts of the world.

▲ **LONG FACE**

This polar bear's skull is longer and more slender than that of other bears. Like other carnivores, it has prominent, dagger-like canine teeth and meat-slicing carnassials at the back.

▲ **PRIMITIVE BEAR**

The spectacled bear is grouped separately from other bears. Apart from the giant panda, it is the most primitive bear. Its short muzzle and the unique arrangement of teeth in its jaw give it a more rounded head shape than its fellow bears.

▲ **GIANT PANDA**

The giant panda has the most massive skull in relation to its size of all living bears. Its round face and head are the result of large jaw muscles needed to grind tough bamboo stems.

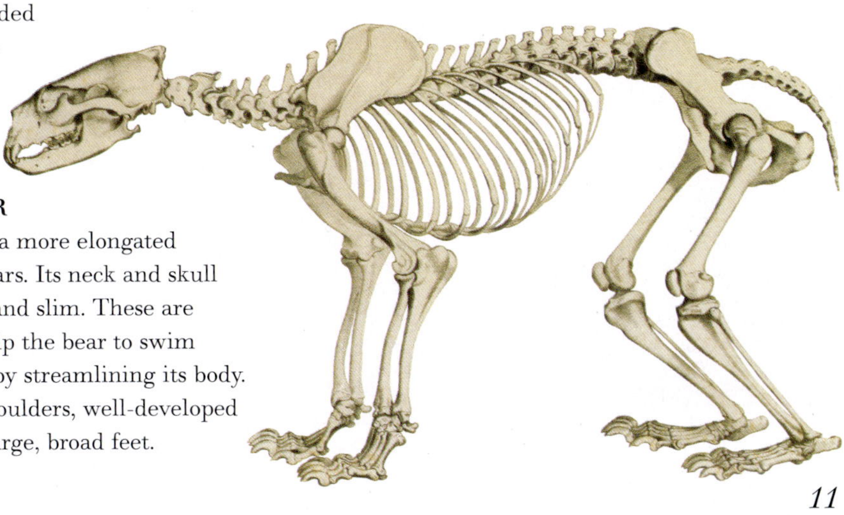

► **SLENDER SWIMMING BEAR**

The polar bear has a more elongated body than other bears. Its neck and skull are relatively long and slim. These are adaptations that help the bear to swim through the water by streamlining its body. It also has lower shoulders, well-developed hindquarters and large, broad feet.

Strong Muscles

Bears are the bully-boys of the animal kingdom. Their strength is mainly in the muscles of their legs and shoulders. Unlike cats and dogs, which run on their toes for speed, bears walk on the flat soles of their broad feet, just as humans do. What bears lack in speed they make up for in strength. Their powerful, mobile limbs can be put to good use digging, climbing, fishing and fighting. They will attack others of their own kind and defend themselves ably from enemies. In a fight, a bear can do considerable damage and survives by sheer brute force. Male bears are generally much larger than females of the same species.

▲ **CLAWS DOWN**
The sun bear has particularly large, curved claws for climbing trees. It spends most of the day sleeping or sunbathing in the branches. At night it strips off bark with its claws, looking for insects and honey in bees' nests.

◄ **PUTTING ON WEIGHT**
This grizzly bear is at peak size. Most bears change size as the seasons pass. They are large and well-fed in autumn, ready for their winter hibernation. When they emerge in spring, they are scrawny with sagging coats.

BEOWULF
An Anglo-Saxon poem tells of the hero Beowulf (bear-wolf). He had the strength of a bear and went through many heroic adventures. Beowulf is famous for slaying a monster called Grendel. Here, Beowulf as an old man lies dying from the wounds inflicted by a fire-breathing dragon.

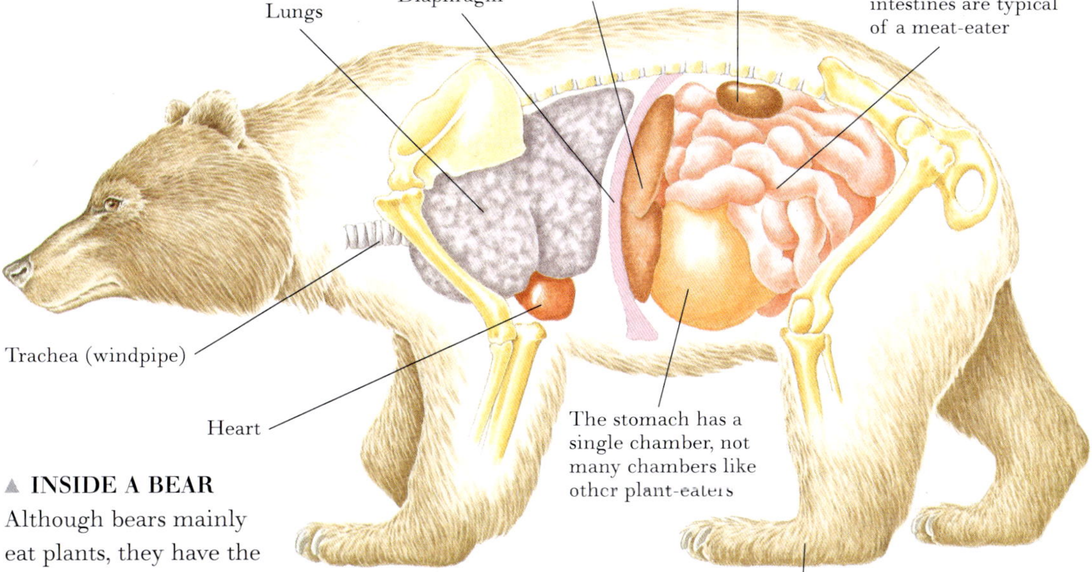

▲ **INSIDE A BEAR**

Although bears mainly eat plants, they have the relatively short intestines more characteristic of a meat-eater, rather than a long gut like a cow. This makes it hard for them to digest their food. Curiously, the bamboo-eating giant panda has the shortest gut of all. Because of this it can digest no more than 20 per cent of what it eats, compared to 60 per cent in a cow.

▲ **SHORT BURSTS**

Bears are not particularly agile and swift, but they can run fast over short distances. The brown bear can charge at 50km/h (lions reach about 65km/h) and does sometimes chase its food. A bear at full charge is a frightening sight and quickly scares away enemies.

▲ **SWEET TOOTH**

The sun bear's long slender tongue is ideal for licking honey from bees' nests and for scooping up termites and other insects. Like all bears it has mobile lips, a flexible snout and strong jaws.

Warm Fur

All bears have thick fur all over their bodies, including the face. Ground-living bears even have fur between the pads of their feet, while the soles of tree-climbing bears are naked. A thick coat insulates well in the cold, but can cause overheating in summer, so many bears moult. Most bears' coats tend to be brown, black, cinnamon (reddish-brown), grey or white and some have face and chest markings. The most strikingly marked bear is the giant panda. Its startling black and white pattern blends in with the shadowy bamboo of its mountain home. This is particularly so at dawn and dusk, the panda's most active times. In winter, against snow, black rocks and trees, the panda is almost invisible.

▲ **WHITE BEAR**
The Chinese call the giant panda *bei-shung* (white bear). It is considered to be basically white with black ears, eye patches, legs and shoulders. Sometimes the black areas have a chestnut reddish tinge.

BERSERKIR
Among the most feared of Viking warriors were berserkirs. They were named after the bear pelts or bear-skin shirts they wore. Berserkirs worked themselves up into a frenzy before battle. We still use the Norse word berserk to mean crazy or wild. This walrus ivory berserkir is part of a chess set from the island of Lewis off the west coast of Scotland. The berserkir is shown biting his shield and clasping his sword in the rage of battle.

▲ **VELVET COAT**
The sun bear has relatively short fur, a bit like velvet. It is generally black with a grey to orange muzzle and pale feet. Some bears have white or pale orange-yellow, crescent-shaped markings on the chest.

▲ **BEAR FOOT**

The broad, flat paw of a brown bear has thick fur on the upper surface and some between the toes, too. Ground-living bears, such as the brown bear, use their claws to hold on to a salmon or to catch and kill a young deer.

▼ **SILVER TIP**

The grizzly (American brown bear) gets its name from the way in which the long hairs of the shoulders and back are frosted with white. This gives the bear's coat a grizzled appearance. Brown bears, like the grizzly, are usually a dark brown colour, but they may also be any shade between light cream and black.

▲ **FACIAL MARKINGS**

The spectacled bear is generally dark brown or black. It has an unmistakable, spectacle-like pattern of white or yellowish hairs around the eyes and across the nose. These markings may extend to the chest.

► **LONG HAIR**

The sloth bear has long, black shaggy fur. The longest hair is between the shoulders. The black fur can be tinged with brown or grey. There is a white, or yellow to chestnut-brown, patch in the shape of a U or Y on the bear's chest. Chest markings may act as a warning sign when the bear stands up.

Life on the Ice

The polar bear is perfectly adapted to life in the frozen Arctic, where winter temperatures can drop to -50°C. Beneath its skin lies a thick layer of fat. The bear's entire body, including the soles of the feet, is covered in insulating fur made up of thick hairs with a woolly underfur. Each hair is not actually white, but translucent and hollow. This acts like a tiny greenhouse, allowing light and heat from the sun to pass through, trapping the warm air. Sometimes, such as in zoos, the hairs are invaded by tiny algae and the polar bear's coat has a green tinge. In the wild, the fur often appears yellow, the result of oil stains from its seal prey. Beneath the fur the skin is black, which absorbs heat. This excellent insulation keeps the polar bear's body at a constant 37°C.

RESPECT FOR THE ICE BEAR
The polar bear is the most powerful spirit in Arctic cultures. The Inuit believe that a polar bear has a soul. It will only allow itself to be killed if the hunter treats it properly after death. It is forbidden to hunt another bear too soon. Time must be left for the bear's soul to return to its family. Some Inuit offer a dead male bear a miniature bow and arrow, and a female bear a needle holder.

SEA-GOING BEAR
Polar bears are excellent swimmers. They must swim frequently for their icy world is unpredictable. In winter, the Arctic Ocean freezes over. But with the arrival of storms and warmer weather the ice breaks up. Then the bear must swim between ice floes in search of seals. The thick layer of fat below the skin and dense, insulating fur allow a polar bear to swim in the coldest seas without suffering. In such cold water, a human being would be dead in a few minutes.

▲ **COOLING DOWN**

Polar bears are so well insulated they are in danger of overheating on warm days. To keep cool, they lie flat out on the ice. At other times they lie on their backs with their feet in the air.

▲ **WARM BEARS**

The insulating fur and fat of a polar bear is so efficient that little heat is lost. In fact, if a scientist were to look at a polar bear with an infra-red camera (which detects heat given off by the body), only the bear's nose and eyes would be visible.

► **LAZY DAYS**

Polar bears are most active at the start of the day. During summer, when the ice melts and retreats, bears may be prevented from hunting seals. Then they rest, living off their fat reserves and eating berries.

▲ **BEAR SLUMBERS**

A polar bear, like a human, sleeps for seven or eight hours at a time. This helps the polar bear to conserve energy and heat. Polar bears are not at risk of attack when they are sleeping, so they do not have to hide like other animals. Most often, polar bears find a sheltered area to protect them from the cold polar winds.

Using Brain and Senses

All bears are very intelligent. Size for size, they have larger brains than other carnivores, such as dogs and cats. They can remember sources of food and are very curious. Bears use their brains to find food or a mate, and to stay out of trouble. They are mainly solitary plant-eaters, however, so they have little need to think up hunting tactics or ways of communication. They rely on smell (the part of their brains that detects scent is the largest of any carnivore), with small eyes and ears compared to their head size. Bears often appear short-sighted, although they have colour vision to recognize edible fruits and nuts. People can easily chance upon a bear and take it by surprise, prompting the startled animal to attack in self-defence.

▼ **SCENT MARKING**
An American black bear cub practises marking a tree by scratching. When it is older, the bear will leave a scent mark to indicate to other bears that the territory is occupied.

◀ **TEMPER TANTRUM**
A threatened bear puts on a fierce display. First it beats the ground or vegetation with its front feet. Then it stands up on its back legs to look larger. This is accompanied by a high-pitched snorting through open lips or a series of hoarse barks. The display of aggression finishes with snapping the jaws together.

▲ FOLLOW YOUR NOSE

A brown bear relies more on smell than sight. It often raises its head and sniffs the air to check out who or what is about. It can detect the faintest trace of a smell, searching for others of its kind.

▼ GETTING TO KNOW YOU

Polar bear cubs rub against their mother to spread her scent over themselves. Smell allows a mother and cubs to recognize each other. They also communicate with sounds. Distressed cubs make low-pitched snores that develop into high whines.

BRUNO THE BEAR

Aesop's Fables are a set of tales written by the ancient Greek writer Aesop. One fable features Bruno the bear. He is shown as stupid and easily deceived. Bears were considered slow-witted because they sleep a lot. But Bruno was kind, unlike the cunning Reynard the fox. Bruno cared about others and forgave those who played pranks on him. Bruno was the forerunner of characters such as Winnie-the-Pooh.

▲ SNOWY SCENT TRAIL

Even in the Arctic, polar bears can pick up the trails of other bears and follow them. There are few objects around to use as scent posts for polar bears, so trails may be marked by the dribbling of urine on the ground.

Finding Food

Most bears eat whatever is available at different times of the year. They have binges and put on fat in times of plenty, then fast when food is scarce. Brown bears are typical of most bears in that they eat an enormous variety of food, from grasses, herbs and berries to ants and other insects. They also catch salmon and rodents, and on rare occasions hunt down bigger game, such as caribou, seals and birds. Only polar bears eat almost entirely meat, especially young seals. In summer, however, as the ice melts and they are unable to hunt, polars supplement their diet with grasses and berries. All bears, even the bamboo-loving panda, also scavenge on the carcasses of prey left by other animals. They are attracted to easy food, and rummage through rubbish thrown out into dustbins and left behind in campsites.

GOLDILOCKS
The famous story of Goldilocks and the Three Bears *was first told in 1837. In the story, the bears and their ability to organize themselves properly are used to make a strong moral point. In contrast, Goldilocks is shown as a foolish, unthinking girl, who gets a nasty fright.*

▶ **HUNTING DOWN A MEAL**
This American black bear has caught a white-tailed deer fawn. Both black and brown bears are successful hunters. They are able to ambush large animals and kill them by using their considerable bulk, strong paws and jaws. The size of the bear determines the size of its prey. Large brown bears may prey on moose, caribou, bison, musk ox, seals and stranded whales. Black bears are smaller than brown bears and take smaller prey, such as deer fawns, lemmings and hares. Roots, fruit, seeds and nuts, however, form up to 80 per cent of brown and black bears' diets.

▶ **WALRUS CITY**

Polar bears arrive on the northern coast of Russia each year to hunt walruses that have come there to breed. Enormous adult walruses shrug off attacks, but the young walrus pups are more vulnerable.

▲ **BEACHCOMBING**

Brown bears are attracted to beaches beside rivers and in estuaries. They overturn stones to feed on aquatic creatures, such as crabs and crayfish, that are hiding underneath.

▲ **FRUIT LOVERS**

An American black bear snacks on the ripe berries of a mountain ash tree. It carefully uses its incisors (front teeth) to strip the berries from their woody stem.

▲ **INSECT EATERS**

Two sloth bear cubs learn to dig up termites. The sloth bear uses its sickle-shaped claws to break open ant hills, bees' nests and termite mounds. Because it feeds mainly on termites, it has an ingenious way of collecting its insect food. First it blows away the dust. Then it forms a suction tube with its mouth and tongue, through which it can vacuum up its food.

Focus on

The giant panda specializes in eating bamboo, which forms over 99 per cent of its diet. Bamboo is plentiful and easy to harvest. Pandas consume the sprouts, stems and leaves. Digesting bamboo, however, is hard work. This is because the giant panda has a simple stomach and a short gut more characteristic of a carnivore. Most herbivores, such as cows, have several stomachs and very long intestines with bacteria inside that break down the plant tissue. The panda does not, so it must feed on huge quantities of bamboo (up to 40 kg) every day in order to keep going. Much of the leaf and stem matter passes through undigested.

MEAT AND TWO VEG
Whilst pandas spend most of their time eating bamboo, they do sometimes supplement their diet with meat when they can get it. They catch rats and beetles in the bamboo stands, and have been known to scavenge at leopard kills. But they make clumsy hunters and easy prey is scarce. In contrast bamboo is very abundant.

EARLY RISER
The giant panda is most active in the early morning and late afternoon. It spends 16 or more hours a day feeding and sleeps for up to four hours at a time. Most of the water a panda needs comes from bamboo. If it is thirsty, it scoops out a hollow by a stream. When this is full, the panda drinks all it can.

Bamboo Bears

ESSENTIAL FOOD

Umbrella bamboo and arrow bamboo are the giant panda's favourite food. Pandas also eat 28 other species of bamboo. They favour the leaves over the stems because these are the most nutritious parts and are easiest to digest. Pandas also eat other plants such as juniper, vines, holly and wild parsnip.

FEEDING ALL YEAR

Since bamboo is green and nutritious throughout the year, even in winter, the panda has a continuous supply of food. Unlike other bears, whose food is scarce at certain times, the panda has no need to hibernate, even when snow is covering its mountain home. A thick fur coat protects it from the cold.

TABLE MANNERS

A panda usually feeds sitting upright on its haunches. This leaves its forelegs free to handle bamboo. It manipulates the long stems of bamboo using its extra thumb (actually a modified wrist bone) on its front paws. It strips away the woody outer covering with its teeth. Then it pushes the stem at right angles into the corner of its mouth. Here, the centre part is crushed by large back teeth and then swallowed.

Climbing Trees

Trees provide food for some bears and a place of safety for others. Sun bears, sloth bears and spectacled bears climb trees regularly in search of food, such as fruits, seeds and nuts, as well as birds' eggs. Black bears are also agile tree-climbers. Polar bears rarely encounter trees. Those that move into the forest during the summer months, however, rest in hollows dug among tree roots to avoid the heat. Brown bear cubs climb trees to escape danger, but adult brown bears are too heavy to be good climbers. A female sloth bear will carry her small cubs into a tree on her back, unless she is escaping from a leopard since they can also climb trees! Most bears also use trees to mark their territory. They scratch the bark and rub on scents to tell other bears they are there.

▸ BEAR DANGER
A mother American black bear sends her cubs up into a tree while she stands guard at its base. If the danger is from an adult brown bear, the female will flee and return later for her cubs when the grizzly has gone.

◂ TREE-TOP HOME
The sun bear seeks out termite and bees' nests, and will rip away bark to get at insects hidden underneath. Although the sun bear feeds mainly on insects it also eats ripe fruit and preys on small rodents, birds and lizards.

▴ BELOW THE BARK
An adult, cinnamon-coloured American black bear is able to climb into a tree with ease using its short but sturdy claws. It can also lift bark to lick out insects with a long tongue.

▲ UP A TREE

Giant pandas stay mainly on the ground, but they will climb trees occasionally. They do so to sun themselves or to rest. Female pandas sometimes head up a tree to escape males, while males climb trees to advertise their presence.

▼ SAFE HAVEN

Black bears are normally found in forested areas. They have favourite trees located along trails where bears and other animals regularly pass. The bear marks the tree base with its scent and climbs into the branches where it is safe from brown bears.

Short, sturdy claws on a black bear's feet make tree climbing easy

▲ HIGH SCHOOL

Black bear cubs stay close to their mother both on the ground and in a tree. They watch and learn from her how to climb and find food among the branches.

▼ TREE HOUSE

Spectacled bears pull branches together to make a feeding platform. From here the bears feed mainly on tough plants called bromeliads. They also eat fruits, nuts, and honey and may take mice, rabbits and insects.

Gone Fishing

Brown and black bears sometimes overcome their reluctance to be with other bears when there is plenty of food available. This happens regularly on the rivers of the northwest coast of North America. Here, thousands of salmon come in from the sea and head upriver to spawn. The bears fish alongside each other at sites such as rapids where the water is shallower and the salmon are swimming more slowly. An uneasy truce exists between the bears, although isolated fights do occur. The salmon runs take place at different times of the year, but the most important are those in the months leading up to winter. The bears catch the oil-rich salmon to obtain the extra fat they need for the long hibernation ahead.

▲ **STRIPPED TO THE BONE**
Having caught a fish, the bear holds it firmly in its forepaws. Then it strips away the skin and flesh from the bones.

◄ **SALMON LEAP**
Salmon sometimes jump right into a bear's mouth. The bear stands at the edge of a small waterfall. Here the salmon must leap clear of the water to continue their journey upriver. All the bear needs to do is open its mouth!

◀ **COME INTO CONFLICT**
Sometimes the uncertain truce between bears breaks down and they fight for the best fishing sites in the river. Young bears playfight, but older ones fight for real. An open mouth, showing the long canine teeth, is a warning to an opponent. If the intruder fails to back down it is attacked. Fights are often soon over, because the bears are quick to return to the abundant source of fish.

▲ **FISHING LESSON**
Bear cubs watch closely as their mother catches a salmon. The cubs learn by example and will eventually try it themselves. It will be a long time before they are as skilful as their mother.

▲ **A SLOUTHE OF BEERYS**
A group of bears is called a sloth, so brown bears on a salmon river are a sloth of grizzlies. The term "a slouthe of beerys" was used in the Middle Ages. It came from people's belief at the time that bears were slow and lazy.

Focus on

1 A polar bear uses stealth and surprise to catch its prey. The bear approaches its target slowly, moving silently across the ice and swimming between ice floes. Its broad paws act like paddles and propel it effectively through the water. Its hind legs trail in the water and steer like a ship's rudder. A polar bear can also swim underwater to reach its prey, erupting from the sea to surprise seals resting on the edge of the ice.

Because it lives in one of the harshest and most unpredictable places on Earth, the polar bear is constantly on the look out for food. There may be four or five days between meals. Their most common prey is the ringed seal, although harp, bearded and hooded seals, or young walruses, are also taken. In spring, a bear searches for seal nursery dens. It breaks through the roof of snow and ice to reach the seal pups inside by jumping down with its powerful forelegs. During the rest of the year, a bear searches for holes in the ice where seals come up to breathe. While its eyesight and hearing are similar to ours, a polar bear's sense of smell is far superior. It can smell a den or breathing hole from a kilometre away.

2 A polar bear will stand or lie motionless for hours beside a good place for hunting. This may be at a seal's breathing hole or den. In spring, cracks form in the ice where it begins to melt. At points along the ice, seals emerge to rest or bask in the sun. A bear sniffs out suitable hauling-out places and lies in wait. Often, the bear's patience is not rewarded, for only one in fifty attempts to catch prey are successful.

Seal Hunting

3 A harp seal pops up to breathe through a hole in the ice. Before surfacing it would have looked for signs of danger. A bear usually lies on its stomach with its chin close to the edge of the ice. This conceals it from sight until the seal surfaces.

4 When a seal surfaces the bear scoops it out with a paw or grasps it around the head with its teeth and flips it out on to the ice. A powerful bite to the head crushes the seal's skull or breaks its neck and kills it instantly. The bear eats the fat and internal organs, but not the meat unless it is very hungry. It eats quickly since the smell of the kill might attract other bears.

5 Arctic foxes often follow polar bears across the ice in order to take advantage of their leftovers. Polar bears feed mainly on the seal blubber (fat), leaving behind most of the meat and bones. An average bear needs to eat 2kg of seal fat a day to survive. They have huge stomachs, enabling them to eat much more. After the meal, a polar bear cleans its fur by swirling in the water or rolling in the snow.

Winter Shutdown

Black bears, brown bears and pregnant female polar bears hibernate. They do so because food is scarce, not because of the cold. Scientists argued for years whether bears truly hibernate or merely doze during the winter months. Now, according to recent research, the hibernation of bears is thought to be even more complete than that of small mammals. During hibernation, a brown bear's heart rate drops to about 35 beats per minute, half its normal rate. American black bears reduce their blood temperature by at least one degree. They do not eat or drink for up to four months. Bears survive only on the fat that they have stored during the summer. A bear might have lost up to 50 per cent of its body-weight by the end of the winter.

▲ **FAT BEARS**
Before the winter hibernation a bear can become quite obese. Fat reserves make up over half this black bear's body-weight. It needs this bulk to make sure that it has sufficient fat on its body to survive the winter fast. In the weeks leading up to hibernation, a bear must consume large quantities of energy-rich foods, such as salmon.

Did you know? Some hibernating bears sleep for 5½ months non-stop.

◄ **HOME COMFORTS**
A brown bear pulls in grass and leaves to cushion its winter den. American black bears and brown bears sleep in small, specially dug dens. These are usually found on the sunny, south-facing slopes of mountains.

▲ SCANDINAVIAN REFUGE
A hole dug by a brown bear serves as its winter den in this Swedish forest. Bears spend winter in much stranger places, such as under cabins occupied by people, under bridges or beside busy roads.

▲ SNOW HOUSE
Female polar bears leave the drifting floes in winter and head inland to excavate a nursery den. They dig deep down into the snow and ice, tunnelling about 5m into the ground. Here they will give birth to their cubs. In severe weather, a male polar bear rests by lying down and allowing itself to be covered by an insulating layer of snow.

▲ READY FOR ACTION
If disturbed, a bear wakes easily from its winter sleep. Although it is dormant, a bear's body is ready for action. It is able to defend itself immediately against predators, such as a hungry wolf pack.

▲ WINTER NURSERY
In the early winter, one bear enters a den, but three might emerge in spring. Female polar bears, like most bears, give birth (usually to twins) while hidden away in their dens. The tiny cubs are born in the middle of winter, in December or January.

A Solitary Life

Bears do not like other bears. They prefer to be alone. When two bears meet there is sometimes a fight, but usually it is just a shouting match and display. Young bears have playfights, not serious contests but rehearsals for battles later in life. Bears will break off hostilities when food is plentiful. Brown bears tolerate each other at fishing rivers and polar bears scavenge together at whale carcasses and rubbish dumps. But bears have to be constantly wary of other bears. Cannibalism (eating members of the same species) is more common in bears than in any other mammals. Male bears will fight and kill cubs. Deadly fights between adult male bears may end in one killing the other and then feeding on the loser's body.

▲ **CANNIBALISM**
Adult American male black bears and male polar bears are cannibals. They eat mainly younger bears. However, this sort of behaviour is thought to be relatively rare.

◄ **ICE DANCE**
Like two ballet dancers, young male polar bears play at fighting. They use exaggerated lunges and swipes with their paws and jaws. They do not hurt each other, but they must learn to fight well. Later in life as fully grown adults they will compete with other males for females during the breeding season. Fights between well-matched individuals can be violent and often bloody.

THE JUNGLE BOOK

Rudyard Kipling's famous story The Jungle Book *was first published in 1894. A young boy named Mowgli is brought up by wolves. He is befriended by Baloo the bear and Bagheera the panther who teach him the law of the jungle. The tiger Shere Khan plots to kill the man-cub.*

▲ FRIENDS AT THE FEAST

Brown bears gather to catch salmon at a popular spot in Alaska. If they get too close to each other, the bears will contest their fishing rights. Usually the larger bears succeed in fishing the best sites.

▲ TRAGEDY ON THE ICE

A mother polar bear stands over her cub, which has been mauled by a male. Male polar bears usually kill cubs for food. The female attacks and tries to drive the male away. Males are much bigger than females, but a female with cubs is a fierce opponent.

▲ FISHING BREAK

Young brown bears take a break from learning to fish and practise fighting instead. They fight by pushing and shoving at each other, using their enormous bulk to overcome their opponent. They also try to bite each other around the head and neck.

Meeting a Mate

Brown and black bears mate in the summer between May and July. Males use their well-developed sense of smell to track down females in heat. Brown bears often group together near rich food sources in the mating season, so males have to fight each other for the right to mate. Courtship between male and female bears is very brief, but when the male finds the female is ready to mate, he tries to isolate her from other males. The female is pregnant for from six to nine months. The length varies, because, no matter the time of mating, all the cubs are born at roughly the same time in the new year. This is because female bears delay the development of the fertilized egg in their wombs until late summer.

▲ **COURTING COUPLE**
A male and female brown bear may stay together for over a week during courtship and mate several times. The act of mating stimulates the female to release an egg. The male keeps other male bears away to be sure he is the father of any offspring.

◀ **LOOK!**
Male giant pandas often climb trees to advertise to females they are willing to mate. They wail, yap and bark to attract attention. Their loud calls also attract other males. The most dominant male mates with the female first. Pandas mate in spring.

▲ **COURTSHIP DEADLOCK**
A male has to overcome the female's natural tendency to be wary of him. The two assess each other with some gentle sparring interrupted by brief stand-offs and a lot of sniffing.

Did you know? A female polar bear may rear 8 cubs in her lifetime.

HEAVENLY SCENT

Bears rely on their keen sense of smell to find a partner. An Alaskan brown bear approaches a receptive female, checking her odour track on the ground. If she lets him approach he sniffs her head, body and rear for signs that she is ready to mate. This period of courtship can last for up to 15 days, before the female is finally ready.

ARCTIC ENGAGEMENT

Polar bears mate from late March to late May. They tend to congregate where there are plenty of seals. There are usually more males willing to mate than females. Like brown bears, a successful male tries to keep his temporary partner away from other male bears. They mate many times over a period of a week or more.

MATING TIME

If a female brown bear is receptive, the male places his paw on her back. He mounts and grasps her in a bear hug, and bites the back of her head and neck. Mating is brief, lasting for only a few seconds up to about three minutes. The pair may mate up to 16 times in one day, and this may be repeated over several days. This ensures the female becomes pregnant by that particular male.

Nursery Dens

Most bears give birth hidden away from the outside world in dens. American black and brown bear cubs are born in winter, in January and February, when their mother is closeted in her winter den. A den can be a cave, in a hollow tree, under a tree that has been pushed over by the wind, or in a self-made hollow. Usually, two to three cubs are born, naked and helpless. Newborn bear cubs are small for the size of their mother. This is because a female bear's gestation period is very short. The mother has also to rely only on her fat reserves to build up their tiny bodies. A mother bear is only able to feed her cubs if she has eaten enough food in the months before her winter hibernation.

▲ **BLIND AND HELPLESS**
Ten-day-old brown bear cubs nestle into their mother's fur for warmth. With their eyes and ears tightly closed shut, they are totally dependent on her. The cubs remain in the den until May or June when they are about four months old.

Did you know? Polar bear cubs are no bigger than guinea pigs when born.

◀ **TWINS**
A polar bear mother tends her two young. The family leaves its den between late February and April depending on where they live. The further north they are, the later in the year they emerge.

▲ **TRIPLETS**
This American black bear has given birth to three healthy cubs. Females may have up to four cubs at one time. About the size of a rat and naked at first, the cubs grow quickly. They will leave the den in April or May.

▲ **PANDA BABY**
At Wolong breeding station in Sichuan, China, a baby giant panda is put in a box to be weighed. Giant pandas give birth to one or two cubs in a cave or tree hollow. If twins are born the mother often only rears one, leaving the other to die.

◄ **MOTHER'S MILK**
Three-month-old polar bear cubs suckle on their mother's milk. The milk is rich in fats and the cubs suckle for up to a year. Polar bear cubs are born covered with fine hair.

► **IN THE DEN**
American black bears weigh less than 300g when they are born in late January or early February. Their small size and lack of fur makes them vulnerable to the cold. The mother cleans and dries the cubs, then cuddles them close. The den is lined with branches, leaves, herbs and grasses to make a warm blanket. The mother spends a lot of time grooming her cubs and keeps the den scrupulously clean by eating their droppings.

Focus on

From late October, a pregnant female polar bear digs a snow den. Usually it is on a slope facing south, some distance from the sea. This is where she will give birth to her cubs. The mother warms her den with heat from her body. A tunnel to the nursery chamber slopes upwards so that warm air rises and collects in the chamber, which can be 20°C warmer than outside. She gives birth during the harshest part of winter from late November to early January, when permanent night covers the Arctic. The cubs grow fast. Around March the mother drives two holes through the walls of the den and helps the cubs to emerge for the first time.

1 A bank of snow makes an ideal site for a polar bear's winter den. The pregnant female bear digs into the south side of the snowdrift. The prevailing northerly winds pile up snow on the other side.

2 This etching shows an artist's impression of the inside of a polar bear's snow hole. The female bear gave birth to her cubs about three months ago. The cubs are now strong enough to follow their mother towards the sea where she can hunt and feed.

3 The female polar bear emerges from her winter home for the first time in the middle of March. The den's southward-facing entrance and exit hole faces towards the Sun, which is low on the horizon in early spring. The mother and her cubs are warmed by the Sun's rays.

Snow Homes

4 Sitting upright in a snow hollow, a female polar bear nurses her cubs. She differs from other female bears in having four working nipples rather than six. Her cubs also stay with her longer than most bears. She protects them from male bears until they are three years old. During this time she will teach her cubs how to survive in the cold Arctic conditions and also how to hunt seals.

5 On first emerging, the family remains at the den site for a few days so that the cubs become used to the cold. They play outside in the snow during the short days, and shelter in the den at night and during storms.

6 The cubs' first journey outside can be a long one. They may have to walk up to 22km to reach the sea ice where they will see their first seal hunt. The mother takes great care to avoid adult male bears who might try to kill her cubs.

Raising Cubs

Bear cubs spend the first 18 months to three years with their mother. If something should happen to her, they may be adopted by another mother with cubs the same age. The cubs learn everything from their mother. They learn to recognize the best foods and where and when to find them. They must also learn how to escape danger and how to find a winter den or shelter in a storm. Without this schooling, the young bears would not survive. Mothers and cubs can communicate by calling, particularly if they become separated or if a mother wants her offspring to follow her. During their development, cubs must keep out of the way of large male bears who might attack them.

▲ **SAFE IN THE BRANCHES**
Black bear cubs instinctively know that they should head for the nearest tree when danger threatens. It is easier for a mother to defend a single tree than a scattered family.

▼ **MILK BAR**
A mother brown bear suckles her twins. Her milk is thick and rich in fats and proteins, but low in sugars. It has three times the energy content of human or cow milk. The cubs are small at birth and must put on weight and grow quickly.

LEAVING THE FAMILY

Young bears on their own, such as this brown bear, often become thin and scrawny. Despite being taught by their mothers where and how to feed, they cannot always find food. At popular feeding sites, such as fishing points, they are chased off by larger bears. When the time comes for young bears to look after themselves, the mother either chases them away or is simply not there when they return to look for her.

LONG APPRENTICESHIP

Polar bear cubs are cared for by their mother for much longer than other bear cubs. They need to master the many different hunting strategies used by their mother to catch seals. These are not something that the cubs know instinctively, but rather skills that they must learn.

Did you know? Giant panda cubs are the first to leave their mothers at 18 months old.

FAMILY TRAGEDY

This polar bear cub is the victim in a tragic tug-of-war. A male bear has attacked the cub and its mother is trying to save it. Female polar bears fight ferociously to protect their young, but are often unsuccessful against larger opposition. Attacks like this, starvation, the cold and diseases mean that about 70 per cent of polar bear cubs do not live to their first birthday.

Where in the World are Bears?

Bears are found in a variety of habitats including the Arctic tundra, mountain slopes, scrub desert, temperate and tropical forests and tropical grasslands. Each species of bear, however, has its own preferred environment. The polar bear, for example, inhabits the lands and sea ice bordering the Arctic Ocean. It favours the shoreline areas where the ice breaks up and cracks appear, as this is where seals congregate. Most other bears are less specialized and have the uncanny ability to turn up wherever food is abundant. However, many of the wilderness areas where bears live are under threat. Habitat loss, as more land is cultivated for farmland and forests all over the world are cut down, is a major threat to many bears.

▼ **MOUNTAIN BEAR**
The spectacled bear lives in South America. It lives high in the Andes Mountains of Peru, Ecuador, Venezuela and Colombia. It is found in forests, including rainforests, cloud forests, and dry forests, as well as steppe (grassland) and rocky outcrops.

▲ **IN THE BAMBOO FORESTS**
The giant panda is restricted to areas of abundant bamboo forest. It was once much more widespread across eastern Asia, but now survives in just three provinces of western China – Gansu, Shanxi and Sichuan.

▲ **BROWN BEAR**
The brown bear is the most widespread of all bears. It is found in Europe, the Middle East, across Russia and northern Asia to Japan. North American brown bears live in Alaska and the Canadian Rockies.

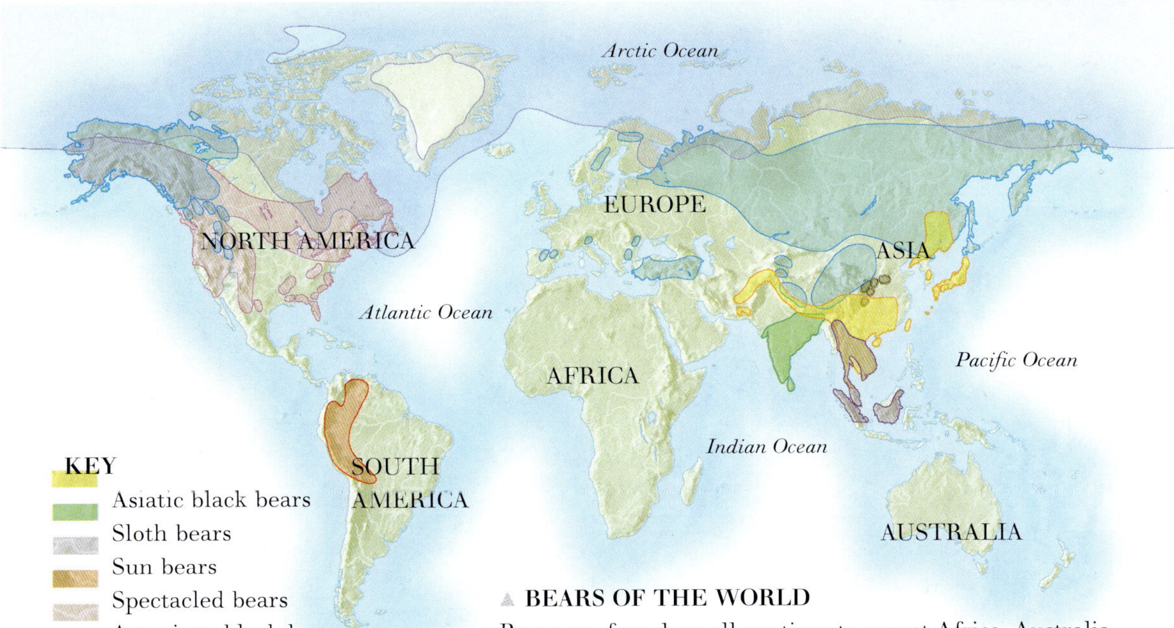

BEARS OF THE WORLD

Bears are found on all continents except Africa, Australia and Antarctica. There were once brown bears living in the mountains of North Africa, but they became extinct (died out) in the 1800s.

ASIATIC BLACK BEAR

The Asiatic black bear lives in mountainous regions over a wide area of southern and eastern Asia. It is found in northern India, Pakistan and on the islands of Japan and Taiwan.

LIFE IN THE FOREST

The sloth bear (*above*) lives in dense, dry forests in India and Sri Lanka. The Malayan sun bear lives in similar lowland but tropical forests of South-east Asia. It is also thought to live in Yunnan in southern China, although no recent sightings have been reported.

Finding the Way

Bears have an uncanny knack of finding their way home even in unfamiliar territory. How they do this is only just beginning to be understood. For long distances, they rely on an ability to detect the Earth's magnetic field. This provides them with a magnetic map of their world and a compass to find their way around. When closer to home, they recognize familiar landmarks. In fact, bears have extraordinary memories, especially where food is involved. For example, a mother and her cubs are known to have trekked 32km to a favourite oak tree to feast on acorns. Five years later, the same cubs (now adults) were reported to have been seen at the same tree.

STARS IN THE SKY

The Great Bear constellation in the northern hemisphere is known to astronomers as Ursa Major. In Greek mythology, it was said to have been made in the shape of a she-bear and placed in the heavens by Zeus. The Great Bear is also worshipped in Hindu mythology as the power that keeps the heavens turning. The Inuit believe these stars represent a bear being continually chased by dogs.

▲ **ARCTIC NOMADS**

Polar bears are capable of swimming long distances between ice floes at speeds of up to 10km/h. They may travel thousands of kilometres across the frozen Arctic Ocean and the surrounding lands in search of prey.

▲ **BAD BEAR**

A sedated polar bear is transported a safe distance out of town. Nuisance bears are often moved this way but they unerringly find their way back.

◄ TO AND FROM THE FOREST
The polar bears of Hudson Bay, Canada, migrate to the forests in summer and return to hunt on the sea ice in winter. On their return journey, they sometimes stop off at the town of Churchill. They gather at the rubbish dump to feed on leftovers, while they wait for ice to reform.

KEY

- Bears return to ice in winter
- Bears come ashore in June and July
- Bears walk north in autumn

▲ REGULAR ROUTES
Polar bears move quickly even on fast, shifting ice floes. A bear moving north, for example, against the southward-drifting ice in the Greenland Sea, can travel up to 80km in a day.

▲ HAVING KNOWLEDGE
Male brown bears live in large home ranges covering several hundred square kilometres. They must remember the locations of food and the different times of year it is available.

Focus on

Bears in some parts of the world are unique in terms of size or colour. Brown bears (grizzlies) on Kodiak, Shukak and Afongnak islands, Alaska, grow to a gigantic size. Across the Pacific, the brown bears of the Kamchatka Peninsula, Russia, are also giants. It is thought they reach their enormous size by including salmon as an important part of their diet.

Elsewhere, brown bears take the grizzly colour to an extreme and have fur resembling streaked hairstyles. American black bears show a very wide variation in colour. Many are not even black, ranging from white to red-brown. These variations may camouflage bears in different habitats. Black bears are invisible in dense forests, but lighter-coloured fur is an advantage in more open places.

BEAR GIANT
Standing on its hind legs, a Kodiak brown bear would tower over a person. It can weigh up to 750kg – almost as big as a North American bison. The average weight for a male Kodiak bear, however, is about 300kg. These bears are so powerful, they can kill and carry an adult moose.

CINNAMON BEAR
Cinnamon-coloured American black bears have a coat that is reddish brown to blond. Bears in the west tend to be cinnamon or honey-coloured, whereas bears in the east are mainly black.

Super Bears

BLUE BEAR
Blue or glacier black bears are found in north-west Canada. They have a blue-grey tinge to their fur. Like all American black bears, no matter what their body colour is, they have a brown muzzle.

RUSSIAN BEAR
The giant brown bears of the Kamchatka Peninsula, eastern Russia, have a varied diet. They eat the seeds of conifer trees, fish for salmon, hunt for seals and birds, and scavenge on stranded whales.

GHOST BEAR
One in ten black bears on Kermodes Island off North America's Pacific coast is snowy white. These bears are not albinos (animals that lack skin colour) or polar bears, but true black bears.

Ancient Bears

There were bears living in the past that were far bigger than today's huge Kodiak and polar bears. The giant short-faced bear, weighing over 1,000kg, was twice the size of a Kodiak. It is the largest known carnivorous (meat-eating) mammal to have ever lived on land. Bears, however, had much smaller beginnings. The first recognizable bear appeared about 20 million years ago. Called the dawn bear, it was about the size of a fox. Early bears split into three groups. The first group, the Ailuropodinae, followed a plant-based diet and of these only the giant panda still survives. The second group, Tremarctinae, included many species of short-faced bears whose only living relative today is the spectacled bear. The third group, Ursinae, includes most of the bears we see today as well as the now extinct cave bears.

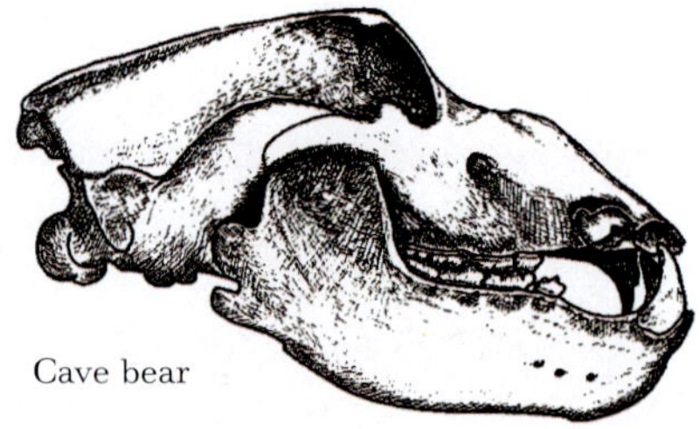

Cave bear

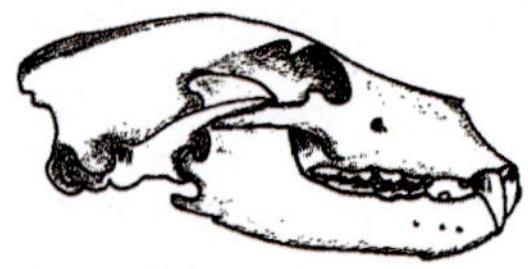

Brown bear

▲ **BIG HEAD**
The ancient cave bear (*Ursus spelaeus*) had a bigger skull than modern brown bears. Its high-domed skull anchored powerful chewing muscles. Enormous back teeth show that it mainly ate plants. It died out about 10,000 years ago.

◄ **DAWN BEAR**
Evolving from an animal that looked part dog, part bear and part raccoon, the dawn bear (*Ursavus elmensis*) was the ancestor of all known bears living today. Twenty million years ago it probably spent most of its time hunting in the tree-tops. By studying its teeth, scientists think the dawn bear supplemented its diet of meat with plant material and insects.

GIANT SHORT-FACED BEAR

The giant short-faced bear (*Arctodus simus*) had much longer legs than today's bears and probably ran down its prey. In profile it resembled a big cat. In its jaws were canine teeth capable of delivering a killing bite and back teeth for shearing meat. It probably hunted ancient camels, bison and horses that once lived on the plains of North America. The only living descendant of the short-faced bear is the spectacled bear (*Tremarctos ornatus*), the sole species of bear to be found in South America.

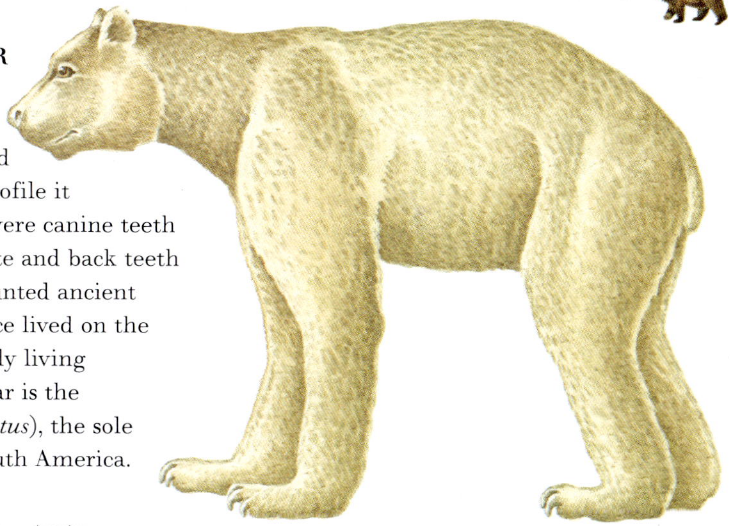

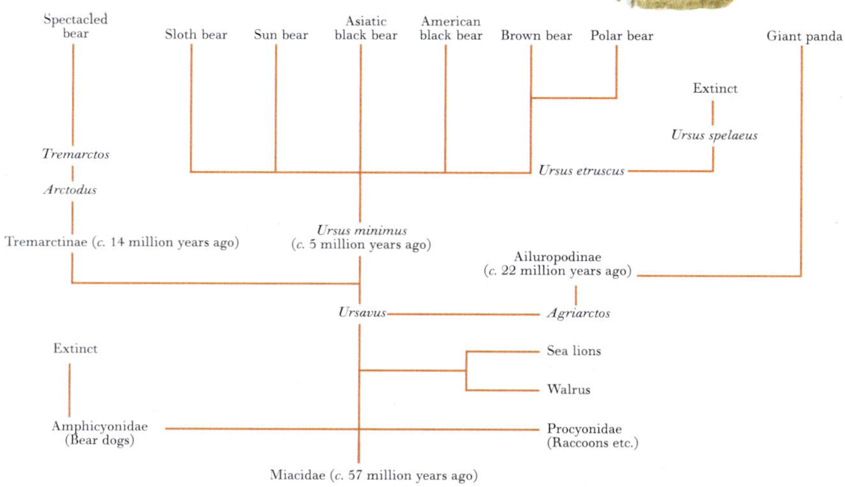

FAMILY OF BEARS

This chart shows how ancient bears gave rise to modern ones. The more ancient bears are at the bottom and modern bears are at the top.

GREAT EUROPEAN CAVE BEAR

About 500,000 years ago, the cave bear (*Ursus spelaeus*) evolved. It was about the size of today's Kodiak bears, with massive front legs, broad paws with claws, and a huge head and large muzzle. Large quantities of bones found in caves throughout Europe indicate that many died during winter hibernation, probably because they failed to put on enough fat to survive until spring.

Family Groups

The eight species of living bears belong to the family Ursidae. They all have the same general appearance, but in different parts of the world, each is adapted to a particular lifestyle. Bears in tropical places tend to be small and spend more time in the trees. Those in northern lands are larger and live mostly on the ground. To help them study bears, scientists divide the eight living species into three smaller groups, or subfamilies. The giant panda and spectacled bear are the sole survivors of two ancient subfamilies, the Ailuropodinae and Tremarctinae. The rest of the bears are grouped together in a third subfamily, the Ursinae. A bear's Latin name reveals how that particular species is grouped.

▲ **SLOTH BEAR**
Very little is known about the origins of the Indian sloth bear (*Melursus ursinus*). Few fossils (remains preserved in stone) have been found for this species of tropical bear. Sloth bears are thought to have evolved during an ice age that started about 1.6 million years ago.

◄ **BROWN BEAR**
Brown bears (*Ursus arctos*) first appeared in China about 500,000 years ago. From here they migrated right across the northern hemisphere into North America and Europe.

▲ **POLAR BEAR**
The polar bear (*Ursus maritimus*) is the most recent bear to have evolved. Its closest relative is the brown bear. In zoos, polar bears and brown bears are sometimes interbred.

▲ GIANT PANDA

The giant panda (*Ailuropoda melanoleuca*) is the sole survivor of the earliest group of bears, the Ailuropodinae. It first appeared around 10 million years ago. Fossils of giant pandas have been found throughout China.

▶ ASIATIC BLACK BEAR

The Asiatic black bear evolved from the same ancestor as most other species of bear, *Ursus minimus*. It is found mainly in the hilly areas of southern Russia, Japan, and southern Asia. This bear is in danger of extinction in some areas, because it is still hunted for food and medicine.

▼ SUN BEAR

The sun bear (*Helarctos malayanus*) is another tropical bear about which very little is known. It too probably evolved during the last Ice Age. Today sun bears are under threat from pressures such as farming. The bears tear out the hearts of oil palms, which ruins the crop.

▼ SPECTACLED BEAR

The spectacled bear (*Tremarctos ornatus*) evolved about 2 million years ago. It is the last remaining member of a group of primitive short-faced bears (Tremarctinae) that were found mainly in the Americas.

Relatives and Impostors

Bears are included in the order Carnivora, a group of mammals that includes cats, dogs and hyenas. The bears' nearest relatives are the raccoon, dog and weasel families. Bears are also probably the closest living relatives of seals and sea lions, members of the order Pinnipedia. The skull of the polar bear, for example, is very similar to that of a seal. Most relatives of bears are meat-eaters, unlike many bears themselves. Some, such as weasels, are loners, much like bears, while others, such as wolves, live in groups and hunt together. A few animals have a strong similarity to bears, but are in fact not related at all. The wombat and koala, two types of marsupial (mammals that raise their young in a pouch), are often confused with bears but are totally unrelated.

▲ **KOALA**
The Australian koala is very similar in appearance to a small bear. It is commonly called the koala bear, but it is not a bear at all. It is a more primitive mammal and is a member of the order Marsupiala. Members of this order, marsupials, have pouches in which their young develop.

◀ **RACCOON**
Raccoons have been given the name washing bear. This may be because of the way they use their front paws to locate prey. They look like they are washing their prey before eating it. Raccoons are not bears, but they are related to them.

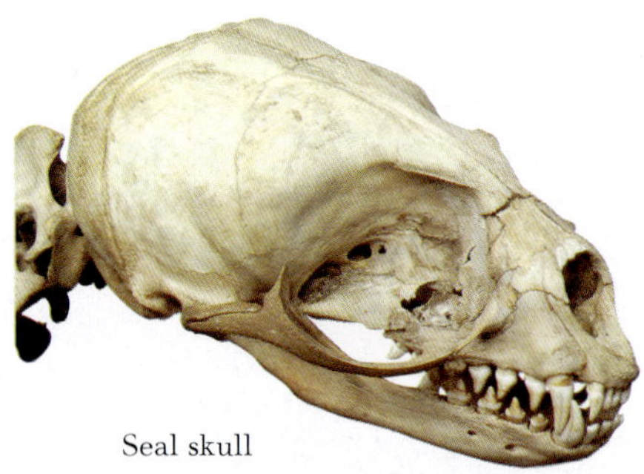

Seal skull

▲ **ARCTIC SIMILARITIES**
Seals and bears are distant relatives. Polar bear and seal skulls have long muzzles, parallel rows of teeth and flattened bones protecting the ear.

▲ THE LONG-TAILED KINKAJOU

The kinkajou is known as the honey bear, but it is not a real bear. It is a South American member of the raccoon family. It spends most of its life in the trees feeding mainly on fruit. It is also very fond of honey. A prehensile (grasping) tail helps it grip on to branches.

▲ DISTANT RELATIVES

Hyenas, unlike bears, live in groups in which the females are the leaders. They have a reputation as scavengers, but are also effective predators. By hunting together they can bring down large prey. Hyenas share a common early ancestor with bears, but are actually more closely related to cats.

▲ NO RESEMBLANCE

The weasel is a solitary carnivore, like the bear. Stoats and weasels share the same ancestor as bears, raccoons and dogs. However, over millions of years, all of these animals have evolved very different characteristics and body shapes.

▶ HUNGRY LIKE A WOLF

The grey wolf is a member of the dog family. It often lives in the same places as brown and black bears. It hunts in packs and sometimes targets the same prey as bears, such as moose and caribou.

Pandas on the Edge

Pandas are the rarest of all bears. Each year their habitat shrinks. There are just 25 pockets of panda forest remaining, each with no more than 50 wild pandas living there. This makes pandas extremely vulnerable. One threat is the flowering of bamboo. When bamboo flowers it dies back. Each species of bamboo flowers at a different time. In the past, pandas would leave their home range and travel to where other species of bamboo were not flowering. Today, pandas are trapped by the surrounding farmland and cannot move from one bamboo stand to another. The Chinese government is desperately trying to conserve the panda. Reserves have been set up to study panda behaviour and establish breeding programmes. The panda is regarded as a national treasure. Anyone killing a panda faces the death penalty.

▲ **FEEDING TIME**
A baby giant panda is bottle-fed at a panda breeding centre in China. By giving the babies a helping hand early in life, it is hoped more pandas may grow to adulthood. They may then be returned to their native bamboo forests.

Did you know? Baby pandas are blind, pink and weigh just 100g at birth.

◄ **CAPTIVE BREEDING**
These are two of the 100 or so giant pandas in zoos and breeding centres worldwide. Unfortunately pandas are slow to have young. Artificial insemination is used in attempts to breed more pandas in captivity.

▲ **SYMBOL FOR CONSERVATION**
The World Wide Fund for Nature adopted the giant panda as its logo. Its rarity and universally popular appeal has made the panda a natural ambassador for all living things facing extinction.

▲ **ON DISPLAY**
This giant panda is far from its natural home. The chances are that it has not been bred in a zoo but taken from the wild. It is, however, safe from poachers who supply a lucrative market in panda skins.

◄ **PANDA TRANSPORT**
Chinese biologists in the Wolong Reserve, Sichuan, try to persuade a giant panda to enter a cage. They will move the panda to the breeding centre for further studies. It is only since the 1980s that scientists have started to understand the panda's biology.

► **BACK TO THE WILD**
Many pandas have been born in captivity, but less than a third of cubs have survived. It was hoped that captive-bred pandas could be returned to the wild, but the results are disappointing. This is because there is not enough vacant habitat to return captive pandas to. There are now 13 panda reserves and these cover about half the remaining habitat.

Focus on

Polar bears come into close contact with people each autumn at the isolated Canadian town of Churchill on the shores of Hudson Bay. The bears are on their way from the forests inland where they spend the summer, to the ice of the bay where they hunt. Often they arrive before the ice forms and cause a lot of trouble while they hang around with nothing to do. Some bears head for town and scare the local townsfolk. Others head for the town dump. The bears are chased away, but frequent offenders are tranquillized and transported somewhere safe. The bears, however, have become a tourist attraction. People come from all over the world to see the congregations of polar bears, which are usually much more solitary in the wild.

LOOKING FOR TROUBLE
This bear has picked up the tantalizing scent of tourists. Many visitors arrive to see the bears each year. They travel about in great buses called tundra buggies, where they are safe from the powerful and inquisitive bears.

OVERSTEPPED THE MARK
A researcher cautiously tests a tranquillized polar bear to make sure it is fully sedated. He holds a gun in case the polar bear is not as sleepy as it seems and attacks. Polar bears at Churchill occasionally threaten people. They are tranquillized and moved a safe distance away or locked up in a trap until the ice refreezes.

Churchill

BEAR BACK
A polar bear keeps cool by rolling in a patch of snow while it waits for the ice to form on Hudson Bay. The days can be warm in the Arctic autumn. Polar bears have thick fur and may overheat if they are not able to cool down.

FAST FOOD
A bear scavenges through the town rubbish dump. This is a favourite rendezvous spot. Household rubbish provides easy food for hungry bears unable to hunt.

FREE FLIGHT
A sedated bear is carried away in a net strung under a helicopter. This is a quick way to move a large animal, but it is also very expensive.

THE SIN BIN
A rogue bear is released from a bear trap. Unfortunately, bears have a well-developed homing instinct and often appear in town again. Persistent offenders are kept in a polar bear jail until the ice refreezes.

▲ **GREAT ESCAPE**
A hiker takes refuge in a tree after surprising a brown bear on a trail in Montana, USA. Adult brown bears do not climb trees. The hiker will have to wait for the bear to lose interest before he can come down.

Bears and People

Bears can be dangerous. They are attracted to food at campsites and rubbish dumps and here they come into contact with people. Fatal attacks by bears are rare. A person walking in bear country is seldom attacked, as long as the bear knows that they are there. Walkers are advised to make a lot of noise, clapping their hands and singing, for example, to warn any bears in the vicinity of their presence. Black bears tolerate people more readily than brown bears or grizzlies. Brown bears, because they have poor eyesight, might mistake people for a threat and charge. Whatever happens, a walker should never come between a mother bear and her cubs, as she will certainly fiercely defend them.

◄ **BEGGING FOR FOOD**
People driving in national parks in the USA offer titbits to black bears, despite warnings not to. The animals have learnt to associate cars with food. To get inside a car bears have been known to break windows or use their claws like a can-opener to slice through metal doors.

▲ **BEAR WARNING**
Signs warn visitors to behave sensibly. Even so, a bear might attack. It acts aggressively at first, chomping its jaws together and hitting the ground. It then charges but usually stops at the last moment.

▲ **TRAPPED IN THE ZOO**
A bear gnaws the bars of its tiny cage in a zoo in Tunisia. Bears are very popular animals, but they are often kept in terrible conditions in small zoos. This concrete-lined cage is totally bare and does not provide an interesting and stimulating home.

Did you know? Since 1900, there have been fewer than 150 attacks on people in the U.S.

▲ **LOOKING FOR FOOD**
The tempting smells coming from a dustbin are irresistible to a hungry bear. Unfortunately, bears begin to associate people with easy food. Harmless scavenging can become dangerous if the bear fails to get the food that it wants.

▲ **DANGEROUS SITUATION**
Bears and people cross paths occasionally. These people are too close to a brown bear for their own safety. If possible, bears should be given a wide berth. They will usually ignore or avoid people, and can be watched safely from a distance.

Bears in Danger

Of the eight species of bears living today, six are considered to be endangered. Only polar bears and American black bears are holding their own, and even they would not survive without considerable protection. Bears face many dangers. Their habitat is shrinking as natural areas are used to provide homes and farmland for people. Cubs are kept as pets but sold when they grow into troublesome adults. Many wild bears do not reach old age because they are shot by hunters. Hides and heads are used as wall hangings and trophies. In many parts of the world bear meat is eaten as a delicacy. Blood, bones and body parts are used in traditional Oriental medicines and as good luck charms. By far the biggest threat to bears is from poaching to supply the medicine trade.

▲ **DANCING BEAR**
A sloth bear is made to dance in India. Despite laws against it, bear cubs are taken from the wild. They are taught to dance using cruel methods and kept in poor conditions.

◀ **BEAR CIRCUS**
Bears have long been a popular act in circuses. Their ability to walk on their hind feet makes them appear almost human. These agile and clever animals are forced to perform tricks such as skipping, riding a bicycle and walking the tightrope. Performing bears are sometimes badly cared for and may be made to work all year round.

▲ ANCIENT PERSECUTION

Bears have been entertaining people for centuries. This carving from the AD300s depicts gladiators fighting bears in the arena. Both brown bears and polar bears were killed to entertain the audience.

◄ HEALING

We can learn a lot from bears. Native Americans discovered that many plants eaten by bears have medicinal properties. The Cheyenne treat diarrhoea with a plant called bear's food and the Crow use bear root to cure sore throats.

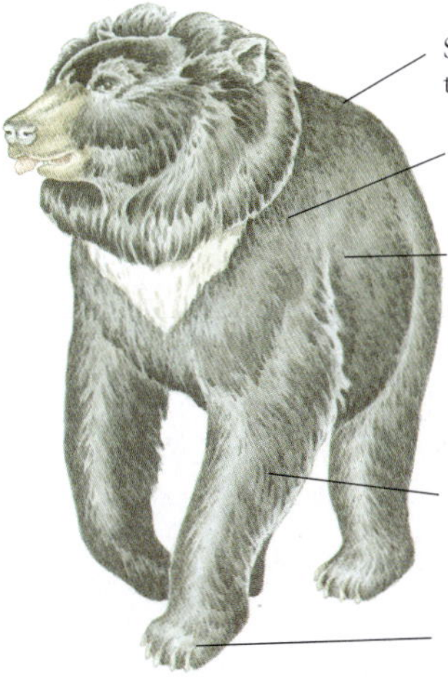

▲ MEDICINE CHEST

Bear organs are important in Oriental medicine. The most valuable part is the gall bladder, said to cure a whole host of ailments including fevers. Many bears are killed for their gall bladder alone.

► MEASURING UP

A scientist takes measurements from a tranquillized bear. A better understanding of the biology and behaviour of bears will hopefully secure them a safer future.

GLOSSARY

adaptation
A change in a living thing to suit a new set of conditions. This change helps the animal (or plant) to survive in its environment.

algae
A group of very simple plants, some of which are only the size of a single cell. Others are much larger. Most types of algae live in water, but they also grow on damp soil and tree trunks.

bacteria
A type of microorganism – a living thing that cannot be seen by the naked eye. Huge numbers of bacteria live in the soil and break down dead material. Millions of bacteria live in the gut and on the skin of many animals, including humans.

canine
A sharp, pointed tooth used for killing and holding prey.

carnassial
A strong, shearing tooth at the back of a carnivore's mouth.

carnivore
An animal that feeds partly or mainly on the flesh of other animals.

conservation
Protecting living things and helping them to survive in the future.

continent
One of the seven main land masses on Earth. The continents are North America, South America, Europe, Africa, Asia, Australia and Antarctica.

cultivation
The preparation and use of the ground in order to grow crops.

diet
The usual types of food eaten by an animal.

digestion
The process by which food is broken down so that it can be taken into the body.

dormant
When an animal lies motionless and inactive, as in sleep.

environment
The surroundings in which an animal or plant lives. It includes both living things (other animals and plants) and non-living things (such as stones, the air, temperature and sunlight).

estuary
The tidal mouth of a large river, where it flows into the sea.

evolution
The gradual development of new types of living things from existing species over very long periods of time.

extinction
When a whole species or larger group of animals or plants disappears, dying out completely.

fable
A story, often with a mystical theme, which is not based on fact.

fertilization
The joining together of a male sperm and a female egg to start a new life.

genetic
Relating to the genes, inside the cells. Genes control the characteristics passed on from parents to their offspring.

grizzly bear
Another name for the brown bear. It is particularly used to mean North American brown bears.

habitat
A place that has certain kinds of animals and plants living there, such as tropical rainforest or semi-desert.

herbivore
An animal that eats plants.

hibernation
A period of sleep during the winter when body processes slow down. Animals hibernate mainly because food is scarce and they might starve otherwise.

Ice Age
A time when large parts of the Earth were covered by glaciers and ice. The last Ice Age ended about 10,000 years ago.

ice floe
A sheet of ice floating in the sea.

incisor
Front tooth used for biting off chunks and cutting up meat.

insulation
A covering, such as a layer of thick fat beneath a polar bear's skin, that prevents heat leaving a warm body to the cold outside.

intestine
Part of an animal's gut where food is broken down and absorbed into the body.

Inuit
A people of the Arctic region, also known as Eskimos. The word Inuit, which means people, is preferred because Eskimo means eaters of raw meat and the Inuit cook at least part of their diet.

mammal
An animal with fur or hair and a backbone, which can control its own body temperature. Female mammals feed their young on milk made in mammary glands in their bodies.

mating
The pairing up of a male and female to produce young. During mating, the fertilization of an egg with a sperm takes place.

migration
A regular journey some animals make from one place to another, because of changes in the weather or their food supply, or to breed.

molar
A broad, ridged tooth in the back of a mammal's jaw, used for grinding up food.

nipple
A teat through which young suck milk from the mammary glands.

playfight
Early preparation for learning how to fight when cubs are older. Playing helps to build up muscles, improve co-ordination and develop good reflexes.

predator
An animal that catches and kills other animals (prey) for food.

prehensile tail
A tail that is capable of grasping and holding on to objects such as branches.

prey
An animal that is hunted for food by another animal (predator).

rainforest
Dense forest that is wet all year round.

scavenge
Feed on the remains of a kill left behind by another animal or on naturally dead animals or rubbish.

spawn
Produce eggs in large numbers to be fertilized.

species
A group of animals that share similar characteristics and can breed together to produce young.

streamline
The rounded, tapering shape that allows water (or air) to flow smoothly around an object.

subspecies
A species is sometimes divided into even smaller groups called subspecies, which are sufficiently distinct to have their own group.

suckle
Suck milk from the breast of a female mammal.

temperate
Mild regions of the Earth that do not experience extreme heat or cold, wet or dry.

territory
An area in which an animal or group of animals live. The borders of territories are marked, so that others of the same species know to keep out.

trail
A path through the forest or open ground that an animal such as a bear uses regularly.

tropical
Warm, wet regions of the Earth, near the Equator.

tundra
A treeless region of the Earth with permanently frozen soil just below the surface.

womb
An organ in the body of female mammals in which young grow and are nourished until birth.

Published by Anness Publishing Ltd,
Blaby Road, Wigston, Leicestershire LE18 4SE
Email: info@anness.com
Web: www.annesspublishing.com

Anness Publishing has a new picture agency outlet for images for publishing, promotions or advertising. Please visit our website www.practicalpictures.com for more information.

Publisher: **Joanna Lorenz**
Managing Editor, Children's Books: **Linda Fraser**
Senior Editor: **Nicole Pearson**
Editor: **Louisa Somerville**
Designer: **Vivienne Gordon**
Illustration: **Julian Baker, David Webb**

ETHICAL TRADING POLICY
Because of our ongoing ecological investment programme, you, as our customer, can have the pleasure and reassurance of knowing that a tree is being cultivated on your behalf to naturally replace the materials used to make the book you are holding. For further information about this scheme, go to www.annesspublishing.com/trees

PICTURE CREDITS
b=bottom, t=top, c=centre, l=left, r=right
Animals Animals: pages 11tr, 38tc, 45c and 55t; Bruce Coleman: pages 4b, 9b, 13tr, 13br, 15tl, 19tl, 20t, 21t, 22t, 27b, 39t, 42b, 43c, 45t, 46t, 48b, 54b, 57t, 60t, 61c and cover; Bridgeman Art Library: pages 31cr and 59b; Ecoscene: pages 8cb, 8bl, 9c and 41br; Mary Evans Picture Library: page 9tl; FLPA: pages 2, 3, 7tr, 7bl, 10t, 10bl, 12b, 14t, 19cl, 19cr, 23t, 24c, 30c, 32-33, 32t, 32b, 34b, 35t, 36-37, 36l, 42t, 50t, 50bl, 52-53, 52cl, 52b, 53bl, 54c, 55b, 56t, 58b, 61t, cover and title page; NASA: page 17br; Natural History Museum: pages 4-5c, 6c and 49cl; Natural Science Photos: pages 18c and 26b; Nature Photographers: pages 11c and 58t; NHPA: pages 12tr, 15b, 24tl, 24b, 29t, 29b, 34t, 35b, 37t, 37b, 38bl, 56c, 57c, 57b, 58c and 61b; Oxford Scientific Films: pages 2, 5bl, 6b, 8cr, 13bl, 16bl, 16br, 18tl, 20b, 21b, 22c, 22b, 25b, 27t, 28t, 28b, 29cl, 29cr, 30b, 31t, 31cl, 33b, 35c, 36b, 38cr, 39b, 40b, 41t, 41bl, 42c, 43t, 44c, 49cr, 49b, 55c, 56b, 59t, cover and title page; Planet Earth Pictures: pages 2, 4tr, 11bl, 14b, 16t, 18b, 19b, 23c, 25t, 26, 30t, 31b, 39c, 44t, 44b, 45b, 46c, 46b, 47t, 47c, 47b, 48t, 49t, 50c, 51cr, 51bl, 51br, 53cr, 54t, 59c, 60b and cover; Spacecharts: page 5tl and 17b; Visual Arts Library: pages 10cr, 15c and 53br; Zefa Pictures: pages 5tr and 51t.

C O N

4-5 Whale Order
6-7 Whales Large and Small
8-9 Whale Bones
10-11 Whale Bodies
12-13 Staying Alive
14-15 Whale Brain and Senses
16-17 Sounds and Songs
18-19 Feeding Habits
20-21 In for the Kill
22-23 Lunging for Lunch
24-25 Swimming
26-27 Ready, Steady, Dive!
28-29 Social Life
30-31 The Mating Game

T E N T S

32-33 Bringing up Baby

34-35 Having Fun

36-37 Whales on Display

38-39 Where Whales are Found

40-41 Migration

42-43 High and Dry

44-45 Grey and Right Whales

46-47 Rorquals

48-49 Sperm and White Whales

50-51 Beaked, Pilot and Killer Whales

52-53 Oceanic Dolphins

54-55 Porpoises and River Dolphins

56-57 Fellow Travellers

58-59 Whale Slaughter

60-61 Whale Conservation

62-63 Glossary

64 Index

Whale Order

Whales and dolphins are among the most delightful creatures on Earth. Like fish, they spend all their lives in the sea and are masters of their watery habitat. But unlike fish, they breathe air, have warm blood and suckle their young. They are more closely related to human beings than fish because they are mammals. Many whales are enormous – some are as big and as heavy as a railway carriage full of passengers. Dolphins are much smaller – most are about the same size as an adult human being. Porpoises, which look much like dolphins, are also roughly the same size as humans.

Although they are much smaller, dolphins and porpoises are in fact kinds of whales, too. All whales, large and small, belong to the major group, or order, of animals called Cetacea. The word comes from the Greek word cetus, meaning sea monster. From the word Cetacea comes the word cetacean, which means any kind of whale.

Heavyweights

The largest of the whales are the biggest animals that have ever lived on Earth. This leaping humpback whale is nearly 15m long and weighs over 25 tonnes – as much as five fully-grown elephants. Some other kinds of whales, such as the fin and blue whales, are very much bigger than this.

Whale Ancestors

More than 50 million years ago, creatures like this were swimming in the seas. They seem to have been ancestors of modern cetaceans. This creature, named basilosaurus (meaning king lizard), grew up to over 20m long. It had a snake-like body with tiny front flippers and traces of a pair of hind limbs.

Baleen Whales

These humpback whales are feeding in Alaskan waters. They belong to the group, or suborder, of whales known as the baleen whales. The baleen, or whalebone, whales are in general very much larger than those in the other main group, the toothed whales. Although the baleen whales include the largest animals on Earth, they feed on some of the very smallest.

Whale in the Sky

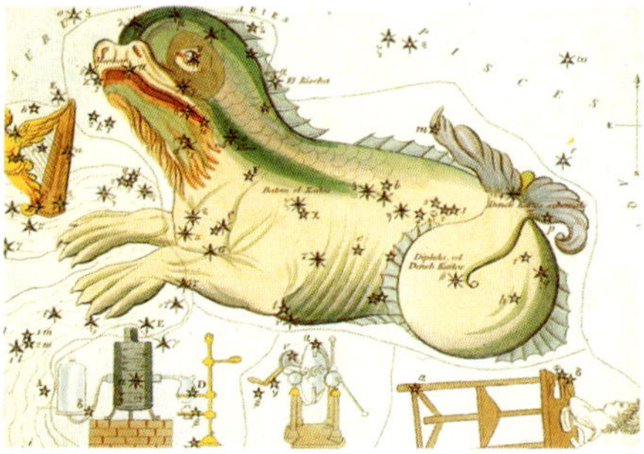

This star map shows a constellation of stars named Cetus, meaning the sea monster or whale. In Greek mythology, Cetus was a monster that was about to devour Andromeda, a beautiful maiden who had been chained to a rock as a sacrifice. Along came the hero Perseus, who killed the sea monster and saved Andromeda.

bottlenose dolphin

basilosaurus

Toothed Whales

A bottlenose dolphin opens its mouth and shows its teeth. It is one of the many species of toothed whales. Toothed whales have much simpler teeth than land mammals and many more of them. The bottlenose dolphin, for example, has up to 50 teeth in both its upper and lower jaws.

common dolphin

Air Breathers

Because they are mammals, all whales and dolphins breathe air. Here, a common dolphin is breathing out through a blowhole on top of its head as it rises to the surface. These dolphins usually breathe several times a minute, although they can hold their breath for five minutes or more when diving.

Did you know? A blue whale can weigh as much as 25 elephants.

Whales Large and Small

The word whale conjures up a picture of a huge creature – and some whales really are vast. Most of the biggest ones belong to the major group of cetaceans called the baleen whales. These whales do not have teeth in their mouth. Instead they have brush-like plates, called baleen, that hang from their upper jaw. They use the baleen to filter their food from the water. One large whale that does not belong to the baleen group is the sperm whale. It has teeth, not baleen, and belongs to the other major cetacean group, called the toothed whales. This group also includes the much smaller dolphins, porpoises, white whales and beaked whales. All these cetaceans have teeth for biting and grasping the prey they feed on.

grey whale

Grey Whale

The massive grey whale can grow up to nearly 15m long, and tip the scales at 35 tonnes or more. It is a similar size to the humpback, sei, bowhead and right whales, but is quite different in appearance. Instead of the smooth skin of the other whales, the grey has rough skin and no proper dorsal fin on its back.

Did you know? Some whales have as many as 3,000 baleen plates in their jaws.

beluga and bowhead whales

Bowhead and Beluga

At 16m the bowhead whale, which has a highly curved jaw, can grow as large as the sei whale. It is closely related to the right whale, which can be slightly bigger, at 18m. The bowhead is famous for its long baleen plates and extremely thick layer of blubber. The toothed whales we call belugas (above left) never grow longer than about 5m. The first part of the word beluga means white in Russian and belugas are also commonly known as white whales.

Sei Whale

At up to about 16m, the sei whale can grow slightly longer than the grey. It looks much like its bigger relatives, the blue whale and the fin. They are all members of the group called rorquals, which have deep grooves in their throat. These grooves allow the throat to expand greatly when the animals take great mouthfuls of water for feeding. A sei whale has up to 60 grooves in its throat, while a fin whale has up to 100.

sei whale

Did you know? The blue whale's tongue weighs as much as an African elephant.

Relative Sizes

Whales come in many sizes, from dolphins smaller than a human to the enormous blue whale, which can grow to 30m or more. In general, the baleen whales are much bigger than the toothed whales. The exception is the sperm whale, which can grow up to 18m.

porpoise

dolphin

narwhal

killer whale

beaked whale

grey whale

sperm whale

right whale

blue whale

Risso's dolphin

Risso's Dolphin

The dolphin pictured leaping here is a Risso's dolphin. It has a blunt snout and as few as six teeth. Most of the toothed whales that we call dolphins are on average between about 2 and 3m long. Risso's dolphins can grow a little bigger – up to nearly 4m long. The biggest dolphin of all, known as the killer whale, can grow to more than twice this size.

Whale Bones

Like all mammals, whales have a skeleton of bones. The skeleton helps give the body its shape and also protects vital organs like the heart. Unlike land mammals, a whale's body is supported by the surrounding water. For this reason, the bones of whales are not as strong as those of land mammals, and are quite soft and spongy. The main part of the whale skeleton is the long backbone, made up of many vertebrae. Joints between the vertebrae give the backbone some flexibility. The backbone provides some body support. However, it acts mainly as a strong anchor for the whale's muscles, particularly the powerful swimming muscles that drive the tail. Instead of a land mammal's limbs, a whale has a pair of modified fore limbs, called flippers. It uses the flippers as paddles for manoeuvring in the water.

Whalebone Corset

This is an advertisement for a whalebone corset. It dates from 1911. Corsets were once worn by women to give them a shapely figure. The corsets, in fact, were not made of the bones of whales, but the baleen plates found in the mouths of the baleen whales. The plates were misnamed whalebones.

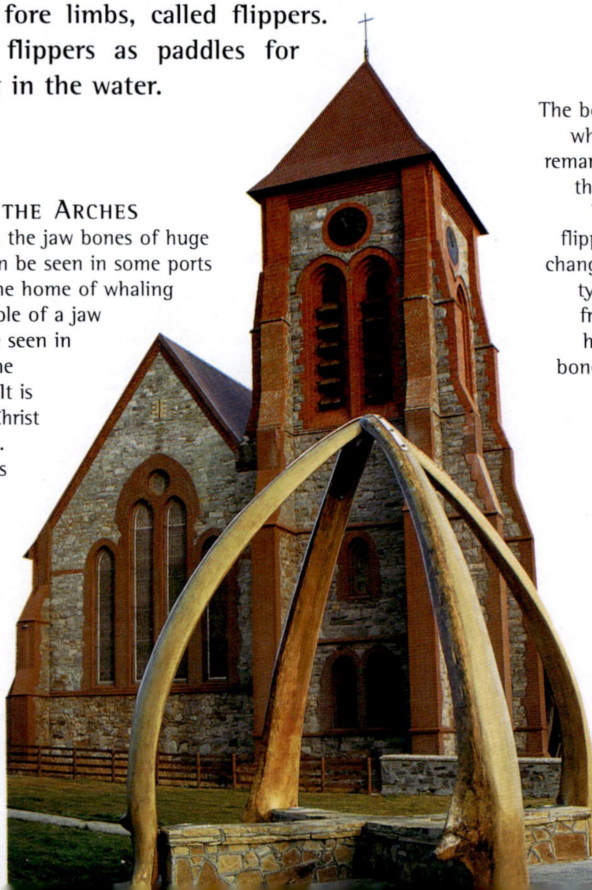

Underneath the Arches

Arches built from the jaw bones of huge baleen whales can be seen in some ports that were once the home of whaling fleets. This example of a jaw bone arch can be seen in Port Stanley in the Falkland Islands. It is located outside Christ Church Cathedral. Nowadays, whales are protected species. Building such arches is now forbidden.

Hands Up

The bones in a sperm whale's flipper are remarkably similar to those in a human hand. A whale's flippers are a much changed version of a typical mammal's front limbs. Both hands have wrist bones, finger bones and joints.

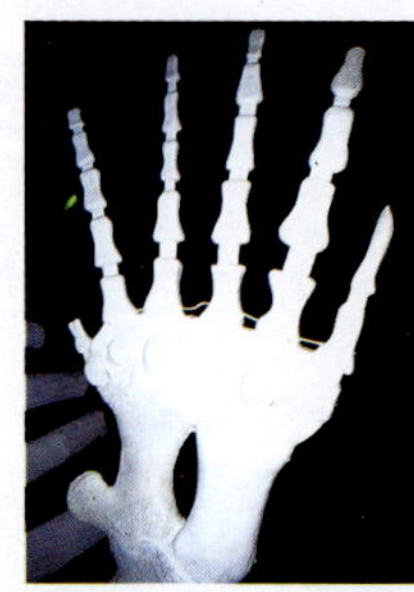

sperm whale flipper

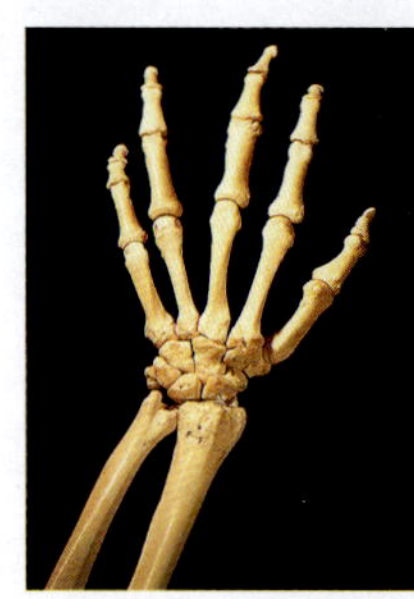

human hand

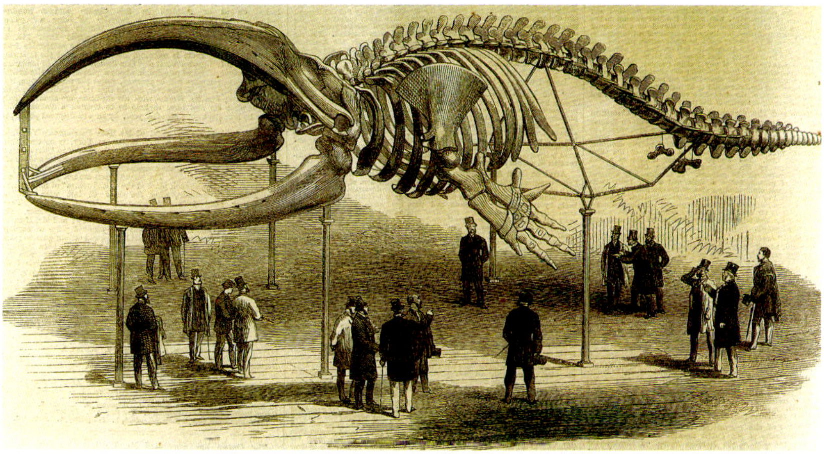

Big Head

This huge skeleton of a right whale was put on display in London in 1830. Its large jaw bones tell us that it is a baleen whale, which needs a big mouth for feeding. As with other mammals, the whale has a large rib cage to protect its body organs. Unlike land mammals, however, it has no hind limbs or pelvic girdle – just a few tiny bones.

Did you know? Whales are very oily and very smelly.

Toothy Jaw

This is the skeleton of one of the toothed whales, the false killer whale. The head is much smaller than that of the baleen whales, and it has powerful jaws studded with teeth. As with other whales, it has a long spine made up of segments called vertebrae. The vertebrae in the whale's waist region are large. This is so that they are strong enough to anchor the animal's powerful tail muscles.

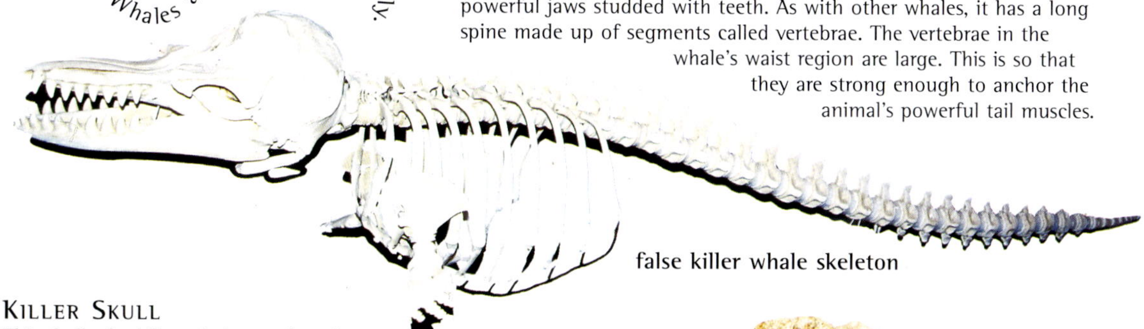

false killer whale skeleton

Killer Skull

This skull of a killer whale was found on the seashore. Both the upper and lower jaws are studded with vicious teeth, which curve inwards and backwards. The teeth are more than 10cm long. When it was alive, the whale would have been a deadly predator, attacking seals, dolphins and sometimes whales that were even bigger than itself.

killer whale skull

Whale Bodies

On the outside, whales could hardly be more different from land mammals. Over many millions of years they have developed quite distinct features that suit them to a life spent mostly underwater. They have long, rounded bodies and smooth, almost hairless, skin. Like fish, whales move about using fins. They have a large head and a very short neck, which most species cannot move. On the inside, whales are similar to land mammals. They have the same body organs, such as heart, lungs, stomach, intestines, kidneys and sex organs. The main difference is that in the big whales, the body organs are much larger than in land mammals.

bottlenose dolphins

Big Mouth

A grey whale surfaces, mouth wide open, to look around. It is one of the baleen whales, and the baleen can be seen hanging from its upper jaw. Baleen whales need a big mouth so that they can take in large mouthfuls of water when they are feeding. Grey whales usually feed at the bottom of the sea.

baleen

grey whale

Jonah and the Whale

This picture from the 17th century tells one of the best known of all Bible stories. The prophet Jonah was thrown overboard by sailors during a terrible storm. To rescue him, God sent a whale, which swallowed him whole. Jonah spent three days in the whale's belly before it coughed him up on to dry land. The picture shows that many people at this time had little idea of what a whale looked like. The artist has drawn a creature with shark-like teeth and a long, curly tail.

Leaping Dolphins

A pair of bottlenose dolphins leap effortlessly several metres out of the water. Their sleek, smooth-skinned bodies glisten in the sunlight. Powerful muscles near the tail provide them with the energy for fast swimming and leaping. Dolphins leap for various reasons – to signal to each other, to look for fish or perhaps just to have fun.

barnacles on humpback whale

whale lice

Lousy Whales

The skin of the grey whale is covered with light-coloured patches. These patches are clusters of lice, about 2 to 3cm long. These ten-legged creatures, called cyamids, feed on the whale's skin. All the large whales have lice, which are often found around clusters of barnacles on the skin.

Hangers On

As a humpback whale pokes its head out of the water here, you can see its throat. It is not as smooth as it might be because it is covered with barnacles. They are able to take hold on the whale's body because it is quite a slow-moving animal. They cannot easily grip the bodies of swifter-moving cetaceans, such as most dolphins. A dolphin sloughs rough skin away as it moves through the water. This also makes it harder for a barnacle to take hold.

Did you know? Whales have whiskers on their faces.

melon-headed whales

Body Lines

A pod, or group, of melon-headed whales swim in the Pacific Ocean. This species is one of the smaller whales, at less than 3m long. It shows the features of a typical cetacean – the body is well rounded with a shortened neck and has only a single fin on its back. It has a pair of paddle-like front flippers and a tail with horizontal flukes. Cetaceans have no outer ears or back limbs.

Staying Alive

Like all mammals, whales are warm-blooded creatures. To stay alive, they must keep their bodies at a temperature of about 36–37°C. The trouble is that they swim in very cold water, often in temperatures of only a few degrees. The water quickly takes heat away from the surface of their bodies. To stop too much body heat getting to the surface, whales have a thick layer of a fatty substance called blubber just beneath the skin. This acts as insulation and holds in the heat.

Like other mammals, whales must breathe in order to stay alive. They breathe through nostrils known as blowholes. Unlike mammals' nostrils, a whale's blowhole is situated on top of the head, not in front. When a whale blows, or breathes out, it sends a column of steamy water vapour high into the air.

southern right whale

In the Warm

A southern right whale leaps out of the water in a magnificent display called breaching. This whale spends the summer feeding in ice-cold Antarctic waters. The sheer size of the whale helps limit the percentage of the body heat it loses to the water.

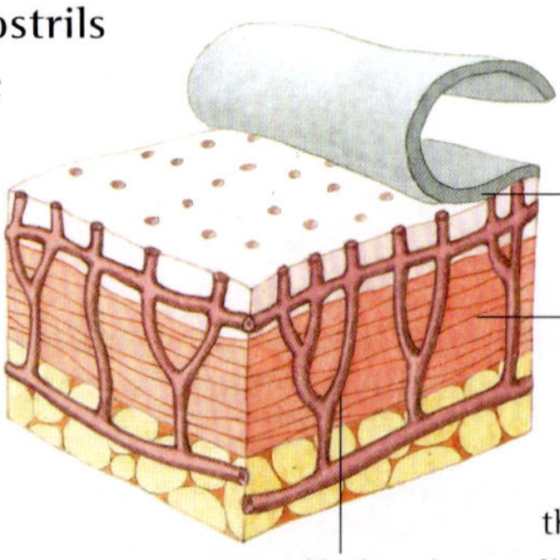

Skin Deep

This is a cross-section of the thick outer layer that covers a whale's body. Beneath a thin skin there is a thick layer of fatty blubber, which insulates the whale from ice-cold water.

Small Species

Two Atlantic spotted dolphins play follow-my-leader. This species is among the smaller cetaceans. They are about the size of a human. Because it is quite small, its body has a relatively large surface area for its size and therefore loses heat faster than its big relatives. This is probably why the Atlantic spotted dolphin lives in quite warm waters.

Atlantic spotted dolphins

Did you know? The temperature of a whale is about the same as yours.

Sky High

In the waters of Frederick Sound, in Alaska, a humpback whale surfaces and blows. The column of warm, moist air shoots high above the water. As it rises, it is cooled by the surrounding air. The moisture in it condenses into a cloud of tiny water droplets.

humpback whale

Deep Diving

Whales feed at different depths. Most dolphins feed on shoaling fish quite close to the surface. They never stay under water for more than 15 minutes. Blue whales can dive deeper, but not for much longer. The sperm whale holds the diving record, being able to descend to about 2,000m and stay under water for up to an hour.

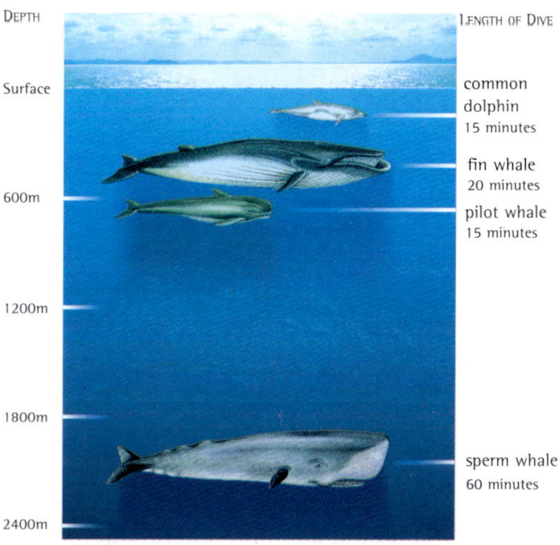

One Blowhole

A bottlenose dolphin rises to the surface to blow. It has only one blowhole, like all of the toothed whales. Here, the blowhole is seen open. When the dolphin dives, thick lips of elastic tissue will close it shut and prevent water entering, no matter how deep the dive or how high the water pressure.

bottlenose dolphin

Two Blowholes

These are the blowholes of a humpback whale. The humpback breathes out through a pair of blowholes, which are located behind a raised ridge called a splashguard. This helps prevent water entering the blowholes when the whale is blowing.

Whale Brain and Senses

Like all animals, the whale controls its body through its nervous system. The brain is the control centre, carrying out many functions automatically, but also acting upon information supplied by the senses. The size of the whale brain varies greatly, according to the size of the animal. However, the dolphins have much bigger brains for their size, almost matching those of humans. Dolphins appear to be remarkably intelligent. Whales have the same five senses as we do, although their senses of taste and smell seem very limited. Hearing is by far their most important sense. They pick up sounds with tiny ears located just behind the eyes.

Tiny Eyes
Compared with its large body, a whale's eyes are tiny. It can see quite well when it is on the surface and often spy-hops, or lifts its head out of the water to look around. But when it dives deep for food, its eyes are useless in the dark water.

Atlantic spotted dolphins

Did you know? A sperm whale's brain is five times the size of a human's.

Close Encounters
A group of Atlantic spotted dolphins swim closely together in the seas around the Bahama Islands. Like most other cetaceans, the dolphins often nudge one another and stroke each other with their flippers and tail. Touch plays a very important part in dolphin society, especially in courtship.

humpback whale

Slap Happy

A humpback whale slapping its tail, or lob-tailing, a favourite pastime for the great whales. It has raised up its tail and is about to slap it on to the water. This will create a noise like a gunshot in the air, but, more importantly, it will make a loud report underwater. All the other whales in the area will be able to hear the noise. Hearing is the most important sense in most whales.

Cupids and Dolphins

In this beautiful Roman mosaic, cupids and dolphins gambol together in the sea. In Roman mythology, Cupid was the god of love, who shot arrows of passion from his bow. Roman artists were inspired by the dolphin's intelligence and gentleness with human beings. They regarded dolphins as sacred creatures.

Brainy Dolphin?

Some dolphins, such as the bottlenose, have a brain that is much the same size as our own. It is quite a complex brain with many folds – but still has a lot fewer folds than a human brain. Although the dolphin has a big brain, it is not necessarily highly intelligent.

bottlenose dolphin and trainer

In Training

A bottlenose dolphin is shown with its trainer at the Institute of Marine Sciences at Roatan, Honduras, in Central America. This species has a particularly large brain for its size. It can be easily trained, has a good memory and is a superb mimic. A bottlenose dolphin can observe other animals and learn to mimic their behaviour in a short space of time. It is also good at solving problems, something we consider a sign of intelligence.

Sounds and Songs

Sounds travel easily in water, and sounds are vital to whales. They use sounds to communicate with one another, and many use sounds to find their food. The large baleen whales use low-pitched sounds for communication. Underwater microphones have picked these sounds up as moans, grunts and snores. The humpback whale is the most vocal of the baleen whales, producing many kinds of sounds and stringing them together to make complicated songs. The toothed whales make higher-pitched sounds. These sounds are picked up as squeaks, creaks or whistles. Whales also use high-pitched clicks when hunting. They send out beams of sound, which are reflected by objects in their path, such as fish. The whale picks up the reflected sound, or echo, and can work out the object's location. This is called echo-location.

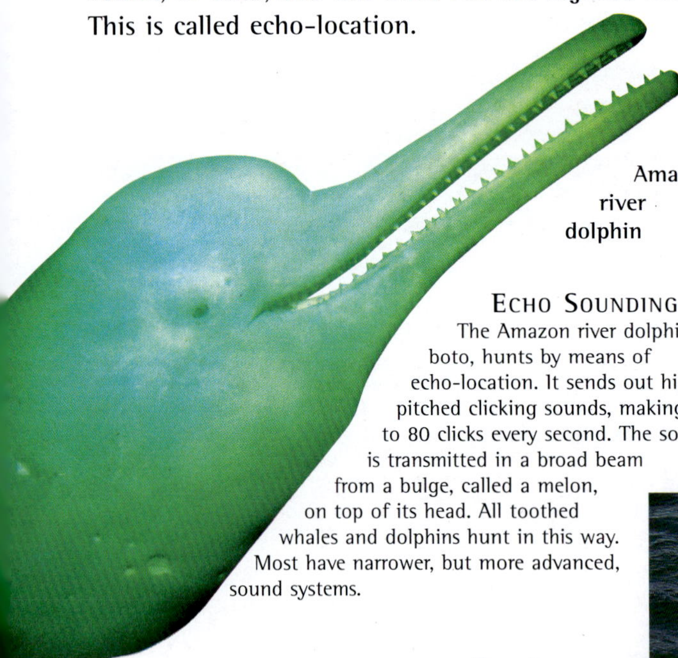

Amazon river dolphin

Echo Soundings

The Amazon river dolphin, or boto, hunts by means of echo-location. It sends out high pitched clicking sounds, making up to 80 clicks every second. The sound is transmitted in a broad beam from a bulge, called a melon, on top of its head. All toothed whales and dolphins hunt in this way. Most have narrower, but more advanced, sound systems.

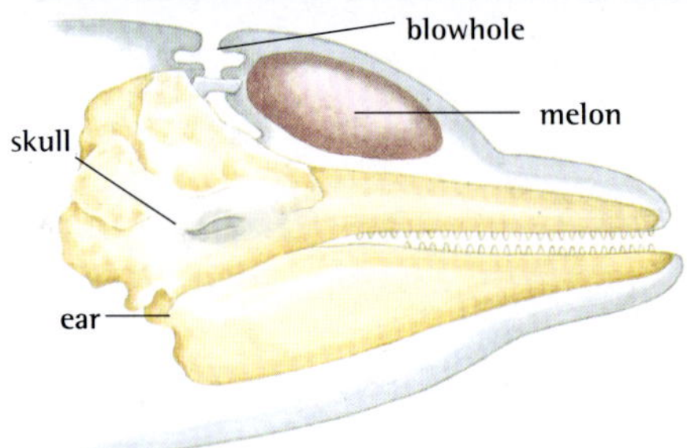

Making Waves

A dolphin makes high-pitched sound waves by vibrating the air in the passages in its nose. The waves are focused into a beam by the melon – the bulge on top of its head. The beam of sound is transmitted into the water.

group of belugas

Sea Canaries

A group of belugas, or white whales, swim in the shallows of a bay in northern Canada. Belugas are noisy creatures whose voices can clearly be heard above the surface. This is why they are sometimes called sea canaries. Belugas also produce high-pitched sounds we cannot hear, which they use for echo-location.

humpback whale

Super Songster

It is the breeding season, and this male humpback whale is heading for the breeding grounds where the females are gathering. The male starts singing long and complicated songs. This may be to announce its arrival and to attract a mate, or it could be to warn other males off its patch. The sound of these songs can carry for 30km or more.

Long Songs

This is a voice print of a humpback whale singing a song. The sounds have been picked up underwater by a hydrophone, or underwater microphone. Prints like these show complex musical phrases and melodies. Humpback whales often continue singing for a day or more, with the animals repeating the same basic song time and time again.

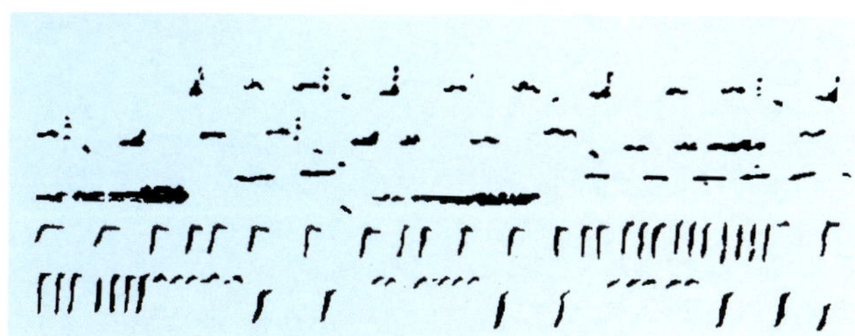

Did you know? A dolphin picks up sounds through its lower jaw.

Sound Echoes

One of the most exciting underwater battles is that of a sperm whale wrestling with a giant squid. The whale can locate a squid more than a kilometre away by transmitting pulses of sound waves into the water and listening. The returning echo is picked up by the teeth in the lower jaw. The vibrations are sent along the jaw to the ear.

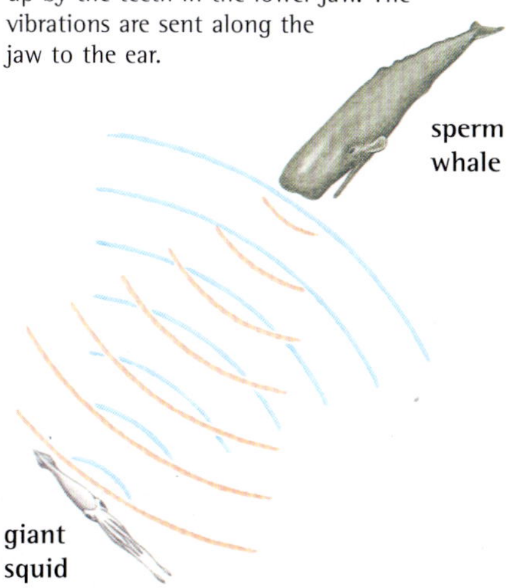

sperm whale

giant squid

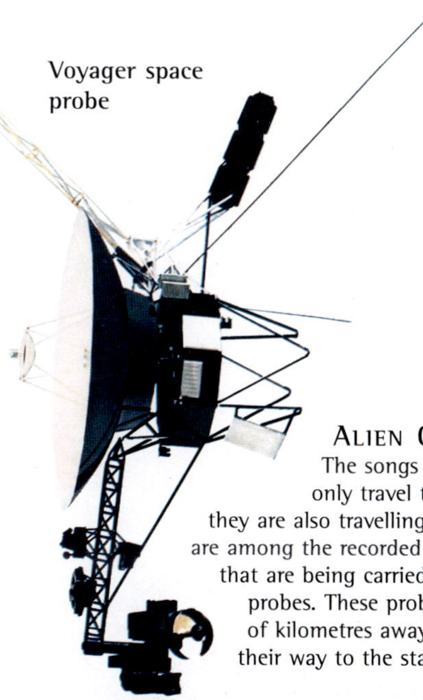

Voyager space probe

Did you know? A blue whale's long moans are the loudest sound of any animal.

Alien Greetings

The songs of the humpback whale not only travel through Earth's oceans, but they are also travelling far out into Space. They are among the recorded typical sounds of our world that are being carried by the two Voyager space probes. These probes are now many millions of kilometres away from Earth and are on their way to the stars.

CRUNCHY KRILL

These creatures are known as krill. Like crabs and shrimps, they are crustaceans. They form the main part of the diet of many baleen whales, especially in Antarctica. Measuring up to 75mm long, they swim in vast shoals, often covering an area of several square kilometres. Most krill are found in Antarctic waters.

FEEDING HABITS

The two main groups of whales eat quite different food. Most of the huge baleen whales feed by taking great mouthfuls of seawater containing fish and tiny shrimp-like creatures called krill or plankton, as well as algae, jellyfish, worms and so on. The whale closes its mouth and lifts its tongue, forcing water out through the bristly baleen plates on the upper jaw. The baleen acts like a sieve and holds back the plankton or fish, which the whale then swallows. The toothed whales feed mainly on fish and squid. They find their prey by echo-location. It is essential for species like the sperm whale, which dive deep into the ocean where it is dark. Some toothed whales hunt in packs.

grey whale

PLOUGHED FOOD

A grey whale ploughs into the seabed, stirring up the sand and ooze. It also dislodges the tiny crustaceans that live there and gulps them down. These crustaceans, called amphipods, are only about 15mm long. Grey whales feed mostly during summer in the Arctic region before they migrate further south.

southern right whale

SKIM FEEDING

With its mouth open, a southern right whale cruises slowly over the surface of the sea. It filters tiny crustaceans, called copepods, out of the water with its fine baleen. It can eat as much as two tonnes of these plankton every day. It needs to eat so much because of its huge size - up to 80 tonnes. Usually right whales feed alone, but if food is plentiful, several will feed cruising side by side.

Succulent Squid

A shoal of squid swim in Californian waters. Squid is the sperm whale's favourite food and is eaten by other toothed whales and dolphins as well. Squid are molluscs, in the same animal order as snails and octopuses. But, unlike octopuses, they have eight arms and two tentacles, and are therefore called decapods (meaning ten feet). Squid swim together in dense shoals, many thousand strong. They are fast swimmers. There are many species of squid, measuring from a few centimetres long up to 20m in the case of the giant squid.

Toothy Smile

A Ganges river dolphin shows its teeth. The front teeth are visible even when its mouth is closed because they are particularly long. In all, the dolphin has more than 100 teeth. Ganges river dolphins eat mainly fish, but they also take shrimp and crab. They usually feed at night and find their prey by echo-location.

Did you know? A blue whale eats nearly 1,000kg of krill in a single meal.

Ganges river dolphin

sperm whale

Hunt the Squid

This sperm whale is hunting. It is the largest toothed whale, notable for its huge head and tiny lower jaw. The sperm whale feeds in deep water on squid. It often hunts the giant squid that live in waters around 2,000m deep. At that depth, in total darkness, the sperm whale hunts its prey by sonar, or echo-location.

beluga whale

Beluga Lunch

Belugas, or white whales, feed on squid and small fish, which are in plentiful supply in their icy ocean home. Unlike common dolphins, belugas do not have many teeth. Scientists think that many squid-eating whales may simply suck prey into their mouths. Many species of beaked whale, which also feed on squid, have no teeth suitable for clutching prey.

The Attack Begins

A hungry killer whale has spotted a sea lion splashing in the surf at the water's edge. With powerful strokes of its tail, it surges towards its intended prey. The whale's tall dorsal fin shows that it is a fully-grown male. The sea lion seems totally unaware of what is happening but, in any case, it is nearly helpless in the shallow water. The killer is scraping the shore as it homes in for the kill.

In For the Kill

Among the toothed whales, the killer whale, or orca, is the master predator. It feeds on a wider variety of prey than any other whale. It bites and tears its prey to pieces with its fearsome teeth and may also batter them with its powerful tail. It is the only whale to take warm-blooded prey. Fortunately, there is no record of a killer whale ever attacking human beings. However, wildlife watchers and photographers often report killer whales spy-hopping when they approach, as though sizing them up for a meal. As well as fish and squid, a killer whale will hunt seals, penguins, dolphins and porpoises. It may even attack large baleen whales many times its size. Killer whales live in family groups, or pods, which often go hunting together. Co-operative hunting greatly improves the chance of success. But killer whales will also go hunting by themselves when they chance upon a likely victim, such as a lone sea lion.

Terrible Teeth

Suddenly the killer's head bursts out of the water, and its jaws gape open. Its vicious teeth are exposed, ready to sink into its sea lion prey. The killer whale may have fewer teeth than most toothed whales, but they are large and very strong.

No Escape

Now the killer snaps its jaws shut, clamping the sea lion in a vice-like grip. With its prey struggling helplessly, it slides back into deep water to eat its fill. Killer whales sometimes almost beach themselves when they lunge after prey but, helped by the surf, they usually manage to wriggle their way back into the sea.

Lunging for Lunch

The humpback whale is a filter-feeder, taking in great mouthfuls of water and filtering out the food in them through its baleen plates. Usually it scoops up the water as it lunges forwards and upwards. This way of lunge-feeding is typical of the baleen whales known as the rorquals, which also include the blue, fin, sei and minke whales. All these whales have grooves in their throats, which allow the mouth to expand greatly. This, in turn, enables them to take in many tonnes of water containing their food. Sometimes, before lunge-feeding, humpbacks may round up the fish by blowing a circle of bubbles around them. The bubbles act like a net and stop the fish escaping.

On the Lookout

Poking its head out of the water, a humpback whale spy-hops in the feeding grounds of Alaska. It is looking for signs of shoals of fish, such as anchovy and cod. Here, in the Northern Hemisphere, humpbacks feed mainly on fish. The Southern Hemisphere population of humpbacks feed mainly on plankton, such as krill.

Forward Lunge

Once it has located a shoal, the humpback swims towards it. In the middle of the shoal, it opens its enormous mouth and lunges forwards. The throat grooves in the bottom jaw expand as the water – and the fish – rush in. Next, the humpback uses its tongue and cheek muscles to force out the water through its baleen plates, leaving the fish behind in its mouth.

Upward Lunge

Here, the humpback is using a different technique. It sinks below the surface and then flicks its tail to help it to shoot upwards again. With mouth gaping open, it lunges at the fish from below.

Ring of Bubbles

The surface of the sea is boiling with a ring of frothy bubbles. Unseen, beneath the water, one or more humpback whales are swimming round in circles, letting out air as they do so.

Bubble Netting

The circle of bubbles rises up to the surface from the whales circling under the water. It forms a kind of net around a shoal of fish and prevents the fish inside from escaping. The whales then swim upwards to the surface with their mouths gaping, ready to engulf the netted prey.

SWIMMING

All the whales are superb swimmers. Even the huge baleen whales can swim faster than an Olympic swimmer. All parts of the whale's body help it move through the water. The driving force comes from the tail fin, or flukes. Using very powerful muscles in the rear third of its body, the whale beats its tail up and down. The whole body bends when this happens. The whale uses its pectoral fins, or flippers, near the front of the body to steer with. The body itself has a streamlined shape and is very smooth to help it slip through the water easily with the least drag, or resistance. The body has some flexibility and can change shape slightly to keep the water flowing smoothly around it. Little ridges under the skin help as well. Some whales also give out tiny droplets of oil from their skin, which further reduces drag.

STEERING

Among whales, the humpback has by far the longest front flippers. Whales use their paddle-like flippers mainly for steering as they travel through the water. The humpback also uses its flippers for slapping the water. Flipper-slapping seems to be some form of communication.

grey whale's tail flukes

TAIL POWER

The tail flukes of a grey whale rise into the air before it dives. Whales move their broad tails up and down to drive themselves through the water.

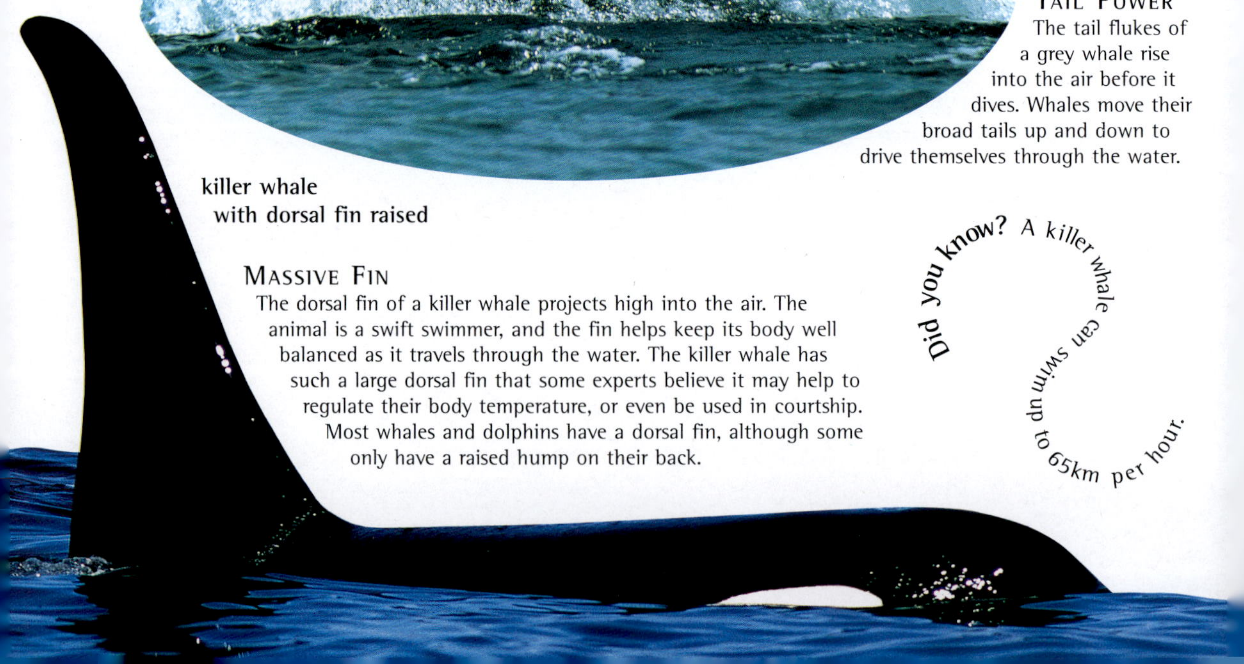

killer whale with dorsal fin raised

MASSIVE FIN

The dorsal fin of a killer whale projects high into the air. The animal is a swift swimmer, and the fin helps keep its body well balanced as it travels through the water. The killer whale has such a large dorsal fin that some experts believe it may help to regulate their body temperature, or even be used in courtship. Most whales and dolphins have a dorsal fin, although some only have a raised hump on their back.

Did you know? A killer whale can swim up to 65km per hour.

Atlantic spotted dolphins

Streamlining

Atlantic spotted dolphins swim without effort in their underwater habitat. Their bodies are beautifully streamlined. This means they are shaped so that they slip easily through the water when they move. The dolphin's body is long and rounded, relatively broad in front and becoming narrower towards the tail. Apart from the dorsal fin on the dolphin's back, and flippers, nothing projects from its body. It has no external ears or rear limbs, and the sexual organs are usually protected by folds of skin.

bottlenose dolphin

How a Dolphin Swims

Like all the cetaceans, dolphins use their tails to propel themselves through the water. They beat their tail flukes up and down by means of the powerful muscles near the tail. The flukes force the water backwards at each stroke, and as the water is forced back, the body of the dolphin is forced forwards. The other fins on the dolphin's body help guide it through the water. They do not supply any swimming power.

dolphin swimming

Smooth Skinned

This bottlenose dolphin is tailwalking – supporting itself by powerful thrusts of its tail. As can be seen, it has an amazingly smooth skin. Unlike most other mammals, it has no covering of hair or hair follicles – the little dimples in the skin from which the hair grows. Its very smooth skin helps the dolphin's body slip easily through the water.

How a Fish Swims

A fish has similar fins to a dolphin and, like a dolphin, it is mainly the tail that provides the power for swimming. A fish's tail has vertical fins, unlike the horizontal fins of the dolphin. The fish swims by beating its tail and body from side to side, not up and down like a dolphin.

fish swimming

Ready, Steady, Dive!

Most whales feed beneath the surface, some often diving deep to reach their food. We can usually identify the species of whale from the way it prepares to dive, or sound. The sperm whale, for example, is one of the species that lifts its tail high into the air before it descends into the ocean depths. It is the deepest diver of all the whales, sometimes descending to more than a kilometre in search of squid. It can remain under water for 60 minutes or more before it has to come up for air. As in other whales, its lungs collapse when it dives. It is thought that the great mass of oil in its huge head helps the whale when diving and surfacing. The oil freezes and becomes heavier on the way down, then melts and becomes lighter again on the way up.

Blowing
Two sperm whales swimming at the surface. The one on the left is preparing to dive. Its head is in the air, and it is filling its lungs with air in a series of blows. The sperm whale's blow is easy to recognize because the spout projects forwards, as in no other whale.

Spurting
Seen from above, the diving whale lashes its tail and accelerates through the water, creating a foaming wake. Its head goes under. If it is going to make a deep dive, the whale may not take another breath for over an hour.

Arching

Now the whale starts the dive, thrusting its bulbous head down and arching its back steeply. The rounded hump on its back rises high into the air. The lumpy knuckles behind the hump become visible as the body arches over.

Fluking

Soon the body disappears with just the tail flukes poking out of the water. The body is now in a vertical position and that is how it remains as the whale dives swiftly into the deep. Descending at speeds of more than 150m per minute, it is soon in darkness, scanning its surroundings by beams of sound for the squid on which it feeds.

Social Life

In our daily lives, we meet, work, play and communicate with many other people. We are part of human society. We are sociable animals. Some kinds of whales are also very sociable and live together in groups. Sperm whales, for example, live in groups of up to about 50. A group may be a breeding school of females and young or a bachelor school of young males. The older male sperm whales spend much of their time alone, except during the breeding season. Beluga whales have a similar kind of organization. Belugas often live in groups of several hundred, as do some kinds of dolphins. The baleen whales are not as sociable. They tend to move about singly or in small family groups. This is probably because of their enormous appetite – they could not find enough food if they lived close together all the time.

beluga whales

Herd Instinct

Beluga whales gather together in very large groups, or herds, and they mostly stay in these herds for life. Many of the animals in this group, pictured in the Canadian Arctic, have calves. These can be recognized, not only by their smaller size, but also by their darker skin colour.

Did you know? Dolphins will nudge a sick member of the group up to the surface, so it does not drown.

Nosey Orcas

Two killer whales, or orcas, spy-hop in Antarctic waters. They rise out of the water together, as if on a signal. They are members of the same pod, which stay together all their lives. The bonds between the animals are very strong. This helps them coordinate their activities, especially when hunting for food.

killer whales

Did you know? Male whales often try to help injured females. Females rarely try to help injured males.

Atlantic spotted dolphins

Staying Close

Two Atlantic spotted dolphins swim with their young. The young have no spots and are smaller. Spots do not start to appear until the animals are about a year old. As with many other species, the young stay very close to their parents most of the time. Spotted dolphins are usually found in quite small groups of less than 20.

Human Contact

A bottlenose dolphin swims alongside a boy. These dolphins live in social groups but lone outcasts, or animals that have become separated from their group, often approach humans.

Amazon river dolphin

Solitary Swimmer

An Amazon river dolphin, or boto, is resting on the river bed. It may spend most of its life by itself, or maybe with just one other animal. This solitary behaviour is typical of the other river dolphins too. They are not sociable animals like most other whales and dolphins.

Pilot Error

Long-finned pilot whales are stranded on a beach in Tasmania. Pilot whales usually live in large groups, and there are strong bonds between group members. This can lead to the disaster of mass stranding. One of the whales may get into difficulties and strand itself on the beach. The other members of the pod try to help it and become stranded themselves.

long-finned pilot whales

THE MATING GAME

Like most other mammals, whales court one another and mate at a certain time of the year. Baleen whales, for example, mate during the late autumn after the whales have migrated to their warm-water breeding grounds. One whale will mate a number of times with different partners. Sometimes several males will attempt to manoeuvre a female into a mating position. Often the males fight each other for the chance to mate, biting, headbutting or lashing their tails at one another. Male narwhals fence one another with their long tusks. But mating behaviour can also be gentle, with the males and females caressing one another with their flippers.

beluga whales

WHITE WEDDING

A pair of belugas begin to take an interest in each other. Male and female belugas spend much of the year in separate groups, only mixing during the mating season in the summer in northern waters. Mating and calving (about 14 months later) take place in shallow bays in the far north.

humpback whale

LOVE SONGS

Like other mammals, whales attract a mate by body language and sound. Amazingly, whales such as this humpback can pinpoint each other's position, and perhaps exchange messages, through the ocean over great distances.

grey whale

MATING BRAWL

Two grey whales court in the winter breeding grounds in the seas off Baja California, Mexico. Usually, a group of males fights furiously for the right to mate with a female, causing a lot of commotion in the water. The female might mate many times with her battling escorts.

Rollover

Courtship for these southern right whales is nearly over. The male (top) has succeeded in getting the female to roll over on her back and is manoeuvring into the mating position. This scene is in the Valdez Peninsula, Argentina, the winter breeding grounds for southern right whales.

The Fabulous Unicorn

In the breeding season, male narwhals fight one another using their long tusks, which often break. In ancient times, when these small whales were little known, people found these tusks and wondered what kind of beast they came from. This may have led to the idea of the unicorn, usually shown as a horse-like creature with a long spiral horn (like the narwhal's) on its forehead.

southern right whales

Belly to Belly

A pair of southern right whales mate, belly to belly. The male has inserted his long penis into the female to inject his sperm. Usually, the male's penis stays hidden in the body behind a genital slit. It will be nine months or more before the female gives birth to a single calf.

Big Baby

A sperm whale calf snuggles up to its mother, almost as if the two are stuck together. A calf might measure up to 4.5m long when born, nearly 15 months after mating took place. The mother feeds it for a year or more, leaving it alone only when diving deep for food.

sperm whale and calf

Did you know?

Female sperm whales can mate at the age of 8. Males cannot mate until they are nearly 20.

Bringing Up Baby

Birthday
A bottlenose dolphin gives birth to her baby. The baby is being born tail-first, the usual way. It has been growing inside the mother's body for about a year, and is now nearly 1m long. This birth is taking place near the bottom of an aquarium. In the wild, birth takes place close to the surface so the baby can surface quickly and start to breathe.

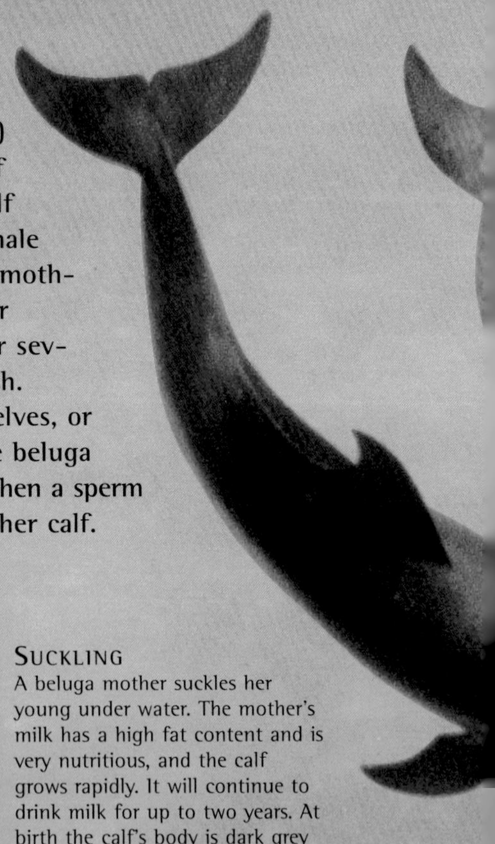

After a successful mating, the female whale becomes pregnant and a baby whale starts to grow inside her body. After about a year, the baby (calf) is ready to be born. By now it can weigh, in the case of the blue whale, up to 2.5 tonnes. The first thing the calf must do is take a breath, and the mother or another whale may help it up to the surface. Soon it finds one of the mother's nipples and starts to suck the thick, rich milk in her mammary glands (breasts). It will continue to suckle for several months until it learns to take solid food such as fish. Mother and calf may spend most of the time by themselves, or join nursery schools with other mothers and calves. The beluga and sperm whales are known for their nursery schools. When a sperm whale mother has to dive for food, an aunt looks after her calf.

Suckling
A beluga mother suckles her young under water. The mother's milk has a high fat content and is very nutritious, and the calf grows rapidly. It will continue to drink milk for up to two years. At birth the calf's body is dark grey in colour, but it gradually lightens as the calf matures.

At Play

A young Atlantic spotted dolphin and its mother play together, twisting, turning, rolling and touching each other with their flippers. During play, the young dolphin learns the skills it will need later in life when it has to fend for itself. The youngster is darker than its mother and has no spots. These do not start to appear until it is about a year old.

Did you know? The calves of baleen whales stop breast feeding after about 9 months – much sooner than toothed whale calves.

Togetherness

A humpback whale calf sticks closely to its mother as she swims slowly in Hawaiian waters. The slipstream, or water flow, created by the mother's motion helps pull it along. For the first few months of its life, it will not stray far from its mother's side.

Having Fun

Dolphins have delighted people with their acrobatic antics for thousands of years. They were a favourite subject for ancient Greek and Roman artists, for example. They leap, spin and somersault, ride the bow waves of boats and go surfing. The spinner and dusky dolphins are particularly lively, hurling themselves high into the air. Some dolphin antics undoubtedly have a purpose, such as sending signals to other dolphins. But often the animals appear to perform acrobatics just for fun during their playtime. In most animal species it is the young that play. In whale and dolphin society, the adults play too. One of the games southern right whales play is sailing. They hang vertically in the water with their heads down and tails in the air. The tails act like sails and catch the wind, and they are blown along.

Atlantic spotted dolphins

Playful Pair
Two Atlantic spotted dolphins jostle one another as they play with a type of coral, called a sea fan. Dolphins spend much of their time playing, especially the younger ones. They seem to make up all kinds of games, using anything they can find, such as pieces of driftwood, leaves, feathers and stones. Sometimes their games last for hours.

Jumping for Joy
A pair of bottlenose dolphins leaps high into the air. They leave the water at exactly the same time, as though they have been rehearsing their act for the photographer. They seem to be jumping just for the joy of it, but their behaviour may have some kind of social function within their family group.

bottlenose dolphins

Porpoising on Purpose

A group of long-snouted spinner dolphins go porpoising, taking long, low leaps as they swim swiftly. As you can see, they churn the water behind them into a foam. Many dolphins practise porpoising, which is a more efficient method of travelling fast on the surface.

Did you know? Killer whales like brushing against each other as they swim at high speed.

long-snouted spinner dolphins

Pacific white-sided dolphin

Riding the Wake

This picture shows a Pacific white-sided dolphin surfing on the waves. This is one of the most active and acrobatic of the dolphins. It is also often seen bow-riding in front of boats. Many other species of dolphins also like to ride in the foam and waves left in the wake of passing boats.

Did you know? The rough skin on a porpoise's back may be for giving calves piggy-back rides.

dusky dolphin

Did you know? A dolphin may play cat and mouse with its prey before eating it.

Aquatic Acrobat

This dusky dolphin is in a playful mood, throwing itself high into the air. It twists and turns, spins and performs somersaults. It is one of the most acrobatic of all the dolphins. This behaviour is like a roll call – to check that every dolphin in the group is present and ready to go hunting. This behaviour is repeated after hunting to gather the group together once more.

WHALES ON DISPLAY

One of the most spectacular sights in nature is of the great whales leaping out of the water and crashing back to the surface amid a shower of spray. This activity is called breaching. It is common among humpbacks. Some animals have been observed to breach up to 200 times in succession. When one whale starts breaching, others start breaching too. Like other whales, humpbacks put on other displays as well, including slapping their flippers and tail on the surface. These activities make a lot of noise, particularly underwater. They could, like breaching, be some form of signalling. Spy-hopping is another common activity among most whales. It is often done to look for signs of fish to eat.

BREACHING
Propelled by powerful thrusts of its tail, the humpback launches its vast bulk into the air, twisting as it does so. For a creature weighing up to 30 tonnes, this is no mean feat. As breaching ends, the humpback crashes back to the surface with a splash. This time it lands on its back, with one of its flippers up in the air.

FLIPPER-FLOPPING
The humpback swims on the surface, raising one of its long flippers in the air. Then it rolls over and slaps the flipper on the water. It may do this several times in succession, perhaps to warn off rivals. The humpback's flipper-flopping (or flipper-slapping) is particularly noisy because its flippers are so large.

What a Fluke!

The humpback raises its tail in the air during the display known as lob-tailing or tail-slapping. The tail is also exposed when the whale is about to dive, behaviour called fluking. It is easy to tell if a humpback is lob-tailing or fluking. In fluking, the tail disappears below the surface quietly.

Lob-tailing

In lob-tailing the tail is brought down sharply on to the surface of the water in a shower of spray and with a noise like a gunshot. Apart from when it is lob-tailing, the humpback only shows its flukes when it is about to go on a deep dive.

Spy-hopping

The humpback on the right of the picture is spy-hopping. It positions itself vertically in the water and pokes out its head until its eyes are showing. Then it has a good look round. The other humpback here is doing the opposite, poking out its tail, ready to lob-tail.

Where Whales are Found

Whales of one kind or another are found in all the oceans of the world. Some species of whale range widely, while others are found only in a certain area. They may stay in the same place all year long, or they may migrate from one area to another with the seasons. There are whales that stick to shallow coastal waters, and those that prefer the deep waters of the open ocean. Some live in the cool waters of far northern or far southern parts of the world, or at high latitudes. Others are more at home in tropical regions near the Equator or in low latitudes. You will find some kind of whale or dolphin in practically all parts of the oceans. There are even some species living in rivers.

Ocean Wanderer

A humpback whale surfaces to blow while swimming at Cape Cod off the north-east coast of North America. The humpback is found in all the oceans, but in different places at different times. In winter it remains in high latitudes feeding. Then it migrates thousands of kilometres to low latitudes to breed during the summer months.

Muddy Waters

The mud-laden waters of the River Amazon in South America are the habitat of the largest river dolphin, the Amazon river dolphin. Here, one is showing its fine set of teeth. This species ranges along the Amazon and its tributaries, and throughout the Orinoco River system to the north.

killer whale

Did you know? Some dolphins come and go between salt water and fresh water.

Worldwide Killer

Among ice-floes in the Arctic Ocean, a killer whale hunts for prey. Killer whales may be found in all oceans of the world, from the high Arctic to the Antarctic. They live in coastal areas but sometimes venture out to the open ocean. Killer whales also swim inshore among the surf, sometimes even beaching to snatch their prey.

Snow White

These belugas, or white whales, are swimming in Hudson Bay, Canada. They are cold-water creatures, living around the coasts in the far north of North America, Europe and Asia. They venture into estuaries and even up rivers, such as the Mackenzie River in Canada. In winter they hunt among the pack ice in the Arctic.

beluga whales

melon-headed whales

Tropical Melons

A pod of melon-headed whales is shown swimming in the Pacific Ocean. These creatures prefer warm waters and are found in subtropical and tropical regions in both the Northern and Southern Hemispheres. They generally stay in deep water, keeping well away from land.

Did you know? Several species of beaked whale have been so elusive that no one has seen them alive.

bottlenose dolphin

Wide Ranger

A bottlenose dolphin lunges through the surf in the sunny Bahamas. This animal is one of the most wide-ranging of the dolphins, being found in temperate to tropical waters in both the Northern and Southern Hemispheres. It is also found in enclosed seas such as the Mediterranean and Red Sea. Mostly it stays in coastal waters. When bottlenose dolphins migrate to warmer areas, they lose weight. When they return to colder climes, they gain more blubber.

Migration

Grey whales spend the summer months feeding in the food-rich waters of the Arctic Ocean. Many of the females are pregnant. Before winter comes, the whales head south towards Mexico for warmer waters, where the females will give birth to their calves. In the warmer climate, the calves stand a better chance of surviving. Mating also takes place around late summer. When spring comes, the greys begin to head north to the Arctic. Their annual journey between their feeding and breeding grounds is one of the greatest animal migrations, involving a round trip of some 20,000km. The northern and southern populations of humpbacks take part in long migrations too. Most of the other large rorquals and the right whales seem to undergo similar migrations from cold water feeding grounds to warm water breeding grounds. Among dolphins, which live mostly in warmer waters, there is little evidence of such large-scale migrations.

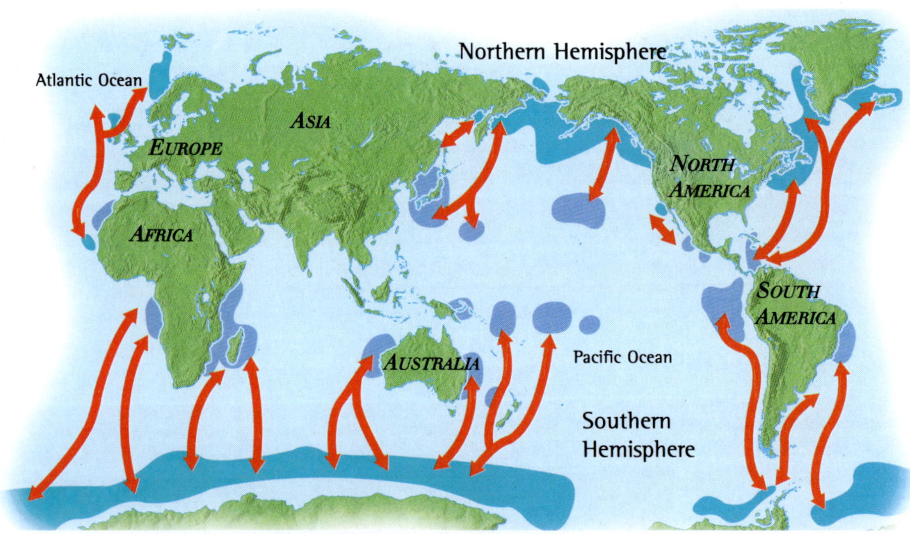

Migration

This map shows the routes taken by humpback whales during their annual migrations between their summer feeding and winter breeding grounds. There are at least three main groups, two in the Northern and one in the Southern Hemisphere. It is thought that the three groups never mix.

Key to Map

breeding grounds

feeding grounds

Summer Feeding

Two Southern Hemisphere humpbacks feed in the rich waters of the Antarctic Ocean during the summer months. This is when the krill and other plankton they feed on thrive. These two whales have just taken large mouthfuls of water, which they will now sieve for plankton using the baleen on their upper jaws.

Southern Hemisphere humpbacks

humpback whales

Did you know? Grey whales make longer migrations than any other mammal.

Winter Breeding

It is early winter in the Southern Hemisphere, and these two humpbacks have spent the past weeks migrating north from the Antarctic. They have arrived at their destination in a shallow bay on the coast of eastern Australia. It is here that one large group of humpbacks have their breeding grounds. Here they will mate and, about 12 months later, the females will give birth.

Right Location

The tail fluke of a southern right whale is thrust into the air as the whale sails. This whale is one of a group of right whales in winter breeding grounds off the coast of Argentina in South America. By summer the whales will have returned south to feed in the Antarctic Ocean.

southern right whale

Mating Greys

In the winter breeding grounds of Baja California, a grey whale surfaces near a boatload of whale-watchers. Grey whales spend about three months in the region, where mating and (two years later) births take place. They take some three months to migrate north back to Arctic waters.

grey whale

Atlantic white-sided dolphin

High and Dry

Dead whales, large and small, are often found washed up, or stranded, on the seashore. Live whales are sometimes found too, particularly open ocean species, such as pilot whales and sperm whales. Some live whales probably strand when they become ill and weak. Others may strand when they lose their sense of direction.

It is thought that whales may navigate, or find their way around, using the Earth's magnetism as a kind of map. Slight changes in this magnetism might cause them to turn the wrong way and head for the shore. Sometimes mass strandings take place, with scores of whales left helpless. This happens particularly among sociable species, such as the pilot whales. It is believed that when one animal gets stranded and becomes distressed, the other whales travelling with it try to help and become stranded too.

Beached Dolphin
This Atlantic white-sided dolphin is stranded on a beach in the Orkney Islands. Strandings are quite common in this species. The dolphins usually travel in quite large groups, so mass strandings can sometimes take place too.

Waiting For the Tide
People come to the assistance of more than 60 stranded long-finned pilot whales on the beach at Golden Bay, New Zealand. People cover the whales to stop them becoming sunburnt and throw water over them to keep their skin moist.

Did you know? When people help stranded whales, the whales often swim back and get stranded again.

Rare Stranding

Marine biologists examine a stranded Stejneger's beaked whale. Beaked whales are among the least known of all the cetaceans. Most of our knowledge about them comes from occasional strandings like this. Note the large tooth protruding from the jaws. Several beaked whales have teeth like this.

Stejneger's beaked whale

Did you know? Whales stranded in Britain belong to the monarch.

In the Shallows

Three belugas, or white whales, are stranded in the shallow waters of a bay as the tide has gone out. They are vulnerable to attack from polar bears when beached like this. Belugas spend a lot of time swimming in shallow water, but do not usually become stranded. If they do, they can usually survive until the tide comes in again.

beluga whales

Big Fin

A fin whale has become beached on a mudflat. As you can see, compared with a human, it is an enormous creature. Some fin whales can grow to 25m or more and weigh up to 80 tonnes. This animal is dead, but even if it were alive, it would be impossible to do anything to help it to return to the water. Its internal organs will collapse when the animal is not supported by the water. However, scientists can examine stranded bodies to learn new information about whales.

Grey and Right Whales

Grey whales and the three species of right whale, including the bowhead, are all filter-feeders with baleen plates in their upper jaws. The bowhead has the longest baleen of all, while the grey has short baleen. Unlike most other baleen whales, however, the grey whale feeds mainly on the seabed. It is also noted for its long annual migration along the Pacific coast of North America from the Arctic to Mexican waters. The grey whale is found only in the Northern Hemisphere, but there are populations of right whales in both hemispheres. Right whales were named by whalers because they were the right whales to catch. They swam slowly, they could be approached easily and they floated when dead. Right whales yielded large amounts of oil and baleen.

Did you know? The bowhead whale has longer baleen than any other whale – over 4m long.

Mottled Mammal

The distinctive long, narrow head of a grey whale breaks the surface. Its closed blowholes are in the middle of the picture. The head is covered here and there with clusters of barnacles and lice, which are also found over many other parts of the body. This, together with lighter body patches, gives the whole animal a mottled appearance.

bowhead whale

White Chin

A bowhead whale thrusts its head out of the water, exposing its white chin patch. It is the only whale to have a white chin. This is covered with numerous black patches. The skin is smooth, with no growths like those on the skin of the northern and southern right whales.

Did you know? We know a lot about grey whales because they stay in shallow waters near the coast.

Lively Lob-tailing

Near the coast of Argentina in South America, a southern right whale is lob-tailing. In seconds, its tail will come crashing down on the surface with a resounding smack that will echo off the cliffs on the shore. The noise will be heard by other whales, many kilometres away. Right whales often lob-tail and also do headstands, waving their tails in the air.

southern right whale

Beard and Bonnet

A southern right whale cruises in the South Atlantic. Two features make this whale impossible to mistake. One is the deeply curved jawline. The other is the southern right whale's beard and bonnet. These are large growths on the whale's chin and nose, which become infested with barnacles.

southern right whale

northern right whale

Hairy Monster

There are two populations of right whale – the northern and southern right whale. The northern right whale, pictured here, lives in the North Atlantic. Whalers used to refer to the crusty hard skin on its head as its bonnet or rock garden. Barnacles and whale lice develop on this skin and it can grow to an enormous size. The right whale is also the hairiest of all whales, retaining more of its hair after birth than any other cetacean. It even grows facial hair in places where humans do!

grey whale

Did you know? You can tell a grey whale by its unique long, narrow head.

Bristly Jaws

A grey whale opens its mouth, showing the baleen plates on its upper jaw. The baleen is quite short, stiff and coarse. The whale uses it to filter out the tiny creatures it digs out of the seabed when feeding. Grey whales are not shy and sometimes swim up to the boats of whale-watchers.

RORQUALS

The rorquals are a family of baleen whales that includes the largest creature that has ever lived on Earth, the blue whale. They are named after the grooves that they have on the throat – the word rorqual means a furrow. All rorquals, except the humpback, have a long streamlined body with a sharp nose and a dorsal fin set well back. They can all swim fast, reaching speeds up to about 30kph. The humpback is a slower swimmer with a chunkier body. It has long, knobbly flippers and a pronounced hump in front of the dorsal fin. The humpback is famous for the songs it sings. The minke whale is the smallest rorqual and is also the most common. Bryde's whale lives mainly in warm tropical and subtropical waters. The other rorquals are found in cold waters as well, often venturing into polar waters in the summer.

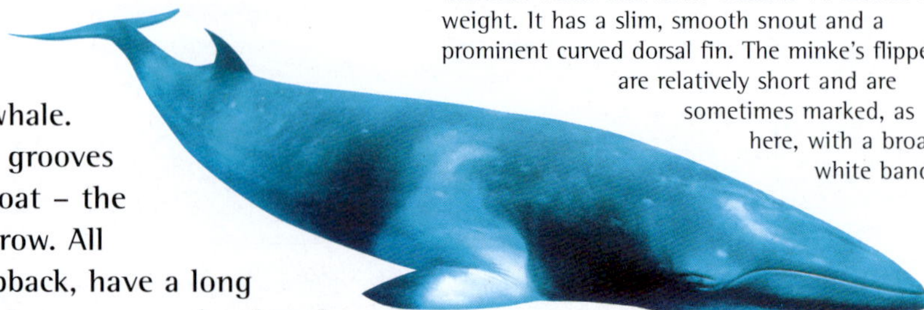

minke whale

TINY MINKE

The minke whale is the smallest rorqual. It grows to only about one-third of the size of the blue whale and never exceeds 10 tonnes in weight. It has a slim, smooth snout and a prominent curved dorsal fin. The minke's flippers are relatively short and are sometimes marked, as here, with a broad white band.

Did you know? Humpback whales can live to age 95.

WHALE WITH A HUMP

This picture of a humpback whale shows the feature that gives it its name very well. Its small dorsal fin sits on top of a pronounced hump on its back. This profile view of the animal also shows the prominent splash guard on its head in front of the blowholes. Note also the many knobs on the head and throat.

humpback whale

KNOBBLY FLIPPER

A humpback whale swims on the surface, with one of its flippers up in the air like a boat sail. The flippers of the humpback are by far the most distinctive of all the whales. They are sturdy and very long – up to a third of the length of the whale's body. The flippers have knobs along their leading (front) edge.

blue whale

Big Gulp

A blue whale feeds in Californian waters. It has taken in a mouthful of water containing thousands of the tiny shrimp-like creatures it feeds on. The grooves on its throat that allow its mouth to expand can be clearly seen. A blue whale typically has between about 60 and 90 of these grooves.

Did you know? A humpback whale may sing for 30 minutes at a time.

Dripping Flukes

A blue whale fluking, with its tail flukes rising out of the water before the animal dives. Among the rorquals, only blue and humpback whales expose their flukes in this way before diving. The humpback's tail flukes are quite different. They are knobbly at the trailing (rear) edges and have white markings on the underside.

blue whale

Did you know? A blue whale's heart is about the size of a small car.

Sei Whale

The sei whale can be found in most oceans. It feeds in the cool Arctic or Antarctic waters during the summer and migrates to warmer waters in the winter to breed. With a length of up to about 18m, it is slightly larger than the similar looking Bryde's whale.

Sperm and White Whales

The sperm and the white whales are two families of toothed whales. The sperm whale and dwarf and pygmy sperm whales have an organ in their head called the spermaceti organ, which is filled with wax. The wax may help the animals when they dive and may play a part in focusing the sound waves they use for echo-location. All three species spend most of their time in deep water. The sperm whale is similar in size to the large baleen whales. It has teeth on only its tiny lower jaw and no dorsal fin, just a hump. The two white whales, the beluga and the narwhal, also have no dorsal fin. The beluga has up to about 20 teeth in each of its jaws, but the narwhal has only two. In the male narwhal, one of the teeth grows into a long spiral tusk, measuring up to 3m.

Baby Eyes

The eye of a sperm whale calf. Like all whales, the sperm whale has tiny eyes compared with those of most other mammals. But this does not matter because when the whale dives to feed, it descends deep into the ocean where light never reaches. It depends on its superb echo-location system to find its prey.

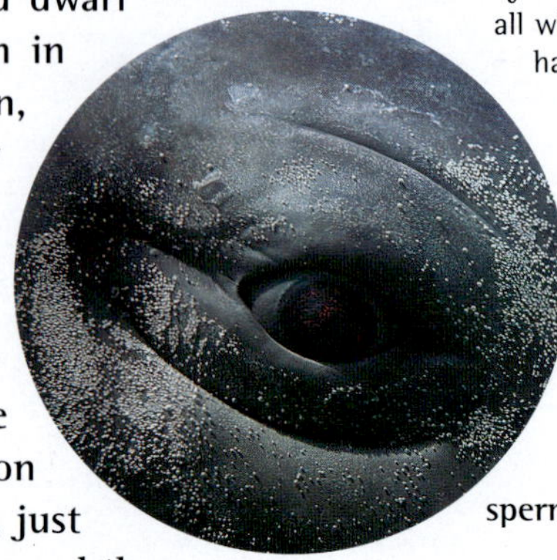

sperm whale

Did you know? Perfume is made from foul-smelling wax made in sperm whales' guts.

Did you know? A sperm whale can dive as deep as 3,000m in search of squid.

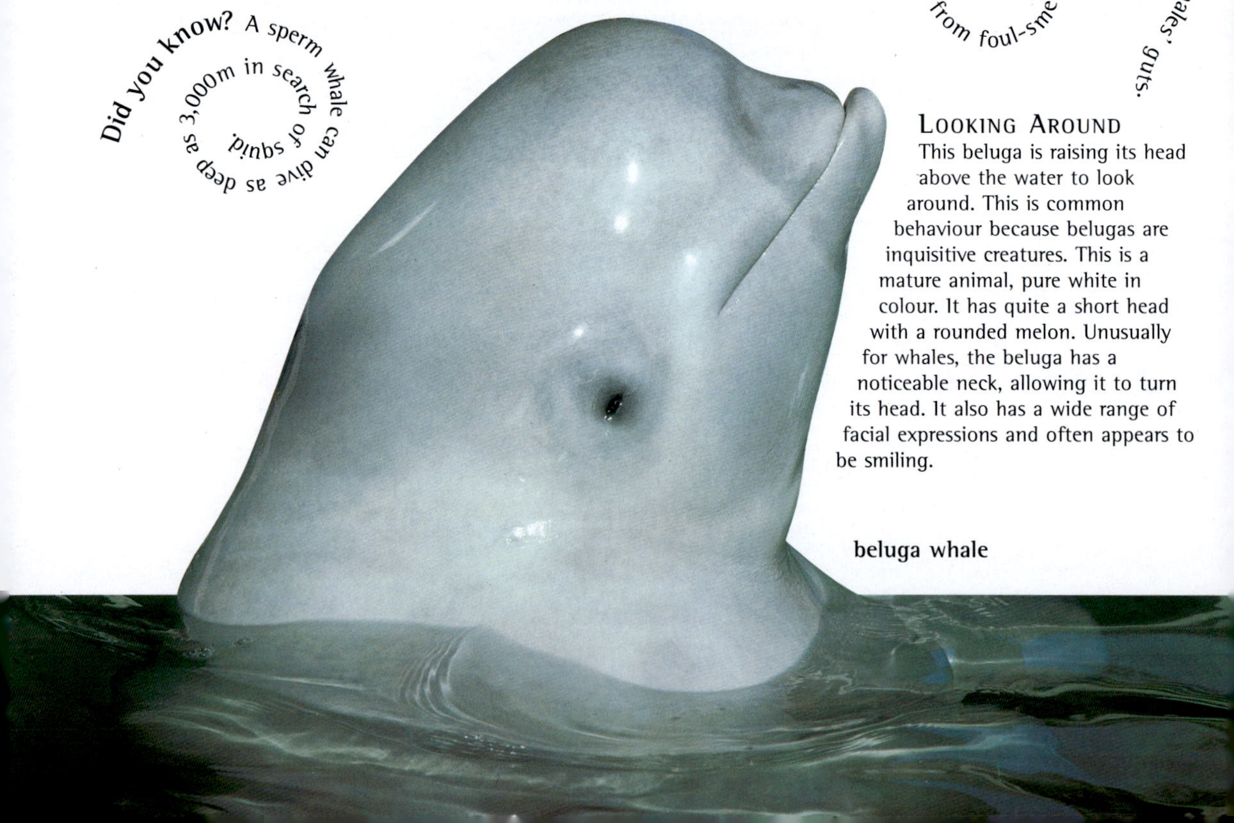

Looking Around

This beluga is raising its head above the water to look around. This is common behaviour because belugas are inquisitive creatures. This is a mature animal, pure white in colour. It has quite a short head with a rounded melon. Unusually for whales, the beluga has a noticeable neck, allowing it to turn its head. It also has a wide range of facial expressions and often appears to be smiling.

beluga whale

COW AND CALF

This sperm whale cow is swimming with her calf. Cows suckle their young for at least two years, staying in a nursery group with other cows and calves. This picture shows the unique body shape of the sperm whale, with its huge blunt snout. The sperm whale does not have a dorsal fin, just a triangular lump on its back.

sperm whales

HIGH SOCIETY

This pod of belugas is swimming in Arctic waters off the coast of Canada. Belugas are usually found together in such pods, because they are very social animals. Note the typical body characteristics, including broad stubby flippers and the lack of a dorsal fin.

beluga whales

LONG IN THE TOOTH

In freezing Arctic waters a male narwhal comes to the surface to blow, its long tusk raised. The tusk has a spiral shape and can be up to 3m long. It is one of the narwhal's two teeth. A small number of males produce twin tusks.

tusk

narwhal

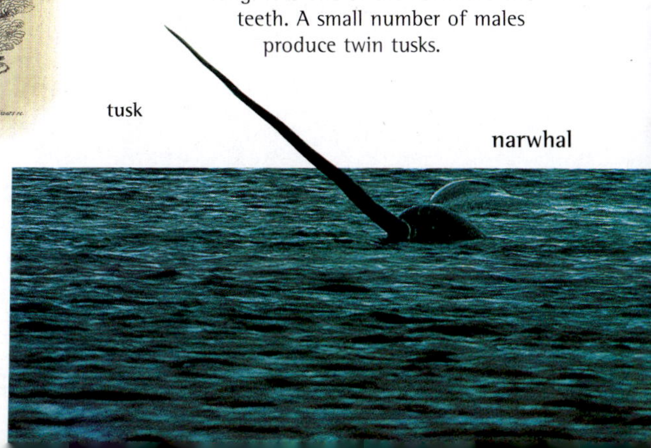

COLOUR, SHAPE AND WEIGHT

The narwhal's stocky body is much like that of the beluga, the main difference being in the colour. Whereas the beluga is white, the narwhal is mostly a mottled dark and light grey. The narwhal and beluga are almost identical in size and body shape, growing up to about 5m long and weighing up to 1,500kg.

Beaked, Pilot and Killer Whales

The beaked whales are named after the beak they have, which is rather like that of many kinds of dolphin. They are much bigger than dolphins and, unlike dolphins, they have hardly any teeth. Most have just two teeth near the front of the lower jaw. Beaked whales are found mainly in the deep ocean, far from land, and little is known about the way they live. Pilot and killer whales, however, are much better known. They are in fact part of the dolphin family and are, like many dolphins, sociable animals that tend to live in quite large groups. Because pilot whales and killer whales are mostly black, they are often called blackfish. The killer whale is the largest and best known of the family and is one of the fiercest predators in the ocean. It can be found worldwide, but prefers cooler waters. Most of the blackfish prefer warmer waters.

A Telling Tail

A killer whale lob-tails by extending its tail into the air and slapping the water with it. A killer whale's tail is black on top but mainly white underneath, with a distinct notch in the middle. Note also the pointed tips of the flukes.

killer whale

Into the Breach

A killer whale leaps high into the air while breaching in Alaskan waters. The whale may twist and turn before it falls back to the surface with a resounding splash. Look at this killer whale's broad paddle-shaped flippers. The size and shape of the flippers and the dorsal fin mark this specimen as a male.

killer whale

Did you know? Killer whales have never been known to attack humans in the wild.

pilot whale

Cruising Pilot

Here is a head-on view of a short-finned pilot whale. The animal has a broad, bulbous head, and is for this reason sometimes called the pothead whale. The flippers are long and sickle-shaped, and the curved dorsal fin rises from a broad base. It prefers tropical and subtropical regions. The long-finned pilot whale is very similar to the short-finned species, but with slightly longer flippers. This species is found mainly in the Southern Hemisphere in cool, as well as warm, waters.

White Lips

A pod of melon-headed whales are swimming closely together. This is typical behaviour for the species. One of the animals is spy-hopping, and shows its distinctive melon-shaped head. Note its white lips.

melon-headed whales

Sleek Performer

The superbly streamlined body of the killer whale is displayed to perfection as it comes out of the water while performing at Sea World in California. The picture also shows its characteristic oval white patch behind the eye, the white chin and the white patch at the side. There is a greyish saddle patch behind the tall dorsal fin.

False Teeth

A false killer whale spy-hops, taking a good look at what is happening around it. It is showing a fine set of teeth. False killer whales have as many as 20 teeth in both the upper and lower jaws. The false killer does not look much like the killer whale and is much smaller. It has no white patches on its body, and its head is more slender.

false killer whale

Lone Swimmer

A beaked whale swims alone. Most beaked whales spend a lot of time by themselves or with just one or two companions. They prefer deep waters, and some species rival the sperm whale when it comes to diving.

beaked whale

Oceanic Dolphins

Dolphins are the most common of all cetaceans. They are swift swimmers that often perform leaps and other acrobatics as they travel close to the surface. They have sleek, streamlined bodies with, usually, a prominent dorsal fin. Their predominant body colour is dark grey on the back and white or pale grey on the belly. Many dolphins have contrasting stripes along the sides. About half the dolphin species have a long beak, and have as many as 250 teeth in their jaws. The other half have short beaks and fewer teeth. Dolphins can be found in most oceans, but they do not usually venture into the cold waters of far northern or far southern regions. Most species are highly sociable, some travelling together in groups of hundreds.

Did you know? There are about 32 species of dolphin.

southern right whale-dolphins

Striking Stripes

The distinctive black and white striped bodies tell us that these two animals are southern right whale-dolphins. The back is jet black, while the beak, forehead, belly and flippers are white. Note that this species has no dorsal fin.

Porpoising Dolphins

This group of common dolphins is porpoising – taking long, low leaps out of the water. These dolphins have a prominent beak and yellow or tan-coloured markings on their sides, leading from the eye. Behind the dorsal fin, the markings become grey. The dark skin on the upper back looks rather like a saddle. This is why it is sometimes called the saddleback dolphin.

Did you know? Dolphins make loud noises when hunting to panic fish into bunching together.

common dolphins

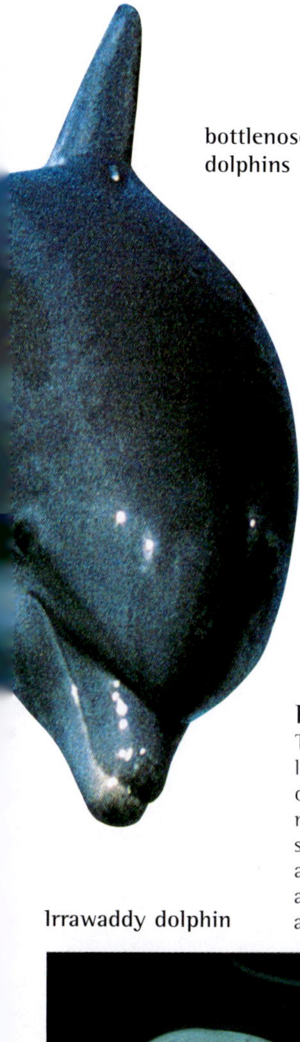

bottlenose dolphins

Great Leapers

These two bottlenose dolphins are launching themselves with great energy several metres into the air. Their bodies are mainly grey in colour, darker on the back and paler on the underside. The head of the bottlenose dolphin is more rounded than that of most other beaked dolphins.

Risso's dolphins

Blunt Heads

A group of Risso's dolphins is easy to recognize by their blunt heads and tall dorsal fins. Their bodies are mainly grey on the back and sides, paler on the underside. The colour becomes paler with age, and some old adults are nearly all white. The older animals are nearly always heavily scarred.

Pale Face

This close-up shows one of the oddest looking of the dolphin species, the Irrawaddy dolphin. It has a rounded head and a distinct neck, rather like the beluga, with which it is sometimes classified. Its flippers are large and curved. It is found in rivers and estuaries, as well as coastal waters from south of India as far as northern Australia.

Irrawaddy dolphin

Dolphin Rescue

An old Greek tale tells of a famed poet and musician named Arion. After a concert tour, sailors on the ship that was taking him home set out to kill him for his money. They granted his request to sing a final song. Then he jumped overboard. He did not drown because a dolphin, attracted by his beautiful song, carried him to the shore.

Porpoises and River Dolphins

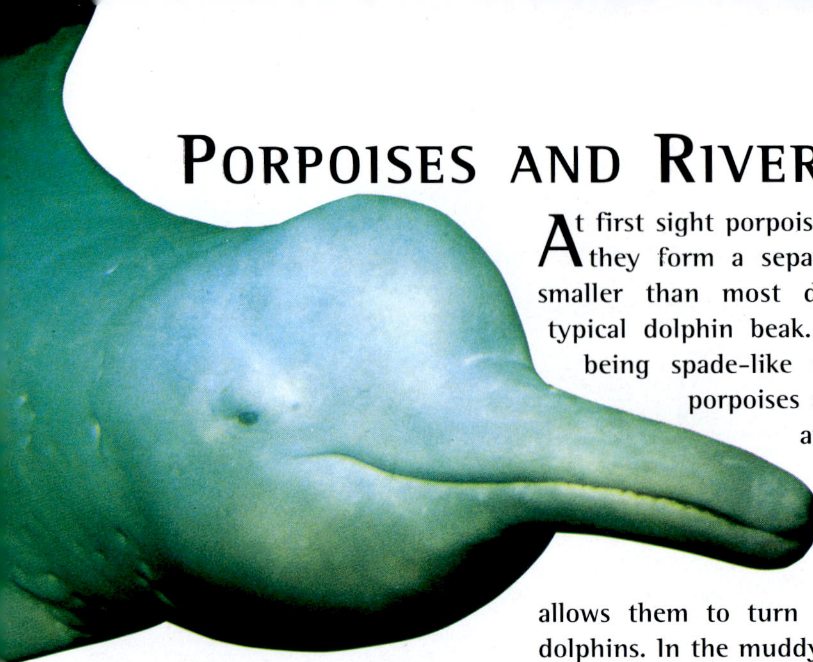

Amazon river dolphin

At first sight porpoises look rather like dolphins, yet they form a separate cetacean family. They are smaller than most dolphins and do not have a typical dolphin beak. Their teeth are different too, being spade-like instead of cone-shaped. Most porpoises are shy creatures that stay well away from boats. The rare river dolphins also form a separate family. They have a long slender beak and a rounded forehead. Their flexible neck allows them to turn their head, unlike the oceanic dolphins. In the muddy waters where they mostly live, they use echo-location rather than their poor eyesight to find the fish, shrimps and other creatures they feed on. All these dolphins are river dwellers, except the La Plata dolphin, which lives in coastal waters near the La Plata estuary in South America.

Beaked Boto

The Amazon river dolphin, or boto, has the typical long beak of the river dolphins. It also has a pronounced melon on its head, which can change shape. Its body colour can vary from pale bluish-grey, as here, to pink. It has no proper dorsal fin, just a fleshy ridge along its back.

Resting Porpoise

A Dall's porpoise rests on the surface of the water, displaying the body features of its species. It has a stocky black body, with a large white patch on the sides and belly. The upper part of its dorsal fin and the trailing edges of its tail flukes are white as well. Unlike most other porpoises, which are quite shy, the Dall's porpoise loves to bow-ride fast boats.

Dall's porpoise

Rare Sneezer

A Yangtze river dolphin, or baiji, thrusts its head out of the water. Like all river dolphins, it has tiny eyes and very poor eyesight. Its blowhole is circular and its blow sounds a bit like a sneeze! This dolphin is one of the rarest of all whales and dolphins, with a population that may be as small as 150 individuals.

Did you know? The harbour porpoise is rarely seen in harbours.

Yangtze river dolphin

Fast and Furious

Dall's porpoises are the most energetic of the porpoises. Their swimming is fast and furious. They kick up fountains of spray as they thrust themselves through the surface of the water. Unlike other porpoises, they often swim with moving boats.

Dall's porpoises

Noisy Snorter

The harbour porpoise is usually seen from a distance, and seldom comes near boats. It is noted for its noisy, snorting blow. The general body colour is dark grey on the back with paler patches on the flanks. Its belly is white, and it has black flippers and lips.

harbour porpoise

Did you know? Dall's porpoises are one of the fastest marine mammals – travelling up to 35 knots.

porpoise

Dolphin or Porpoise?

Porpoises are close relatives of dolphins and may easily be mistaken for them. However, they belong to a different family and have different body features. For example, dolphins have a beak, while porpoises do not. We know relatively little about porpoises as they are very shy creatures. Scientists can take advantage of strandings such as this one to study them.

polar bear

Fellow Travellers

Whales are not the only mammals that are at home in water. Other aquatic mammals include otters and many species of seal. The otter spends most time on land and has the furry coat of a typical land mammal. Seals are much better adapted to life in the water, with a sleek, streamlined body and flippers instead of legs. They have some fur, but it is the thick layer of fatty blubber under the skin that keeps them warm in the water. Blubber is a very good insulator and stops body heat escaping. It also insulates against the cold air, when seals return to land to breed. Other mammals that are at home in the water are the dugong and the manatee. Often called sea cows, these creatures have a bulky seal-like body. They are found in rivers and coastal waters in tropical and subtropical regions.

Bear at Sea

The polar bear spends most of its time drifting on the pack ice in the Arctic Ocean. It often takes to the water to hunt the seals it feeds on. Instead of the whales' layer of blubber, the polar bear has a thick furry coat to give it protection from the Arctic climate. Here, temperatures can fall to -50°C and below.

Did you know? Whales were probably descended from a 4-legged land mammal called a mesonychid.

Californian sea lion

Fin-Footed

The Californian sea lion is one of the most graceful swimmers. It swims using powerful strokes of its front flippers. Its body is much more adapted to the water than an otter's, with its paddle-like flippers. Its body is partly hairy, but is partly smooth and streamlined.

Furry Swimmer

The otter spends most of its time in the water, but is equally at home on land. Its body is that of a land mammal, being four legged and furry. However, it is also adapted for life in the water. The otter's legs are short, and its toes are webbed, which make efficient paddles. Its body is streamlined, and its fur waterproof.

otter

whale shark

Whale-like

This creature is called a whale shark. However, it is not a whale, but the biggest fish of all – a harmless member of the shark family. It lives in warm seas throughout the world. The whale shark is huge, measuring more than 15m long when fully grown and weighing many tonnes. It feeds on plankton, which it takes in through its gaping mouth. Then it sieves out the plankton from the water through a special gill structure.

Did you know? The largest whale shark ever caught weighed 15 tonnes.

dugong

Sea Cow

A dugong swims in the Pacific Ocean, just off Australia. Unlike the seals, which leave the water to breed on land, dugongs spend all their time in the sea. They have no hind limbs, but a tail, similar to that of a whale. The alternative name for the creature – sea cow – is a good one because the animal feeds on sea grasses.

walrus

Whale-horse

With its whiskers and long tusks, the walrus (whose name means whale-horse) is unmistakable. It is a large mammal of the seal family. Like the true seals, it has no external ears and it swims by means of its rear flippers. It feeds mainly on the seabed, using its whiskers to locate buried clams and its armoured snout to grub them out. The walrus excavates clams that are buried deep in the mud by squirting a high pressure jet of water from its mouth into the clam's burrow.

Whale Slaughter

The baleen whales and sperm whale are so big that they have no natural predators, although killer whales may occasionally attack them. Until a few hundred years ago, the oceans teemed with hundreds of thousands of them. Humans had hunted them for centuries, but did not take enough to affect whale populations. Then in the 15th and 16th centuries, whaling grew into a huge industry and whale slaughter began on a large scale. Whales were killed mainly for their blubber, which could be rendered down into oils for candles and lamps. The industry further expanded following the invention of an explosive harpoon gun in the 1860s. By the 1930s, whale factory ships working in the Antarctic were taking up to nearly 50,000 whales a year. In 1988, commercial whaling was banned.

Whale Soap
The sperm whale was once the prime target for whalers because of the large amounts of oil it contained in its massive head. This tablet of soap was made from a waxy substance, called spermaceti, from the whale's forehead.

Deadly Struggle
Whalers row out from a big ship to harpoon a whale in the early 1800s. It was a dangerous occupation in those days because the dying whales could easily smash the small boats to pieces.

Did you know? Whale blubber was made into lipstick and other sorts of make-up.

Fin Whaling
A modern whaler finishes cutting up a fin whale. A few whales are still caught legally for scientific purposes, but their meat ends up on the table in some countries. The fin whale used to be a favourite target for whalers because of its huge size. Among whales, only the blue whale is bigger.

Pilot Massacre

Every year in the Faroe Islands of the North Atlantic, people round up and kill whole pods of pilot whales. It is a traditional practice, which conservationists have been unable to stop. As the slaughter takes place, the sea turns red with the blood of the dying mammals.

Did you know? Early whalers killed their prey by throwing harpoons from rowing boats.

Killer Net

This striped dolphin died when it was caught in a drift net. Because it could not untangle itself, it was unable to rise to the surface to breathe, and so it drowned. Many tens of thousands of dolphins suffer the same fate each year because hundreds of kilometres of nets are cast into the oceans.

striped dolphin

Did you know? In the 1800s baleen was used to make umbrellas.

'Whale Tale'

***Moby Dick** is an adventure story that was written by Herman Melville in 1851. The one-legged Captain Ahab searches for a great white whale (a sperm whale) called Moby Dick. Eventually he harpoons Moby Dick, but he and all but one of his crew lose their lives.*

Whale Conservation

If full-scale whaling had continued to the present day, there is no doubt that many of the great whales would now be extinct or close to extinction. Even today, only a few thousand blue whales, right whales and bowhead whales remain. Because they are slow breeders, it will take a long time for their numbers to recover. Some of the other whales, such as the grey and the humpback, appear to be recovering well. These two whales are favourites among whale-watchers because they are so approachable. Whale-watching has become increasingly popular and has made people more aware of what remarkable creatures whales are and why they must be protected. Dolphins and killer whales can also be seen performing in aquariums, although the benefits of keeping these creatures in captivity are not certain.

Whale Recovery

This chart shows the effect of whaling on three types of whales. By the middle of this century, the blue, bowhead and grey were close to extinction. Then whaling of these species was banned. Now populations have started to recover.

grey whale

Grey Greeting

A grey whale rises to the surface near a tourist boat off the Pacific coast of Mexico. It is winter, and the greys have migrated to these breeding grounds from the far north. Because these animals stay close to the shore, they are easy to reach by boat.

Did you know? The first whale sanctuary was set up in 1945.

Humpback Spectacular

A humpback whale gives a demonstration of breaching to whale-watchers. It has hurled its 30-tonne bulk into the air, belly up, and will soon crash back to the surface in a shower of spray. Out of all the behaviour whale-watchers come to see, this is by far the most spectacular.

Friendly Flipper

When most people think of dolphins, they usually have in mind the bottlenose dolphin, pictured here. One dolphin character, called Flipper, (played by several different dolphins), starred in a series of television programmes and films. These helped to focus attention on how intelligent these mammals are and how vulnerable they are in the modern world.

bottlenose dolphin

Did you know? You can adopt your own whale by contacting your own local whale and dolphin society.

Whale-watching

A boatload of whale-watchers sees the tail flukes of a humpback whale disappear as the animal starts to dive. The boat is cruising off the New England coast of the United States where some populations of humpbacks feed during the summer months.

whale-watching off the New England coast

Performing Killer

A killer whale leaps high out of the water at a dolphinarium, drawing applause from the huge crowd watching. It is doing what comes naturally, but at the command of its trainer instead. In the wild, the killer whale is a deadly predator, but in captivity – with all meals provided – it proves docile and friendly.

killer whale

Did you know? Some countries continue to hunt whales.

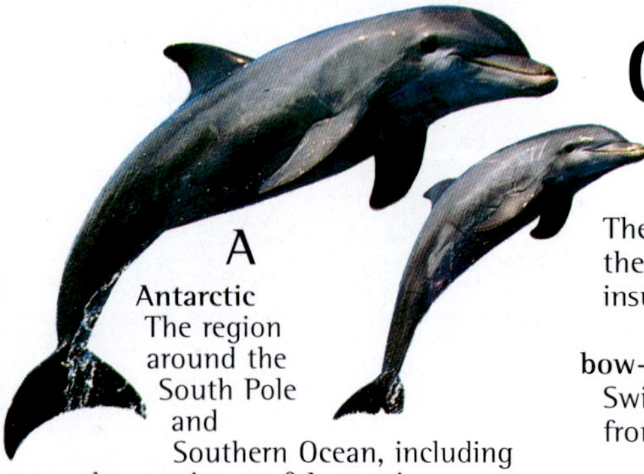

GLOSSARY

A

Antarctic
The region around the South Pole and Southern Ocean, including the continent of Antarctica.

Arctic
The region around the North Pole.

B

baleen
A tough and flexible material, which forms comb-like plates in the upper jaw of baleen whales.

baleen whale
A whale that has baleen plates in its mouth instead of teeth.

beak
The protruding jaws of a whale or dolphin.

blow
The cloud of moist air that is blown from a whale's blowhole when it breathes out.

blowhole
The nostril of a whale. Baleen whales have two blowholes, toothed whales have one.

blubber
The layer of fatty tissue beneath the skin of a whale that acts as insulation against cold water.

bow-riding
Swimming on the bow wave in front of a moving boat.

breaching
Leaping out of the water and falling back with a great splash.

bull
A male whale.

C

calf
A baby whale.

cetacean
A whale, dolphin or porpoise, all of which belong to the animal order Cetacea.

cow
A female whale.

crustacean
A creature with a hard body that lives in the sea. Shrimp and krill are crustaceans.

D

dolphin
A small, toothed whale that has cone-shaped teeth.

dorsal fin
The usually triangular fin on the back of a whale's body.

E

echo-location
The method toothed whales use to find their prey. They send out pulses of high-pitched sounds and listen for the echoes produced when the pulses are reflected by objects in their path.

extinction
When a species of living thing dies out.

F

flipper
A whale's paddle-like forelimbs.

flukes
The tail of a whale.

fluking
Raising the flukes into the air before diving.

K

krill
Tiny crustaceans that are the main food for many of the baleen whales.

L

lob-tailing
Raising the tail into the air and then slapping it down on the surface of the water.

M

mammal
An animal that has warm blood and breathes air. Female mammals feed their offspring on milk from their mammary glands.

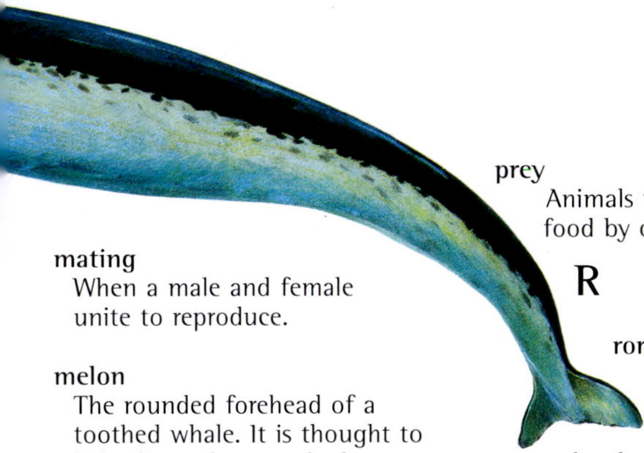

mating
When a male and female unite to reproduce.

melon
The rounded forehead of a toothed whale. It is thought to help direct the sounds the animal uses for echo-location.

migration
The regular journey taken by some animals from one region to another and back at different times of the year.

P

pectoral fin
An alternative name for flipper.

plankton
Tiny sea creatures and plants. They form the basic foodstuff for all life in the oceans.

pod
A group of whales.

polar region
The area around the North or South Pole, where it is very cold.

porpoise
A small toothed whale with spade-shaped teeth.

porpoising
Leaping in and out of the water while swimming fast.

predator
An animal that hunts other animals (prey) for its food.

pregnant
When a female animal has a baby developing in her womb.

prey
Animals that are hunted for food by others (predators).

R

rorqual
A baleen whale with grooves in its throat. The grooves allow the throat to expand when the animal is taking in water when it is feeding.

S

school
Another name for a group of whales.

species
A particular kind of animal. All animals of the same species look alike and can reproduce with one another.

splashguard
A raised area in front of the blowholes of some whales. It helps prevent water entering the blowholes when the whales breathe.

spout
Spout is another word for blow.

spy-hopping
Poking the head out of the water so that the eyes are above the surface.

stranding
Coming out of the water on to the shore and becoming stuck, or stranded.

streamlined
Shaped to slip through the water easily without much resistance.

T

tail fin
Another name for a whale's flukes.

temperate
A climate in which the weather is not too hot and not too cold.

toothed whale
A whale that has teeth and not baleen plates in its mouth. Toothed whales include sperm whales, dolphins and porpoises.

tropical
The climate in the Tropics, the region on either side of the Equator, where the seas are always warm.

W

whale
A cetacean. Commonly the term is applied to the large whales, such as the baleen and sperm whales.

whalebone
A popular name for baleen, but baleen is not bone.

whaling
Hunting whales for their meat and blubber.

INDEX

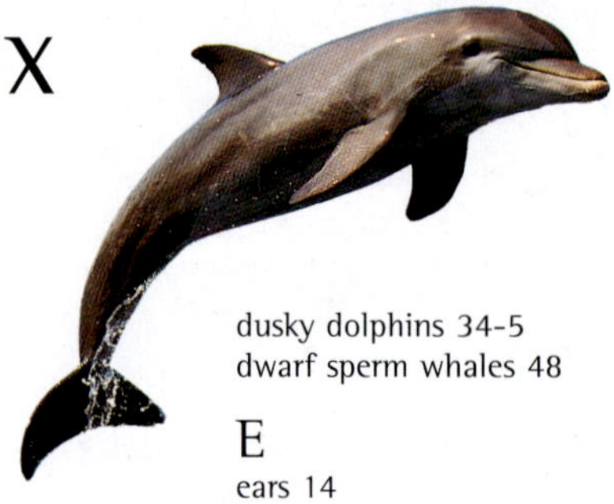

A
Amazon river dolphins 16, 29, 38, 54
aquariums 60
Atlantic spotted dolphins 12,14, 25, 29, 33-4
Atlantic white-sided dolphins 42

B
baleen plates 6, 8, 44-5
baleen whales 4, 6-7, 9-10, 16, 18, 22-4, 28, 30, 58, 60
barnacles 11, 44-5
basilosaurus 4
beaked whales 6, 43, 50-1
beaks 52, 55
beluga whales 6, 16, 28, 30, 32, 39, 43, 48-9
birth 31-2, 40-1
blowholes 5, 12-13, 26, 44
blubber 12, 39, 56, 58
blue whales 4-5, 7, 13, 22, 32, 46-7, 60
bones 8-9
bottlenose dolphins 5, 10-11, 13, 15, 25, 29, 32, 34, 39, 53, 61
bowhead whales 6, 44, 60
brain 14-15
breaching 12, 36, 50, 60
breathing 5, 12, 26
breeding 17, 30-3, 40-1
Bryde's whales 46-7
bubble netting 22-3

C
calves 29, 31-3, 40, 48-9
Cetus 5
common dolphins 5, 52
conservation 60-1
corsets 8
courtship 30-1
Cupid 15

D
Dall's porpoises 54-5
displays 36-7
diving 13, 26-7
dugongs 56-7
dusky dolphins 34-5
dwarf sperm whales 48

E
ears 14
echo-location 16, 18-19, 48, 54
eyes 14, 48, 54

F
false killer whales 9, 51
feeding 13, 18-23, 26, 32, 40, 44
fin whales 4, 7, 22, 43, 58
fins 10, 24-5
fish 18, 22-3, 25, 32, 36, 54
Flipper 61
flippers 8, 24, 30, 36, 46
fluking 27, 37, 47

G
Ganges river dolphins 19
grey whales 6, 10-11, 18, 24, 30, 40-1, 44-5, 60

H
harbour porpoises 55
harpoons 58-9
hearing 14-15
humpback whales 4, 11, 13, 15-17, 22-4, 30, 33, 36-8, 40-1, 46-7, 60-1
hunting 16, 20-1, 28

I
Irrawaddy dolphins 53

K
killer whales 7, 9, 19-21, 24, 28, 35, 38, 50-1, 58, 60-1
krill 18, 22, 40

L
La Plata dolphins 54
lice 11, 44-5
lob-tailing 37, 44, 50
long-finned pilot whales 29, 42, 51
long-snouted spinner dolphins 35
lunge-feeding 22-3

M
mammals 12, 56
manatees 56
mating 30-1, 40-1
melon 16, 48
melon-headed whales 11, 39, 51
migration 38-41, 44
milk 32
minke whales 22, 46
Moby Dick 59
muscles 8-9, 11, 24-5

N
narwhals 30-1, 48-9
navigation 42
northern right whales 45
nursery schools 32

O
oil 26, 44, 58
orcas *see* killer whales
organs 10, 43
otters 56

P
Pacific white-sided dolphins 35
pilot whales 29, 42, 50-1, 59
plankton 18, 22, 32, 40, 57
play 34-5
pods 28, 49
polar bears 56
porpoises 4, 6, 54-5
porpoising 35, 52
pygmy sperm whales 48

R
right whales 6, 9, 40, 44-5, 60
Risso's dolphins 7, 53
river dolphins 54
rorquals 7, 22, 40, 46-7

S
saddleback dolphins 52
sailing 34, 41
sea lions 20-1, 56
seals 56
sei whales 7, 22, 47
senses 14-15
short-finned pilot whales 14, 51
sizes 7
skeleton 8-9
skin 10-12, 24-5
social life 28-9
songs 17, 30, 46, 53
sounds 16-17, 30, 44, 52
southern right whale-dolphins 52
southern right whales 12, 18, 31, 34, 41, 44-5
space probes 17
sperm whales 6-7, 13-14, 17-19, 26-8, 31-2, 42, 48-9, 58
spermaceti organ 48, 58
spinner dolphins 34-5
spy-hopping 28, 36, 51
squid 17-20, 26-7
Stejneger's beaked whale 43
stranding 29, 42-3
striped dolphins 59
swimming 24-5, 46

T
tails 8-9, 15, 24-7, 30, 34, 36-7, 41, 44, 47, 50
tailwalking 25
teeth 6, 9, 19, 43, 48, 50-2, 54
temperature 12
toothed whales 5-7, 9, 16, 18
tusks 30, 48-9, 57

W
walruses 57
wax 48, 58
whale sharks 57
whale-watching 60-1
whaling 58-60
white whales 6, 16, 48-9

Y
Yangtze river dolphins 54

Peter C. Harper

Photographing Nature

A PRACTICAL GUIDE

ROBERT HALE • LONDON

I
Cameras and Equipment

Camera equipment is fearfully expensive — prices inspire bank robbery — so it's best to do it right the first time. Here's a list to help you.

The camera body

Choose a 35-mm single lens reflex camera from a reputable manufacturer. These cameras allow you to see clearly what you are going to take before you take the picture, as well as reducing a host of problems with exposure control when using different lenses and attachments. The big five are Canon, Minolta, Nikon, Olympus, and Pentax. These cameras are well made and reliable and come with a complete range of lenses and accessories. Service agents and parts are readily available should these cameras break down or be damaged. I do not recommend fully automatic cameras (see **Auto focus cameras** later in this chapter) as these, unless they have a manual override, have more control over you than you have over them. You must be able to operate your camera manually when the need arises. Larger format cameras do not necessarily produce better pictures than good 35-mm cameras, and they generally weigh enough to break your back as well as your bank balance. Moreover, they do not have the range of accessories available for most 35-mm cameras.

If you are unfamiliar with photographic equipment, take a friend who knows about cameras with you and explain to them the sort of camera you want so that they can advise you before you enter the shop. Bear in mind that whatever you decide on, the choice must ultimately be yours. Don't nail your friend to the wall if you decide later that you've made the wrong choice. Avoid 'super buys' of unfamiliar makes, and insist on some form of written guarantee with whatever you purchase.

Important features your camera should have include the following:

Aperture control priority: This allows you to set the aperture and the camera selects the appropriate shutter speed. Aperture control priority enables you to control what is to be rendered sharp on the film. This is important for subjects such as landscapes and close-ups.

Shutter priority: Camera shake is the bane of many beginners. With shutter priority you can avoid it easily by accurately determining the shutter speed, and allowing the lightmeter to select the appropriate aperture setting for you.

Many newer types of camera have shutters which range steplessly from their slowest to fastest speed (i.e., if an exposure of exactly 1/139 sec. is required, that is what the shutter uses), and it is not always easy to determine the true speed of the shutter during exposure (for example, with aperture control priority cameras). Shutter priority eliminates the guesswork.

Moreover, by selecting a rather slow shutter speed, such as 1/30 or 1/60 sec., you can follow moving objects and create a blurred background as you do so. This is called *panning.*

With nature photography there will also be times when you want to set the shutter at 1/2000 or 1/4000 sec. for 'stopping subjects' in motion, or to minimise camera shake when using big telephoto lenses.

Manual control: You must have the ability to switch everything off and be totally creative when you want to. Manual control may not appear important to the beginner initially, but as you become more confident in your photography there will be many times when you will want to do it your way, and not the way the camera tells you to.

Note: Manufacturers of newer cameras offer the four above alternatives in one camera body. The nature photographer should be grateful for this, because it offers both ease and flexibility in approaching a subject. Sometimes

you will need to take a 'grab' shot automatically, when time is of the essence; at other times, with something like a landscape, you can turn off the camera automatics and do everything manually.

Depth-of-field preview button: This allows you to see what you have in focus in front of and behind your subject before you trip the shutter. It is inadvisable to buy a camera without one.

Through-the-lens (TTL) metering: Most modern cameras now have this facility — the camera measures the light coming through the lens.

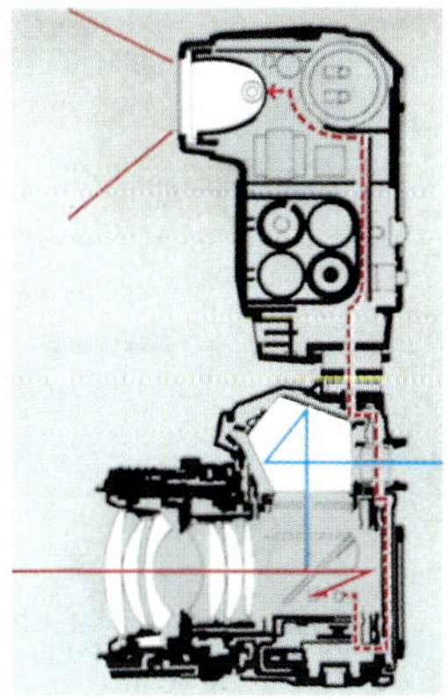

Through-the-lens flash (TTL flash). It is a joy to use a flash matched to your camera. The camera is able to monitor the amount of light falling on the film by electronically turning off the light generated by the flash when the film is correctly exposed, so there are no more guide numbers or complicated sums to confuse you. TTL flash means everyone has the opportunity of taking successful flash pictures. A word of caution, however — do not use your flash on the camera. Buy a TTL flash cable that permits you to place the flash off the camera. This allows more creativity, and better pictures, but more about flash techniques later.

Exposure (value) or compensation dial (EVD). This innocent-looking dial baffles many camera buffs. The EVD allows you to increase or decrease your exposure value by up to 2 exposure values on either side of the correct exposure as determined by your lightmeter. You are, in effect, overriding your lightmeter to achieve a different exposure from the one it thinks you ought to have. The EVD is normally calibrated in increments of 1/3 EV so that small adjustments to the exposure can be easily made when required. In this way you can take a picture using the normal exposure, with the EVD set at zero, and then *bracket* that exposure by turning the EVD to -1 which will *reduce* the amount of light falling on your film by one stop. This will result in *underexposure* (the picture darkens). By turning the EVD to +1 or +2, we are *increasing* the amount of light falling on the film. This means *overexposure* (the picture fades with increasing light). Most modern cameras will provide a visual warning inside the viewfinder when you are using the EVD. This is to prevent you forgetting to return it to zero after you have finished with it.
Note: You can, of course, reproduce the effects of the exposure compensation dial by altering the film speed dial, or by ignoring your lightmeter while you alter the shutter speed or f/stop. I prefer to avoid tampering with the film speed dial, however, because it is too easy to leave it set at the wrong place.

Delayed action release: This is useful for avoiding camera shake without the use of a cable release.

Exposure lock button: This allows you to expose correctly a subject which might not necessarily be in the centre of the picture. If your subject is off centre in the photograph, you can take a lightmeter reading from it, lock that exposure with the exposure lock button, and then recompose the picture the way you want it. The result will be a correctly exposed subject.

Tripod bush in the centre of the camera base: Cameras with the tripod bush near the end of the base plate can swing on a tripod and clout you in the eye when you least expect it. If your camera is so constructed, you can purchase an adaptor plate which will centralise the bush.

Facility to change the focusing screen: Most quality cameras will allow you to remove the standard split-image focusing screen and to replace it with a clear matt screen. Many split-image focusing screens are annoying

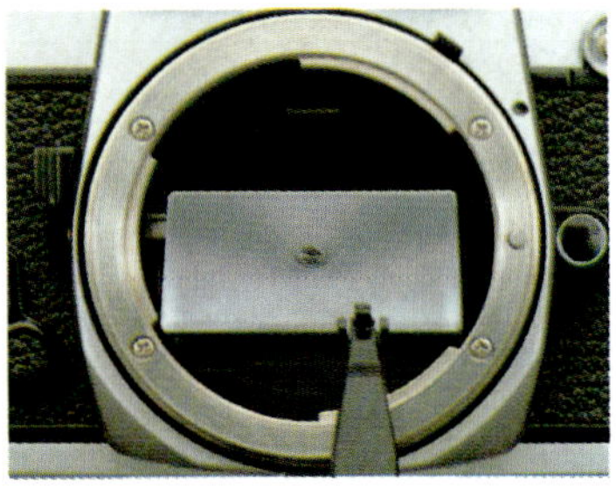

to use with slow lenses and telephoto lenses which limit the amount of light entering the viewfinder. The result is that the split image will sometimes partially dim or darken completely, making focusing difficult. People wearing glasses will find split-image focusing screens soul destroying.

Eyepiece correction lenses: If you wear glasses you may find focusing difficult at the best of times. Camera manufacturers make correction eyepieces which replace the ones fitted to your camera. These should be readily available at your camera stockist.

On a personal note, watch where you put your glasses. I have on three occasions set up my camera, removed my glasses, taken a breathtaking photograph, and congratulated myself by sitting or standing on the glasses. It's sudden death to your feeling of well-being. I can still function optically without glasses, but you might not be able to.

Note: Once you have bought your new camera, never let the salesperson put a film in it. Listen to what they have to say and ask as many questions

as you need, then take the camera home and learn everything about it *before* you put in the first film. Above all, read the instruction book very carefully — nothing is more important than this.

Modern cameras are clever and can do things that will surprise and delight you. Only when you know what everything is and how it works, should you load your first film. Your instruction book shows you clearly how to do this. Whatever you do, make sure both sides of the film are properly engaged in the wind-on sprockets and that the rewind knob is turning each time you advance the film. These are simple but vital checks. People who try to get more pictures on their film by closing the back without the film properly loaded are, in my opinion, plain stupid. You will not find professional photographers taking such risks.

Auto focus cameras

These hi-tech cameras do everything except make tea. They automatically load the film to the first exposure, wind it on, focus, calculate the correct exposure — which they can automatically bracket if necessary — and finally, when the film is done, they automatically wind it back into its cannister.

The potential for these cameras is unquestionably very exciting, and it is clear that the big five manufacturers will be at each other's throats trying to outdo one another in this brave new world of computer-driven cameras (compare the Minolta 9000 with the Canon EOS 620, and Nikon's 801, for example). I expect major advances in the field of auto focus cameras to escalate in the forseeable future.

So what does this mean for the nature photographer? Good news, I think. Provided that an auto focus camera has the important features already outlined above for conventional camera bodies: a depth-of-field preview button, aperture and shutter priorities, manual control override, provision for a cable release and a centrally placed tripod bush, etc., you might well consider one of them. I suspect they will need a bit more care than other cameras — I would not recommend their use in Antarctica, for example, and you would need to be especially careful in corrosive environments such as near the sea. The need to carry plenty of spare batteries at all times would be mandatory — battery failure would be disastrous.

Remember also that with auto focus cameras you will need to have compatible lenses as well. All in all, it represents a major commitment of cash and expectation, so study the pool very carefully before you plunge in. I have recently added a Nikon 801 to my photographic gear in the firm belief that I should fully exploit the auto focus camera.

Lenses

It has been said that the first step to creative photography is the discerning choice of lenses. Most retailers sell their cameras with standard 50-mm lenses fitted to them. This results in the purchaser buying the wrong equipment to start with because you should *not* buy a standard 50-mm lens with your camera. It is the last thing you need. Choose a 50-mm or 55-mm close-up (referred to as either a macro or micro) lens. These lenses will do the same job as the standard lens and have greater resolving power: they will, at the same time, also enable you to explore the fascinating world of the very small. Macro lenses contain their elements in groups that can shift position in relation to each other. This means that they alter their optical configuration as you focus from infinity to very close. They remain sharp over their entire focusing distance, something a standard fixed-element lens cannot do nearly as well.

But what exactly is a good lens? Essentially it is one that produces sharp

images, has good contrast, and faithfully renders true colours — all with a minimum of image distortion. We can define a lens's sharpness as its ability both to produce fine detail (its *resolving* power) and to reproduce subtle differences in shade where they meet along a common border (its *contrast*). Poor contrast comes from light seepage into the dark areas on the film which will reduce the overall image sharpness. The ability of a lens to reproduce colours faithfully (its *colour balance*) probably causes a few grey hairs for the designers, and also for their computers. The lens manufacturer must take into account how the lens handles, how heavy it is, whether it is compact, durable, and of a practical design. He must also make sure a customer can afford to buy it.

It follows from all this that you should buy your lenses with great care. Try to avoid mixing and matching lenses from several manufacturers, because probably they will *not* match; the colour balance may vary quite considerably from one to another.

The next lens to choose depends on what sort of photography you are interested in. Assuming that you haven't any idea to begin with, I suggest you start with a 100 or 105-mm tele lens. Good examples are the Canon FD f 2.8/100 mm or the f 2.5/105 Nikkor. Both are sharp, versatile lenses. If you can afford it, you might consider a macro lens in the 90 to 105 mm range of focal lengths. With these you can photograph small things at life size, and because they allow more working distance between you and your subject, there are no lighting hassles or possibilities of scaring the subject to death by being too close.

Your third lens should be a wide-angle, because these lenses offer you many opportunities to take vistas of landscape as well as interesting close-up pictures. The f 2.8/24 mm lens (picture angle 84 degrees) is an excellent choice because it is light in weight and gives better quality pictures at wider apertures than most 28-mm lenses that I have tested. The 35-mm wide-angle (62 degrees) is frequently used as a standard lens by many photographers, because it has a greater depth of field than the 50-mm lenses, making it ideal for grab shots. Further, its picture angle matches the coverage of most electronic flash units.

In time, your complete lens kit might look something like this:

f 2.8/24 mm
f 2.8/35 mm or f 1.4/35 mm (expensive, but fast and superbly sharp)
f 2.8/55-mm macro
f 2.5 to f4, 90 to 105-mm macro
f 2.8/180-mm ED or f 4/200-mm tele
f 4.5/300 IF/ED tele

These lenses, or their equivalents, are made by the major camera and lens manufacturers.

I have not found a need for any lens larger than the 300 mm. Telephoto lenses with internal focusing (IF) and low-dispersion glass elements in their construction (e.g. Nikon and Pentax ED lenses and Canon L lenses) are easy to use, as well as lighter and closer-focusing than those with conventional helical focusing. They also represent a major advance in optical quality in that the calcium fluorite lens elements keep red and blue light in better focus, resulting in excellent colour correction so that you get sharper, brighter pictures.

I have tested many lenses from many manufacturers over the years and, in my opinion, the above selection should satisfy most photographers who want top-quality lenses to cover a reasonably wide spectrum of focal lengths. You can confidently photograph most things with them.

The list above cannot cater for every situation, however. If you are about to begin your photographic career with a strong interest in birds, then you should aim to obtain a telephoto lens as soon as you can. You might follow the already mentioned macro lens with an f 2.8/180 mm or f 4/200 mm, and then buy a 300-mm or a 600-mm telephoto for photographing species such as wading birds which feed in flat open places where concealment for the photographer can be difficult. Under these conditions you will need a photographic hide.

Zoom lenses: Zoom lenses are convenient but I have been wary of them. My bias stemmed from repairing the earlier models which were simply not up to the quality a nature photographer requires. They lacked contrast, colour fidelity and sharpness, and were very prone to *flare*, a term given to unwanted reflections off the many glass elements (up to 17 or so) which make up the average zoom lens. These and other mechanical failings made them lenses to avoid.

Times have changed, however. Low dispersion glass and computer designed optics and handling have combined to produce some zoom lenses which are satisfactory for most needs of the nature photographer. However, if quality is an essential prerequisite, as it is for me, only the top-of-the-line zoom lenses are worth a second look. These include the Tokina AT-X series of lenses, some of the top Tamron ones and, of course, those produced by the big five. I haven't tested other independent makes of zoom lenses, but those mentioned above are certainly as good as and often better than their equivalents made by the big five. Check prices carefully. To get you started, consider this mix of zoom and fixed focal-length lenses:

f 2.8/24–40-mm zoom	(Tokina AT-X240 or its equivalent)
f 2.8/50 or 55-mm macro	(made by the big five)
f 2.8/80–200-mm zoom	(Tokina AT-X828; Tamron Mod. 30A)
f 2.5/90-mm macro	(Tokina AT-X M90; Tamron Mod. 52B)
f 4.5/300-mm IF/ED	(made by the big five)

One last plea: if you want to do close-up photography use macro lenses and avoid zoom lenses with macro facilities — zoom lenses cannot do everything (see Close-up Photography).

Mirror lenses: Avoid mirror lenses because they have a rather large, fixed aperture and produce optical effects which you might find bizarre. Although mirror lenses can produce acceptably sharp small objects at a distance, they have difficulty coping with large ones. The result, as I found to my cost, was a sea elephant portrayed as a fat-jowled cigar, blurred beyond recognition, and adorned with strange doughnut-shaped rings all over it, except for a small zone of sharpness on its nose where I had focused the lens.

Teleconverters/extenders: These are small auxiliary lenses composed of up to seven elements, which are attached between a telephoto lens and the camera body. Essentially they increase the focal length of the master lens, either doubling it, or increasing it by 1.4x. If, for example, you attach a 2x teleconverter to a 300-mm lens it is converted into a 600-mm lens. The 1.4x converter will change the same lens to one with a focal length of 420 mm.

There is a catch, however. The 2x converter reduces the maximum aperture of any lens it is attached to by two f/stops, whereas the 1.4x reduces the maximum aperture by only one stop.

Teleconverters should be attached only to your best fixed focal length telephoto lenses and not to zooms. Used in this way they become handy devices to give your lenses more flexibility when you need it. My preference is for the 1.4x teleconverter, because you lose only one stop while still preserving a good quality image.

Note: I would stress that beginners should start with the shorter focal-length lenses, which are easy to handle and are forgiving when the inevitable

ice-cream. It is best to remove them for most photography, especially close-ups with flash, where flare off their flat surfaces might ruin a good shot.

Polarising filter: Invaluable for bright sunny days, polarising filters heighten colour, reduce reflections from non-metallic surfaces, and produce dramatic skies. Excellent for landscapes. If your wide-angle lens takes a different-sized filter to your other lenses, fit a polarising filter to it first, because you will probably want to start using such a filter for landscapes before trying its effects with other lenses. There are two types: *circular* for cameras that meter off the film plane and *linear* for all other types of cameras.

An amber filter: On heavily overcast days a bluish cast will sometimes appear in colour transparencies. Amber filters, such as the A2, will correct this by compensating for the blueness. The results definitely warrant slipping one of these filters into your camera bag.

Soft-focus filters: These are special filters designed to produce slight to severe flare (but not blurring) in your photographs. They are extremely effective for portraiture and close-ups of flowers, and when you wish to create landscapes with foggy atmospheres. There are two types: No. 1 which has mild effects, and No. 2, which generates a very 'foggy' picture image.

If you really become keen on soft-focus effects, it is useful to know that some camera manufacturers produce special lenses with up to three adjustments for soft-focus. This sounds interesting, doesn't it?

Finally, *never* screw more than one filter at once on to a lens. Dark, shadowy corners (vignetting) may appear on your transparencies, particularly with wide-angle lenses. You will also greatly increase the chance of flare with two filters on your lens.

Tripods

How sharp your image is depends on the lens's resolution, the resolution of the film you are using, and the stability of your camera. Loss of sharpness is invariably due to camera movement during exposure — you can frustrate the best of intentions of camera and film manufacturers with an unstable camera position. It follows that a good tripod is as important as the film in your camera. Both are essential.

Buy a tripod that collapses completely to the ground, such as the Velbon VEF-3 or equivalent. These tripods have independently adjustable legs, which allow you to set them up on uneven surfaces. Watching someone trying to hang their camera upside down from the centre pole positioned between the legs of a conventional tripod is great fun. They look like a feeding flamingo.

The 'pan and tilt' heads available for tripods are easier to handle than the small ball and socket types. They also offer more support for your camera, particularly with larger lenses. Make sure that the head has large, easy-to-use knobs that can readily be tightened and released. A quick-release shoe that comes away with the camera can be handy for photographers in a hurry.

Some tripod manufacturers make a leather shoulder harness enabling you to carry your tripod more comfortably. You could easily make such a harness.

If hand-holding your camera, use the focal length of your lens as a guide to minimum shutter speed (i.e. 55 mm=1/55 sec., 105 mm=1/105 sec., and so on). It is an old and useful rule of thumb. I generally avoid shooting anything slower than 1/30 sec. without using a tripod. Some people show off by announcing that they can hold their cameras rock steady at a quarter of a second. Just say, 'how interesting', and set up your tripod.

If you are stalking birds and mammals, you might find a shoulder stock or rifle grip convenient for stabilising telephoto lenses of 300 mm and longer. A shoulder stock is a device that functions like a shotgun stock. They aren't easy to buy off the shelf, so if you want one you might have to make it. You will need a long cable release to attach the camera's shutter release to your trigger handle.

Cable releases

Buy a relatively short one of about 30 cm in length. Many are black in colour and a dream to lose, so attach a piece of fluorescent orange tape to it. Cable releases any longer than about 30 cm can suddenly flip in front of the lens at the precise moment you are taking a long exposure

of the last tiny orchid the cows have left you in the field. Other versions of this frustrating occurrence include the 'camera strap blues', and the 'strange case of the waving hand' which happens when you have shaded your lens with your hand and assumed your hand is not included in the picture. All photographers will soon become acquainted with the famous Murphy's Law: 'If anything can go wrong, it will'.

Film

Books have been written on the type of film you should use. Because I have not used negative film in recent years, I must confine my brief remarks to that which I use a lot of, colour slide or transparency film.

I always carry Kodachrome 25ASA and Fujichrome 50ASA Professional in my camera bag and rarely anything else. I have used Kodachrome for

2
Designing a Photograph

Previous pages:
DESIGN IN FOAM
Fujichrome 50, 90-mm lens, 1/15 sec. at f 11.
An image was created by a foamy stream of peaty water eddying into a large pool. It produced an interesting natural pattern to which I have added a rock and moss to provide a focal point for the picture.

Once you have decided what to photograph, the next question is how to photograph it. This is where your skill at designing a photograph comes in. The art of seeing things in a balanced and co-ordinated way is called *composition.*

THE RAPE FIELD
Fujichrome 50, 35-mm lens, 1/30 sec. at f 11.
A simple example of composition and colour. Notice how the eye is drawn across the field of yellow flowers to the right, before being directed left over the undulating land to the dark pine trees. The trees act as a visual full stop to reading the photograph.

LAND AND SKY
Kodachrome 25, 35-mm lens, 1/30 sec. at f 8. Pola filter.
Keep it simple. *This photograph uses the contrast between the warm colour of land and cool blue of the sky. A polarising filter has heightened the effect of the clouds. Note how the frame is divided into horizontal thirds with two-thirds sky, one-third land.*

Opposite:
THE GOD CLOUD
Kodachrome 25, 105-mm lens, 1/500 sec. at f 8. Pola filter.
It is late summer and the air is alive with thunder. This huge cumulo-nimbus cloud towers over rocky outcrops, totally dominating the darkened landscape below. No photographer could afford to miss it.

The subject

To be successful, a photograph must look and feel right. Selecting a subject to photograph is the first step. This isn't always easy because potential pictures are everywhere. The sky, with towering clouds, can create powerful images, or the ground might be alive with tiny sundews or flowers, all waiting to be discovered.

I remember, for example, being drawn to a knot of people standing on a small bridge within the Yuys Hir Nature Reserve, Wales, in May 1986. Their eyes were glued to their binoculars as they gazed intently into the trees, obviously bird-watching. Being interested in birds, I quietly asked where the spotted flycatcher was hiding and bent down to fetch my

RUSTIC ABSTRACT
Kodachrome 25, 105-mm macro lens, 1/30 sec. at f 5.6.
This unusual shot of peeling paint on a corroding steel bridge was taken in Wales. Notice how the rust stains create a three dimensional effect to the picture.

binoculars from my pack. My attention was immediately rivetted to the bridge itself, where the black paint was rusting away to reveal a marvellous texture of orange on black. I'm sure the spotted flycatcher watchers thought I was beyond help, but I had found my subject.

Selecting a subject requires you to look at a scene and ask, 'Why do I like this scene? What is so eye-catching about it?' Be highly specific when answering these important questions. Try to isolate those special qualities of the scene that attract you. It may be the colours, or a specific colour in relation to a neighbouring one; it may be a group of trees, or just one particular tree that attracts your attention. Whatever it is, discover it. Once you have isolated a subject for your picture, the next step is what to include along with it.

Designing the frame

To become a good photographer, you must know what you are doing when you place a camera to your eye. You are designing something on a piece of film. The viewfinder is framing that image: everything between its four walls will appear in the picture. So consider carefully the subject of your picture, and the background you are including along with it. Don't include too much background. In this way your subject receives the priority it deserves. Carefully scan the edges of your picture before you release the shutter.

Tips for beginners

A good idea for beginners is to punch out the unwanted film from a rejected slide and use the mount as a simple viewfinder with which to frame up potential pictures. Movie directors are wedded to their viewfinders for exactly the same reason. Prune out everything that might be ambiguous in the final picture. If this means that you have to climb over the untidy fence that was going to appear as a messy foreground, then do it. *Above all, keep the message simple.* Any photograph that requires an explanation is a failure. No amount of words will save it.

If you are just starting out in photography, it's a good idea to find a subject about as big as you are — something you can walk around easily and use a 36-exposure roll of film to explore it with. Photograph it from every angle. Avoid placing your subject dead centre in the frame unless

Overleaf:
SCOTTISH COUNTRY LANE
Ektachrome 100, 105-mm lens, 1/30 sec. at f 2.8.
One of my favourites. Taken on a windy spring day in the Scottish Borders. The line of stately trees stood like protective matrons guarding the road. Although a brisk wind prevented me from getting everything as sharp as I would have liked, the scene just had to be photographed. This picture was taken with the lens at full aperture and although the light was poor I was well pleased with the result.

Opposite:
RATA FLOWERS IN A MATAI TREE
Fujichrome 50, 180-mm macro lens, 1/125 sec. at f 8.
The colourful Rata flowers high in this Matai tree first attracted my attention. A 180-mm telephoto lens enabled me to use the moss-festooned branches to design an unusual backdrop for the flowers.

REMARKABLE MORNINGS
Kodachrome 25, 180-mm lens, 1/60 sec. at f 5.6.
The jagged Remarkable Mountains, overlooking the resort town of Queenstown, figure in most pictures of this scenic area of southern New Zealand. Here I was interested in the strong compositional lines of the mountain slopes which fall steeply to the shores of Lake Wakatipu. The winding road and the morning haze add depth to the picture.

you specifically want to emphasise symmetry; to place it slightly off centre may greatly improve the balance of the photograph. The old rule of dividing your frame into thirds and placing your subject on the intersection of these imaginary vertical and horizontal lines is a good way to begin.

Ask yourself constantly whether your picture has a horizontal or vertical format. Try both before you commit yourself. Most beginners never get away from the horizontal format so let's avoid this from the outset.

The first shots on your 36-exposure film will be easy, your last ones will not. Probably all your imagination will be required to complete the film. Ask to have your processed film returned unmounted, so that you can clearly see the differences between your first and last efforts. You will notice a big change between frames one and thirty-six. Exercises such as this will give you a valuable insight into the way you perceive your environment, and should greatly improve your photographic awareness.

POND WEEDS
Kodachrome 25, 105-mm macro lens, 1/15 sec. and 1/500 sec. at f 11.
Two pictures which were taken within moments of each other. They illustrate the importance of walking around your subject. The first shot is strongly polarised, to strengthen the Pond Weed (Potamageton cheesmanii) *colours by minimising the sun's reflections off the water. The second picture, taken into the reflected sun without any filters, was underexposed by one stop in order to highlight the dark and light areas.*

Another approach, widely used in photographic schools, is to start photographing something, but each time you take a picture move closer. One of the greatest failings of fledgling photographers is staying too far away from their subject. *Get close: fill the frame.*

Another common fault is taking photographs from eye-level, like everyone else. Nothing is quite so boring. Be different and alter your viewpoint. Get your knees wet, climb something, do anything to try to create novelty and interest. Getting down to ground level is especially important for small plants, flowers, insects, etc.

Walk around your subject and carefully watch the changes in light. Do a lot of walking. Remember you are working with light, nothing else. The word 'photography' comes from classical Greek, and literally means 'writing with light', so, do not take the picture until you have positioned yourself to take advantage of the best light.

A word of caution — be careful not to disturb grass, sand, or snow that you might wish to use as a foreground. Most photographers will admit that somewhere during their career they have ruined an excellent foreground by walking all over it in their search for the best camera position.

WHEELS: photo by Andy Aanensen
Ektachrome 64, 135-mm lens with extension rings, 1/60 sec. at f 16.
Rather than take a conventional picture of these tiny fungi, Andy has carefully dislodged them and photographed his subjects from below. A fine picture resulted.

Telephoto lenses

If you need to get closer, walk, or try a telephoto lens. Tele lenses tighten perspective, and enable you to be highly selective. If you have a busy subject full of many things, use your telephoto lens to scan it carefully. Several pictures may reveal themselves where at first you thought there was only one.

Focusing is critical with telephoto lenses, because at first sight it looks as though they have a very limited depth of field but this is not correct. As Heather Angel (*The Book of Nature Photography*) states: 'The depth of field is a function of the relationship between lens aperture and the image size. It remains constant for a given image size and aperture, regardless of the focal length. Hence the depth of field is increased by *decreasing* the image size, or conversely, decreased by *increasing* the image size.' Think about it.

Make sure to use your camera's depth-of-field preview button to check carefully that all you need in focus, is in focus before you take the picture.

WHITE HERON: photo by Andy Aanensen
Kodachrome 25, 300-mm lens.
An excellent example of composition. Note the careful placement of the bird and the weed. Simple yet highly effective.

PENGUINS MEETING PEOPLE
Kodachrome 25, 180-mm lens, 1/250 sec. at f 8. Hand held.
Sixteen Adelie Penguins and three people meet for the first time. Repetition of colour and the attractive shapes of ship, iceberg, and boat all contribute to an exercise in picture composition.

Wide-angle lenses

Wide-angle lenses have just the opposite effect to telephoto lenses. They push back your scene to create an effect of depth, space and airiness. They are especially useful for landscapes, or when you are surrounded by a close, yet interesting, foreground and you want to create a bit of room in your photograph. Wide-angle lenses have great depth of field. With your camera on a tripod, and using a slow speed and small aperture (something like

WATER RUSHES
Fujichrome 50, 90-mm macro lens, 1/30 sec. at f 5.6.
In the calm of early morning, this sort of picture is easy. Keep the message simple by not showing too many of the rushes.

f 16 or f 22), you will find that everything from about one foot to infinity will be within the zone of sharpness. By lowering the camera viewpoint you can create wonderful pin-sharp vistas of flowers or grass swards which sweep away from the foreground into the distance. When taking scenes with an ultra wide-angle lens, such as the f 2.8/24 mm, watch your horizons. *Keep your horizons horizontal, as nothing looks worse than a sliding horizon.* Because a 24-mm wide-angle lens encompasses some 86 degrees it is very easy to get a sloping horizon, particularly if you have a dark promontory reaching out on one side of the picture. Such features can create optical illusions for the unwary photographer. I use a plain matt focusing screen with etched vertical and horizontal lines (the Nikon Type E screen) which ensures that my horizons are indeed straight.

Wide-angle lenses produce very effective close-ups, provided that there are no straight lines at the picture's extremities. These will be distorted by the angle of view.

Keys to good composition

Remember that the eye tends to group things of similar size, shape and colour together whenever we read a photograph. Watch for interesting patterns, such as rows of windows on a building or ripples in a lake. If you have two areas of interest in your picture, look for ways of linking

Overleaf:
THE PINNACLES, W. AUSTRALIA: photo by Newell Grenfell
Kodachrome 64, 35-105-mm zoom lens, 1/60 sec. at f 8.
Imposing pillars of ancient limestone form the Pinnacles in Nambung National Park, midway on the coast between Perth and Geraldton. Formed more than 25,000 years ago, the surrealistic objects create for the visitor the illusion of being in the midst of a vast field of ruins, or a natural Stonehenge untouched by man.

ROAD TO THE MOUNTAINS
Fujichrome 50, 180-mm lens, 1/60 sec. at f 5.6.
Strong lines of perspective help us race down the highway to the hills beyond. Taken early on a winter's morning from the middle of the road — before the traffic arrived. Ten minutes later, the mist and the dramatic light, were gone.

them together to form an interesting conclusion. The top of a picture can be appropriately linked to the bottom by a waterfall for example, particularly if you capture the motion of the water by using a slow shutter speed. In the resulting picture the water looks like spun glass. Two people in a picture can be linked together by ensuring that they are looking either at each other or at something inside or outside our field of view.

FALLING WATER
Kodachrome 25, 50-mm lens, 1/125 sec. at f 5.6 and 1/8 sec. at f 22.
How to make a waterfall look like water. Use a slow shutter speed and a tripod.

Opposite:
KAMANAWA (Nertera depressa)*: photo by Andy Aanensen*
Kodachrome 25, 55-mm macro lens, 12 secs. at f 11. Natural light.
You don't need a flash for superb close-ups provided that your subject remains still for the required long exposure. Close-ups like this demand a sturdy tripod, even lighting, and a small aperture to ensure a sharp picture. Some films alter their colour balance when subjected to very slow exposure times, but Kodachrome 25 is most forgiving and an excellent film to use under these conditions.

Not everything needs to be incorporated into the photograph. Viewers will fill in for themselves any bits of anatomy or architecture that are implied by the picture but missing from it. We call this *closure*, and it can be effectively used to create interest and tension.

A sense of motion may be critically important to a photograph, depending on the subject. We can either create the feeling of motion by design — directing the way in which the photograph is read; or we can introduce motion by letting the subject do the moving for us, such as panning with a bird in flight or using a slow shutter speed to create movement in leaves rustling on a tree.

Other valuable aids are lines. Horizontal lines are at peace with gravity; they denote calm and repose. Vertical lines are full of protective strength and grandeur, like a strong tree trunk. Diagonal lines are neither of these — they create movement and tension. The eye will follow strong diagonal lines wherever they lead. If they enter the picture, diagonals tend to create lines of perspective, such as an avenue of trees or a row of buildings down a street. Perspective is extremely useful in giving depth to an image. Perspective effects can also be induced by weather conditions. Haze or

3
The Magic of Light

Previous pages:
SUNRISE AT KAIKOURA: photo by Andy Aanensen
Kodachrome 25, 24-mm lens, 1/4 sec. at f 11.
The snow-capped Kaikoura Mountains and rock formations in central New Zealand, photographed in the soft, first light of day. Note how the small aperture and the use of a wide-angle lens combine to produce a sharp image from front to back.

APPROACHING STORM CLOUDS: 1960
Kodachrome 10, 250-mm lens, 1/125 sec. at f 4.5.
The strongly backlit kanuka trees of Little Barrier Island create a tapestry of line and shape beyond which the summer storm clouds gather. An example of an almost abstract landscape which still looks good twenty-nine years after it was taken.

a landscape, accentuating form and giving it a sense of depth and substance. Landscapes glow in the maturing light: pathways to the mountains become shrouded in shadow below a sun-painted background. The photographer walks, stops, ponders, and explores, marvelling at the tapestry of colour and light. As the sun lowers, the landscape fades with haze, and shadows sprawl beneath the sinking sun. Only the magnificence of the sunset remains before nightfall.

In the high Antarctic and Arctic regions, at certain latitudes and times

THE SUN'S RETURN
Kodachrome 25, 300-mm lens, 1/250 sec. at f 5.6.
It had been raining all day. Late in the afternoon the thick clouds briefly parted to provide this dramatic scene.

of year, it is possible to have a sunset and sunrise at the same time. The sun dips low, nearly reaching the horizon before deciding not to go lower. It rises again in a flourish of colour. These are precious moments for the photographer.

Shooting with, and against the sun

Remember that shooting with the sun creates pictures of low contrast and good detail; sidelighting creates strong contrast and shadow. Backlighting is a form of strongly directional light that tends to simplify our images. It deletes detail to produce silhouettes. With backlighting, we can effectively overexpose our pictures slightly, to give them a mood of mellowness.

Shooting into the sun requires care. Lens hoods can help, but flare is a constant danger. Against the sun, zoom lenses and filters generate unwanted rainbows of colour so try to avoid both of these. Use a lens of fixed focal length, no filters, and shade your lens if possible.

Lightmeter failure

If you happen to be out in bright sunshine and your camera's lightmeter fails, either consult the film information leaflet, or remember the handy

TOI-TOIS: A RELATION OF SOUTH AMERICAN PAMPAS GRASS: photo by Andy Aanensen
Kodachrome 25, 200-mm lens, 1/250 sec. at f 5.6.
Backlighting and an attractive subject combine to produce a simple yet effective image. It all comes down to knowing how to look, and what to look for.

Bad weather creates opportunities to photograph vigorous low-key images with plenty of atmosphere. If you are on the coast photographing a violent sea driven by a wailing wind, leave the protective UV filter on your lens. Salt spray can severely damage delicate lens coatings. Make sure you clean your equipment thoroughly as soon as you can, and wash your cleaning cloth to prevent salt on it smearing any clean lens surface.

Avoid getting your camera wet by shooting from a car or hide, or by using an umbrella. Remember that your shutter release has a hole in it to receive a cable release. A direct hit on this by a single drop of rain can slide into your camera, causing mechanical heart failure (yet another example of Murphy's Law). I have stripped and repaired a new Nikkormat camera made totally inert by a single fleck of rust that was embedded in the camera's delayed action gear train. You can avoid this happening by screwing into the shutter release a small soft-touch release button. These drawing pin sized black devices are convenient — and they're great on cold days, when you want to leave your gloves on. But watch where you put them when using a cable release. Gremlins eat them; I have lost at least three.

During wet weather you can of course put your camera in a plastic bag with holes in it for the lens and viewfinder. If you do this, watch for condensation on your camera.

Opposite:
GREY DAY ON LAKE MANAPOURI
Kodachrome 25, 300-mm lens, 1/15 sec. at f 4.5.
Grey sullen clouds nestle in a deep valley overlooking Lake Manapouri in the South Island of New Zealand. Taken in heavy rain, this picture demonstrates that bad weather does not preclude the possibility of taking effective pictures.

HIGH NOON AT LINDIS PASS
Fujichrome 50, 35-mm lens, 1/125 sec. at f 8.
I took this unusual shot in the South Island of New Zealand at noon on 17 June 1987. No filters were used, only a lens hood and tripod. It was a day to exploit the magic of light.

4 Landscapes

MOSSY REMEMBRANCES
Fujichrome 50, 90-mm macro lens, 1 sec. at f 11.
Here old beech trees lie collapsed with age and adorned with moss. Behind their fallen forms a new forest emerges in a penumbra of delicate green. Lack of wind enabled me to take a long exposure and still retain the sharpness needed for this type of picture.

countryside is strong and vibrant, too: the farmer's toil can convert a tilled brown hillside into a brilliant yellow when the rape plants flower. In these ever-changing passages of time lie constant delights for the landscape photographer.

All this action means the photographer must be quick to capture that magic instant when everything is just right. If you see a landscape ready for a photograph, don't pass it by. Stop and collect it. I always grab a few quick shots, and then if the light permits, I consider what else I can do to capture the moment effectively before it's lost.

You should try varying your viewpoints and bracketing your exposures by first taking a picture the way your lightmeter suggests, then using the EVD dial to overexpose by about one-third of a stop (+ 0.3). Finally, underexpose by the same amount (- 0.3). When you come to examine the developed film, check your bracketed exposures carefully. Notice how the slight change in exposure alters the mood of the picture. Which do you like most, and why? If you consistently find that you prefer the slightly underexposed images you might decide to compensate for this by increasing the nominal ASA rating of that particular film. For example, I find my camera gives better pictures on Kodachrome 25ASA film when this film is rated at 32ASA. But this guideline only works effectively in bright light; in dull lighting I revert to 25ASA. Bracketing exposures not only gives you flexibility in exposure times, but it also is a convenient way to calibrate

SEASONAL LANDSCAPES
Kodachrome 25, 35 and 50-mm lenses, both at 1/250 sec. at f 4.
These two summer and winter pictures illustrate the importance of visiting places at differing times of year. Note how the strong side-light highlights the river terraces and how the snow brings life to the mountains.

4 Landscapes

MOSSY REMEMBRANCES
Fujichrome 50, 90-mm macro lens, 1 sec. at f 11.
Here old beech trees lie collapsed with age and adorned with moss. Behind their fallen forms a new forest emerges in a penumbra of delicate green. Lack of wind enabled me to take a long exposure and still retain the sharpness needed for this type of picture.

countryside is strong and vibrant, too: the farmer's toil can convert a tilled brown hillside into a brilliant yellow when the rape plants flower. In these ever-changing passages of time lie constant delights for the landscape photographer.

All this action means the photographer must be quick to capture that magic instant when everything is just right. If you see a landscape ready for a photograph, don't pass it by. Stop and collect it. I always grab a few quick shots, and then if the light permits, I consider what else I can do to capture the moment effectively before it's lost.

You should try varying your viewpoints and bracketing your exposures by first taking a picture the way your lightmeter suggests, then using the EVD dial to overexpose by about one-third of a stop (+ 0.3). Finally, underexpose by the same amount (- 0.3). When you come to examine the developed film, check your bracketed exposures carefully. Notice how the slight change in exposure alters the mood of the picture. Which do you like most, and why? If you consistently find that you prefer the slightly underexposed images you might decide to compensate for this by increasing the nominal ASA rating of that particular film. For example, I find my camera gives better pictures on Kodachrome 25ASA film when this film is rated at 32ASA. But this guideline only works effectively in bright light; in dull lighting I revert to 25ASA. Bracketing exposures not only gives you flexibility in exposure times, but it also is a convenient way to calibrate

SEASONAL LANDSCAPES
Kodachrome 25, 35 and 50-mm lenses, both at 1/250 sec. at f 4.
These two summer and winter pictures illustrate the importance of visiting places at differing times of year. Note how the strong side-light highlights the river terraces and how the snow brings life to the mountains.

your camera's lightmeter for the results you like best. Bear in mind that it may work for one film but not another. Fujichrome 50ASA is best rated at its indicated speed. I find that it responds rather poorly to being underexposed. Experiment with each type of film you try.

Having tinkered with the light, try different lenses on your landscape. Don't expect to use only wide-angle lenses with sweeping vistas. Medium telephoto lenses can produce powerful landscape images too. If you arrived too late to capture that scene clad in autumn gold, when the colours were at their peak, take notes, and plan to be back earlier next year. Planning photographs is an important routine for the landscape photographer.

If you don't see a landscape ready for your camera, keep looking, appraise each bit of country as you round every corner in the road. I find myself frequently backing my car up, sighting down valleys comparing the viewpoints of different lenses in my mind's eye, and asking myself whether the scene will look better in the afternoon when I return home. I'll often stop and take a couple of pictures then, just in case. In the forest, look all around you. The trees overhead can form wonderful tapestries against the sky and the sunlight filtering through the trees can create moments of great beauty. Morning mists on a lake can do likewise, so remember that every scene is unique — it will never be exactly the same again. Ensure that your image is sharp by checking it carefully with the camera's

Overleaf:
PRIMORDIAL LAND
Kodachrome 25, 20-mm lens, 1/30 sec, at f 16. Pola filter.
I wanted to capture this well-known New Zealand landscape of Mitre Peak as it might have appeared long ago, untouched by human hand or influence. I used a small aperture to ensure that the foreground was in focus. Note, in the sky, the effect of the polarising filter: the sun was just out of the right-hand side of the picture.

EARLY MORNING AT LAKE MATHESON
Kodachrome 25, 55-mm lens, 1/60 sec. at f 4.
Taken in the South Island of New Zealand. For scenes like this you need to be up at first light.

EARLY MORNING AT THE LAKE
Fuijichrome 50, 180-mm lens, 1/15 sec. at f 5.6
Early morning mist and sunlight combine to produce a dramatic landscape. The placement of the elements in this picture was made easier by using a telephoto lens.

depth-of-field preview button, and check that the horizon is straight before committing it to film. These things are much easier to check if the camera is on a tripod rather than hand-held.

Foregrounds

In landscape photography what you do with the foreground is important. You can create a fine feeling of distance with a carefully crafted foreground. Overhanging branches can, for example, be used effectively in two ways

AFTERNOON ON LAKE TE ANAU
Kodachrome 25, 20-mm lens, 1/500 sec. at f 11.
The stones in the foreground and at the left of the picture carry the eye away to the mountains beyond, where the sun's stairway of light enables us to return, this time across the waters of the lake. This picture, taken in the South Island of New Zealand, was underexposed in order to achieve the result seen here.

— to soften the stark effect of the picture's frame, and to cut the contrast between a bright sky and dark land by allowing the trees' leaves to drape across the sky in a natural curtain. A film cannot handle too much contrast and still give good shadow detail. The effective use of foregrounds can prevent a sparkling image disappearing into a disappointing silhouette simply because you overlooked the limitations of colour film.

Sometimes you can effectively drop the foreground out of your picture completely, to create a surreal image of landscape that looks as though it's hanging suspended in space. Whatever you do with foregrounds, don't ignore them.

The sky

Landscapes often involve the sky and you need to consider the sky with care. How dominant will it be in your photograph? If the sky is drab and uninteresting, then the photograph must necessarily dwell on the land itself, with the sky reduced as much as possible. Alternatively, there are times of the day and season, particularly in late summer, when enormous cumulo-nimbus clouds tower like cathedrals in the sky, forming magnificent skyscapes which completely overshadow and dominate the land below. Remember that the sky can be just as evocative and just as exciting as the land or sea. It is the skillful integration of an interesting sky with the patchwork of land that is a never-ending challenge to the photographer.

CANTERBURY FOOTHILLS,
Fujichrome 50, 180-mm lens, 1/500 sec. at f 4.
A backlit scene and frontal shadowing can provide an extremely attractive landscape, provided that care is taken with the exposure and that the shadows are not too dominating. The winter sun's gentle highlights help create the mood of this picture taken in the South Island of New Zealand.

SUNRISE
Kodachrome 25, 300-mm lens, 1/125 sec. at f 5.6.
Sunrise or sunset: it makes little difference for both are capable of yielding dramatic pictures.

FIORDLAND REFLECTIONS: photo by Robert Hazard
This marvellous abstract was made by blending sunlit forest with the gently swirling water of Lake Manapouri.

Opposite:
THE TWO TREES
Kodachrome 25, 180-mm lens, 1/15 sec. at f 4.
Attention to composition and exposure makes this image one of my favourites. The vertical format is ideal for these types of pictures.

REFLECTIONS
Kodachrome 25, 35-mm lens, 1/60 sec. at f 8.
Absolute calms in the mountains are rather rare. Shot in early August, this picture of Lake Pearson in the South Island of New Zealand is an example of how beautiful relfections in a lake can be.

Lakes make us rest easy, too. When photographing lakes, you will often need a foreground to provide a visual anchor for your picture. The foreground might be a clump of rushes or trees at the lake edge, a wharf jutting out into the water or a fisherman standing heron-like in his thigh-waders. If the weather is fine and tranquil, the stones beneath the water along the shoreline can prove an interesting introduction to a lake stretching away to the horizon. Gleaming sunlight tracing a fiery path across the surface can add a most attractive highlight to any body of water.

Reflections are, of course, a photographer's dream. They provide symmetry of shape and colour, and allow us to break the 'rule of thirds' to place the horizon on or near the centre of the photograph. Anything reflected in water takes on a new stature — be it a snow-capped mountain or still reeds by the lake-side. If you are photographing floating water plants, which can be a most attractive subject, try using a polarising filter at right angles to the sun to bring out the colour and detail of both the leaves and water, then try shooting the plant leaves into the sun, using the reflected sunlight to transform the floating leaves into silvery motes dancing on the water. You will need to underexpose by at least one stop.

Anything moving across water, bird or boat, will create interesting ripples for the photographer to use with great effect. Any movement, however slight, will distort anything reflected in water, so that a sunlit forest of green trees becomes a medley of coloured shapes in the water. The sun or moon become, not a sphere, but a wobbling stairway of light.

Haze

In many situations you will want to get rid of haze, yet haze can add dimension to a picture by throwing distant hills further into the background and creating a sense of great depth to your image. I remember stopping

EVENING ON THE LAKE
Kodachrome 25, 105-mm lens, 1/60 sec. at f 5.6.
The gentle smoke rising from a homestead across the lake weaves a spell of contentment in this picutre. Without it the total effect would have been lost.

HIGH SUMMER
Kodachrome 25, 105-mm macro lens, 1/30 sec. at f 16. Slightly polarised.
A Canterbury farmer had been busy making hay while the sun shone. I took some time to find the best position to shoot the result of his labour.

late one tranquil afternoon to photograph a sleeping lake. Far across the lake someone had lit a wood stove in preparation for the evening meal. The blue smoke wandered into the sky to obscure the snowy hills nearby, creating an atmospheric gradient in which I could clearly see which hills were closest to me and which were furthest away. At first I was disappointed that the smoke had distorted the crystal clarity of the scene, but as I looked more closely, I could see that the gentle blue haze greatly enhanced the total 'feel', or atmosphere. The call of the crickets in the warm grass seemed louder, and the honking of the swans echoed about the solitude

of the hills. After that experience I always think carefully before reaching for a haze filter.

Note: Speaking of filters, I would suggest that you think carefully about the ones that create phoney colour casts, diamonds of light, and other frivolous things. Why buy them? Why not let the landscape speak for itself, free of gimmickry?

Tripods and the wind

The tripod is undoubtedly the landscape photographer's best friend. With it, you can lower the camera's point of view and use the tremendous depth of field of your wide-angle lens. The scene rolls away, pin sharp, from near your feet to the distant hills.

A breeze is perhaps the greatest nuisance to the landscape photographer.

SHADES OF 2001
Fujichrome 50, 105-mm lens, 1/125 sec. at f 4.
Here, in New Zealand's South Island, the waning sun on a clear autumn day at Castle Hill in Canterbury enabled me to polarise the light, and by tilting the camera so as to accentuate the slope of the ground, I was able to create this strangely evocative landscape.

With slow film, a slow shutter speed, great depth of view, a tripod and a cable release, all ought to go well — unless a capricious wind is blowing. Under these conditions, anything that moves will blur, how much depending on how big the subject is, and how close it is to the camera. If you are faced with a fitful breeze, set your camera up, hold the cable release in your hand, then watch the scene closely. When there is a momentary lull, click. This often requires great patience. I often try for landscapes early in the morning before the wind begins its fickle journey. If you think you might like the blurred motion a breeze creates in a field of wheat, or in a nodding throng of flowers, try a few pictures and see.

People in landscapes

If you use people in your landscape pictures, make sure the viewer can see some of their face, avoid back views. People should be integrated carefully into a scene, so that they don't look as though they've been thrown into it. While the farmer and his handiwork form a vital part of country lore, avoid the trappings of civilization such as cars, buses, filling stations, trash cans or rubbish and conspicuous power lines. Let the countryside, free from human interference, speak plainly through your pictures.

5 Creatures of the Landscape

Opposite:
RED STAG: photo by John Johns
Varicolour 125 negative film, 250-mm lens, 1/60 sec. at f 5.6.
This shot was taken with the camera mounted on a shoulder stock. A positive transparency was subsequently made from the original negative.

appalled at how bad they are, they either end up in a waste bin or are left to fade in an ancient photograph album. I finally managed to take a good photograph of our most recent cat, Caspar, some thirty-five years and nine cats after my first attempt.

To come to photographic terms with your pet you will need a short telephoto lens. The 100 mm or 105 mm is ideal, because its magnification of double the standard 50-mm lens produces photographs that draw the viewer's eye to the subject. Lenses of this focal length also provide a natural perspective and a shallow depth of field. All in all, an excellent lens for portraits of flowers and animals.

Understanding animals

This is absolutely essential and for pets it's easy because you already have the animal's confidence. Wait until it's doing something interesting, even if it's nothing more than looking warm and contented in the morning sun. Put your camera on a tripod, design your picture carefully, check your focusing and depth of field, make certain your shutter speed is fast enough to stop any blurring, wait for the right moment, and take the picture. Bracket your exposures to ensure that at least some will be correctly exposed. Take notes on what you took and how. If your camera has spot metering, take a careful reading of your subject first. This is especially important in those instances where you have say, a black animal sitting in bright sunlight which will create exposure problems.

MACQUARIE ISLAND MALE CHOIR IN CONCERT
Kodachrome 10, 85-mm lens, no further data. February 1967.
Photographing animals can be fun. Note the expressions on the faces of these Sea Elephants as they await direction from their conductor in gumboots.

Remember that most camera lightmeters are calibrated to an 18% reflectance grey card, and will attempt to correctly expose any picture in these terms. This includes a black animal and the bright sunshine. You can have one or the other correctly exposed, but not both. To achieve this you will need to reduce the contrast between the black subject and

its bright background. A cloudy day or open shade will give you a much more pleasing picture. I suggest that you use a fairly fast film, something like Fujichrome or Ektachrome 100ASA, or even faster, such as Kodachrome 200. Incidentally, if you photograph your pets under artificial light inside your home, the pictures will have an orange cast over them. Daylight colour film in general use is balanced for sunlight, not for artificial light. Inside pictures will need a replacement for sunlight, so you will require a flash unit.

The zoo

I have not taken photographs in a zoo because I still recall the trapped look in the eyes of the animals. The newer, open zoos are much more lively and interesting places for the animals as well as the photographer. If you decide to visit, use a short telephoto lens to minimise the signs of captivity, and take care near the primates as some species of monkey delight in peeing on unsuspecting humans, particularly if they're wearing cameras. One of my students had the filter on her lens scratched by a monkey's fingernails!

This caution should be observed out in the open, also. I once took a young Californian lady to see her first penguin. She was captivated with wildlife generally, and the sight of a penguin elicited that wonderful instinct

AUSTRALIAN EMU: photo by Toine Schoenmakers
Fujichrome 50, 80-200-mm zoom lens, 1/60 sec. at f 4.5.
Taken in an open zoo with a zoom lens at its 200-mm setting. A wide aperture was used in order to achieve minimum depth of field so as to ensure that the wire mesh of the cage — and the background — were out of focus.

RED DEER: photo by Andy Aanensen
Ektachrome 400, 1/125 sec. at f 8.
A wild hind and its fawn were photographed in open country — not at a deer farm. Note how the background of dead trees contrasts strongly with the alertness of the animals. The result is a lively picture.

called mothering. She rushed down the beach and swept the hapless little blue penguin into her arms, whereupon it bit her hard on the breast and did something unmentionable down her front. Penguins look cuddly and nice, but they're neither.

Wild animals in the open

Visit the library and read all you can about the animal you are interested in. Ask people who know animals well for assistance. If visiting a reserve or other area where many and various animals occur, take time out to

DINGOES AND AUSTRALIAN RAVENS, SOUTHWEST QUEENSLAND: photo by Graham Pizzey
Kodachrome 64, 400-mm lens, 1/125 sec. at f 5.6.
Surprised at mid-morning near Cooper's Creek in far southwest Queensland, these dingoes (Canis familiaris dingo) *were finishing off the remains of a calf, watched hungrily by Australian Ravens* (Corvus coronoides). *As we crept the vehicle obliquely towards them, it was necessary to forget rules about always using tripods with telephoto lenses, and support the lens as firmly as possible on the vehicle's window frame.*

COSY CA
Kodachro
1/250 sec
These ma
subjects fo
flash has p
together w
eyes.

Opposite:
LITTLE CC
SOUTH A
Graham P
Kodachron
at f 5.6.
A seeming
the presen
(Eucalyptu
sparse wa
beside the
Cooper's C
Australia, a
virtual apa
Corellas (C

Previous pages:
WHITE-FRONTED TERN (Sterna striata) *IN FLIGHT*
Kodachrome 25, 300-mm lens, 1/1000 sec. at f 5.6.
Taken as the bird hovered above its nest, this photograph emphasises the importance of prefocusing and allowing the subject to fly into the zone of sharpness. I used a fast shutter speed to stop the bird's wings.

BLACK-BROWED MOLYMAWK IN FLIGHT
Ektachrome 100, 300-mm lens, 1/250 sec. at f 8.
This NZ Black-browed Mollymawk (Diomedea melanophrys impavida) *was circling the stern of the ship. By prefocusing, I was able to follow the bird until it came sharply into focus within the viewfinder. Note the drops of water hanging from its bill, the interesting pattern on its underwing coverts, and the large webbed feet tucked under the tail.*

Lizards in their habitat can be informative. While returning to camp on New Zealand's Poor Knights Islands late one night, I passed a small Pohutukawa tree that was rustling with activity. Geckos bedecked the branches, drinking nectar from the bright red flower sprays. The geckos were alarmed by my electric torch, however, and only when I remained perfectly still did they resume their nocturnal foraging. I pointed the camera and flash unit at the tree and, in total darkness, fired the flash from about 1.5 m away. The resulting habitat picture was full of interest and illustrated more about the behaviour of these animals than any portrait would have.

Many small, cold-blooded creatures emerge early in the day to bask in the sun, warming their bodies sufficiently to become highly mobile and ready to forage. Until they're warm and active they tend to be lethargic and are relatively easy subjects to photograph.

Seabirds on the wing

A singular blessing of wildlife photography is that everyone can do their own thing. Photographers who are content to crouch in hides might die of seasickness trying to focus on an albatross whizzing past on a really rough day, when a stationary ship is lurching through a vertical arc of some twenty or so feet of ocean swell. It's a bit like being dropped down an elevator shaft at irregular intervals.

If you wish to photograph a bird flying about close to you in a relatively consistent way, such as those following a ship or wheeling about close to their nests, pre-focus the lens, and then follow the bird about with your eye to the viewfinder. If you are lucky and quick, the bird will abruptly loom into view and for a split second will be sharply in focus to receive a rapid 'click'.

If you want to do a lot of this type of photography, I would recommend the purchase of a motor drive. At three to six frames a second, you will race through film at alarming speed. You will need 36-frame film, otherwise plan on spending all day loading your camera. A bulk film magazine fitted to your camera will allow you long moments with your motorised finger on the shutter release button, rattling away like a cine-photographer.

Picture sequences

The Greek Island of Hydra has lots of domestic cats which generally seem content to laze in the Mediterranean sun and watch the camera-toting tourists go by. Sometimes, however, feline lethargy is broken by a border dispute in which male cats contest their territorial rights by yowling at each other. These displays of vocal animosity are great for picture sequences. My two particular cats never stopped yodelling to each other as the black and white cat approached. What interested me was the fact that the cats

took great pains not to look directly at each other, even when they were within a whisker's length. Such photographs form an integral part of studying any animal's behaviour and when no pictures are available the research biologist must often resort to long written explanations which can be wearisome to read. Drawings may substitute, but nothing beats a good picture sequence.

PICTURE SEQUENCES: THE CATS OF HYDRA
Kodachrome 25, 50-mm and 180-mm lenses, 1/250 sec. at f 5.6.
Always be prepared for the unexpected. Two cats suddenly started to sing during a lazy afternoon on the Greek island of Hydra. The black and white animal slowly approached the black one, while both took great pains not to look at each other directly. I rapidly changed lenses expecting fur to fly. None did. They got to within a whisker's length of each other and duetted for about 30 seconds, after which they separated and continued their afternoon as if nothing had happened. Probably discussing a territorial boundary.

Animal portraits

Animal portraits give you a detailed view of an animal's appearance so that you are able to count the number of whiskers on a chin, or the number of feathers in a tail. Portraits of this type are thus useful in positively confirming the identity of any unknown creature. They also enable us to contemplate the animal's personality.

An animal or plant portrait needs to be well exposed with high definition. The old axiom 'always make sure you get the eyes sharp' is a good rule-of-thumb for animal photography. Focus on the eyes, then check with your depth-of-field preview button to ensure that the remainder of the face is sharp. This care in focusing is particularly important if a creature has a long nose or a big bill. My picture of the royal albatross at Taiaroa Head near Dunedin in New Zealand is a case in point. Because the bill

and the flash require the same f-stop. Avoid this by reducing the output of your flash. If your calculations give f 8, reduce this to f 11, and bracket your exposures to be sure.

Should you use a high synchronisation speed, such as 1/250 sec. or higher in dim light, the background may come out too dark through underexposure. The picture will probably look as though you took it at night. To avoid this, use a slower shutter speed.

Opposite:
TUATARA: photo by John Johns
Hasselblad with extension rings, using a 150-mm Zeiss Sonnar at f 16.
A tripod was used, together with two small electronic flash units — open flash and a slave unit. The photograph was taken on Stephen Island and shows a Tuatara which has just shed its old skin.

Freezing the action

Because electronic flashes discharge at very high speeds, 'computer' flashes may have a flash duration of 1/2000 sec. to 1/40 000 sec., depending on how close the subject is. The closer the subject, the shorter the flash duration. This action effectively freezes all movement, and can be useful when photographing small birds flying on to the edge of their nest, or for freezing the motion of falling water. By taking advantage of this high speed flash photography, you will be able to examine things that could otherwise never be stopped in their tracks.

ANCIENT MEETINGS
Kodachrome 25, 105-mm lens, flash.
The notion that Tuataras and burrowing seabirds get along well together in the same burrow is a myth. These two have been skirmishing in a small burrow on the Poor Knights Islands off the northern coast of New Zealand. Confrontations like this are rarely photographed. The total world population of Pycroft's Petrel (Pterodroma pycrofti) *is less than 500 pairs. It is much rarer than the Tuatara.*

7
Close-Up Photography

Previous pages:
THE GOLDEN EDGE: photo by
Toine Schoenmakers
Fujichrome 50, 50-mm macro lens,
1/125 sec. at f 3.5.
A simple composition and backlighting provide the ingredients for a striking picture of an ordinary fern leaf.

Close-up photography is an enchanting realm of nature where the surreal is ordinary and dragons abound. Taking close-ups is a relatively simple procedure to learn. You need:

1. The right equipment.
2. An understanding of depth of field and the importance of correct focusing.
3. An appreciation of how to light something close to the camera.

Necessary equipment

Modern single-lens reflex cameras are ideally suited to close-up photography, because you can see plainly what lies within the frame before you take the picture, and the camera's automatic lightmeter can save a lot of trouble when you are working with outdoor close-ups in natural light. Cameras with metering systems that read the light directly off the film plane are a boon to the close-up photographer, because you can use any lens or attachment and still achieve well-exposed pictures.

The most important piece of equipment for the close-up photographer is undoubtedly the lens. Most major camera manufacturers make special lenses designed for close work, and you should buy one of these in preference to the standard 50-mm lens. Zoom lenses with macro facility seldom produce really good close-ups because, despite advertising claims, no manufacturer has contrived a good lens that will do absolutely everything.

True macro lenses usually come in three groups of focal lengths: 50 or 55 mm, 90 to 105 mm, and 200 mm. In each range it is possible to buy a matching automatic extension ring that allows you to obtain a 1:1 (lifesize) reproduction ratio should you need it. Macro lenses are much more expensive than the standard lens of the same focal length, but they're premium lenses with excellent optical characteristics.

Because of their optical design, macro lenses have a relatively slow maximum aperture. This is of little relevance to the close-up photographer who requires, not speed, but exceptional sharpness, high contrast, and often great depth of field. The last named is often critical in close-up photography, and because of this macro lenses are designed to close down to a small minimum aperture such as f 32, which is very useful where a maximum zone of sharpness is required. This is not always the case, however. There will be many times when a larger aperture allows you to block off a distracting background by effectively blurring it.

RAINDROPS ON GRASS
Fujichrome 50, 90-mm macro lens, 7 secs at f 16 (left), 2 secs at f 16 (right).
A still morning after rain and pictures of everyday subjects can become things of great beauty. A good quality macro lens is essential for this sort of work.
The photograph on the right was taken during a 6.5 Richter scale earthquake in Westland, New Zealand. When I stood up, I discovered the ground was swaying from side to side — like standing on a firmly set jelly! The 'quake lasted about a minute.

With close-up work, I must stress the importance of *correct focusing.* At ultra-close range, the depth of field becomes very shallow, and magnifying your image will also magnify any slight error in focusing. Therefore, focus

THE JOY OF A CLOSE-UP LENS.
Kodachrome 25, 35-mm and 105-mm macro lenses, small apertures, slow shutter speeds.
The Giant Buttercup is a magnificent annual alpine plant whose large white flowers and glossy green leaves appear in spring and early summer. I used the wide depth of field which my 35-mm wide-angle lens offers to take a group of flowers on a single plant, I then focused the 105-mm macro on a single bloom for a portrait. Careful exposure is needed with white flowers. Here I was trying to take advantage of the extraordinary light and shade on the petals which made them appear like exquisite porcelain. I bracketed the exposure times.

very carefully, particularly if you are using larger apertures such as f 4 or f 5.6, where the zone of sharpness may only be a few millimetres.

Close-ups with the 50/55-mm macro lenses

If you buy a 50-mm or 55-mm macro lens, attach its automatic extension ring, and focus on something close to the camera, one thing will become obvious immediately. You'll find the lens barrel ending up only a few centimetres from your subject. This is fine for a subject that doesn't mind

MATT DAISIES ON A RIVERBED
Fujichrome 50, 90-mm macro lens, 1 sec. at f 11.
The attractive colour and texture of these small clumps of matt daisies amongst their background of grey river rocks create a natural design of great beauty. Many people walk past such scenes without even noticing them. The nature photographer is always alert for such a picture.

having an enormous alien object very close to it, but most living things, such as insects, will either flee or hide as fast as they can. You will also notice that with the 50/55-mm macro lens fully extended, the subject can become deformed owing to the accentuated perspective of the lens.

Being very close to your subject creates another problem — one of lighting. Being cheek to jowl with your subject means that you will often find the camera shading it. What to do with the light can be tricky, particularly if your small subject decides to walk away while you are photographing with a one second exposure.

You will not, therefore, be surprised to learn that I seldom use my 55-mm macro lens for ultra close-ups. Let me hasten to add, however, that the 50-mm or 55-mm macro is a superb general-purpose lens, and I always carry one in my camera bag. It is capable of high resolution and high contrast, and is ideal for landscapes and many other photographic subjects.

Opposite:
THE 1000-LEAVED FERN
Fujichrome 50, 90-mm macro lens at 1:1 extension, 11 secs at f 16.
This tiny fern required careful attention to depth of field in order to get maximum sharpness. I used natural light and a long exposure to give life to the water moving slowly under the fern.

The 90 to 105-mm macro lenses

Because of their natural perspective characteristics and excellent performance, these lenses are ideal for close-up photography as well as for general work. The big five and the independent manufacturers each produce a macro lens within the 90 to 105-mm range of focal lengths. With such lenses you can stay a reasonable distance away from your subject and not disturb it unduly. The working distance, which is about twice that of the 50/

RAINDROPS ON BROWN TOP GRASS
Fujichrome 50, 90-mm macro lens, 4 secs at f 11.
Although I got very wet passing through the soaking grass in search of a good picture, I finally found this one. Here I have used green rushes to create a strong vertical element which helps strengthen the composition.

55-mm standard or macro lens, also allows the use of reflector boards or flash to solve problems of low light and camera shadowing.

This book is illustrated by some of the results from my Tokina AT-X f 2.5/90-mm macro lens, which I now use for nearly all of my close-up work. It has a superb optical performance, is easy to use, and the f 2.5 maximum aperture allows more light into the viewfinder for easier

UMBRELLA MOSSES
Fujichrome 50, 90-mm macro lens, 2 secs at f 16.
Umbrella mosses are extremely attractive when wet and glistening with water. The long exposure allowed the water behind this small scene to create an interesting and unusual backdrop.

THE JEWELLED STUMP
Fujichrome 50, 90-mm macro lens, 2 sec. at f 16.
The intricate texture of this old beech tree stump, ravaged with age and festooned with all manner of tiny life, attracted my attention. I positioned my Nikon FA camera about 100 cm from the stump and allowed the gentle forest light and the film to do the rest.

focusing than the slower f 4 macro lenses in the same focal length. Given that the Tokina is also cheaper than its big five equivalents, I can recommend it unreservedly to you.

Alternatives to macro lenses

These include:

- **Close-up attachment lenses:** These one element lenses screw directly onto the front of a lens to allow it to focus more closely. They come in varying magnifications. If you buy some of these, remember to put the strongest lens on first to minimise surface reflections. There is no need to use them on macro lenses.
- **Extension tubes:** These are attached between the camera and the lens; some retain the automatic facility of the lens and light metering system of your camera. If your camera does not have an automatic light metering system, you will have to experiment to find the correct exposure control with the extension tubes you use. Extension tubes are not as quick or as easy as a macro lens, but they will enable you to take close-ups without too much expense.

  Note: By adding extension tubes to your macro lens, you can achieve lifesize magnifications of objects such as small insects and tiny plants. Make sure that any proposed extension tubes work effectively with your lens and camera.

  The Tokina f 2.5/90-mm macro lens comes complete with a matched three-element extension lens which allows me lifesize magnifications of superb quality. Where a longer working distance than my 90-mm Tokina can provide is required, I use an f 2.8/180-mm Nikkor lens with a PN 11 extension tube. Like John Shaw, *The nature photographer's*

CLOSE-UP TECHNQIUE: photos by Ueli Walthert
Kodachrome 25, Canon F1, 100-mm macro lens with bellows attachment, natural light, slow shutter speeds.
A. Southern Rata Flowers (Metrosideros umbellata).
This photograph was taken in morning sunlight. Note the strong contrast. Only careful exposure has prevented the subject highlights from being burnt out through over-exposure.
B. Southern Rata Flowers.
Note the difference when the same shot is taken in bright shade. The contrast is reduced and the true colours can now be fully captured on film.
C. Native Orchid (Thelymitra longifolia).
This picture shows a compromise between the above two approaches. Gentle early morning sun and careful attention to depth of field have created an attractive photograph of a beautiful subject. Contrast has been carefully controlled in order not to lose the delicate colour in the orchid's petals.
D. Native Orchids (Corybas oblongus).
By ensuring that these three attractive orchids were all on the same plane as the camera, focusing and depth of field problems have been kept to a minimum. Note that the background is completely out of focus in order to avoid competing with the main subject for the viewer's attention.
Note: *These flowers are native to New Zealand.*

complete guide to professional field photography, I found this to be an excellent choice for small, easily frightened subjects, or for working near splashing water where the camera might otherwise get wet.

- **Extension bellows:** This is an expanding set of bellows on a focusing rail that allows you to alter the distance between the camera and the lens continuously, which in turn allows you to control the magnification you want. With a 50-mm lens, you can get from about 1/2x to 3x magnification of the image. Most bellows attachments for modern cameras are fully automatic, and frequently have a double cable release that permits you to check depth of field and exposure. Canon, for example, make 20-mm and 35-mm macrophoto lenses specifically designed to fit their automatic bellows. These compact lenses allow magnifications of up to 20X with the 20-mm lens, and up to 10x with the 35 mm. At these magnifications, when the subject can be as close as a finger's width from the lens, focusing is absolutely critical.

OLIVER'S SKINK (Cyclodina oliveri)
Kodachrome 25, 105-mm lens, SB16 TTL flash.
The picture was taken at night on the Poor Knights Islands when this large skink comes out of rock crevices to feed on the forest floor. Because it is active only after dark, this creature should be photographed at night, and not during the day when it normally hides out of sight of predators. This gives such photographs their necessary credibility.
By positioning myself so that the entire animal was on the same plane as the film in my camera, I was sure of a sharp picture. Note the catchlight in the eye and the light reflected off the animal's smooth skin.

Other makers, such as Olympus and Minolta, also provide special macro lenses which are constructed for the same purpose.

- **Reversing the lens:** With an adaptor ring that screws into the front of a lens, it is possible to reverse it on the camera body. The result is magnification capability greater than lifesize, but with loss of automatic metering and automatic aperture control.

The above options are specifically designed for close-up photography. Their drawbacks are that they lack the versatility of macro lenses which can focus to infinity and hence be used as general lenses. Macro lenses are, of course, very expensive, but extension bellows aren't exactly cheap either, so consider your options carefully.

A stick insect and your depth of field

With close-up photography you must know how to use your depth of field effectively for good results. A depth-of-field preview button is extremely helpful because it saves you a lot of guessing or referring to tables which can be tedious and time-consuming.

Opposite:
AUTUMN AT THE WATERFALL
Fujichrome 50, 90-mm macro lens, 3 secs at f 16.
This is one of my favourite images. Two Tree Fuchsia leaves lie beached in a small stream. Their warm colour adds an extra spice to this tranquil scene. It is simple scenes which create the best pictures.

Suppose you are photographing a large stick insect, for example. As long as it is on a plane parallel to the film in your camera, focusing on it will be relatively easy and the insect's entire body should be rendered sharply on film. But if the creature decides to see what you are doing, and totters towards you like a pencil on legs, you will get little of it in focus — perhaps only its head and mouthparts.

Clearly then, where you position your camera is important in close-up photography. Head-on pictures of stick insects require precision focusing to begin with, and careful checking with the depth-of-field preview button to ensure that all you need is sharply represented. For example, if you

Opposite:
COMMON GRASS BLUE (Zizina otis labradus), *VICTORIA: photo by Graham Pizzey*
Kodachrome 25, 105-mm macro lens, 1/30 sec. at f 8.
Described as the commonest butterfly in Australia, this delicately toned beauty mostly passes unnoticed. To secure the best angle and a neutral background the tripod was lowered and cautiously advanced. Soft overcast daylight enhanced the effect.

Lens hoods

Always use a lens hood for close-up photography where the front elements of your lens might be exposed to stray light and reflections, and to ensure the best possible colour saturation. The 50-mm or 55-mm macro lenses have sufficient of their lens housing encircling the lens elements to act as a built-in lens hood.

Filters

In most situations, remove all filters as they detract from the image quality. Polarising filters can be helpful in some situations, but use them sparingly because they cut down the light falling on the film by about one to two stops, and this limits your depth of field.

If, however, you want to try something a little different when photographing something like a small fungus, for example, put a soft-focus filter on the lens and check the viewfinder. Interesting, isn't it?

A plain matt screen

I have already sung the praises of these excellent focusing screens. If you are planning to do a lot of close-up work, a plain screen is absolutely necessary. The split image screen, in terms of the vernacular, will drive you mad.

USING FLASH WITH CLOSE-UPS
Kodachrome 25, 105-mm macro lens with extension tube. Natural light and SB18 TTL flash.
While searching for a group of Sundews (Drosera arcturi), *I happened on this tiny drama. A Crane Fly has trapped itself between two of these sticky, carnivorous plants. I took a few exposures using natural light and 1 sec. at f 22, and then a couple of pictures to demonstrate what can happen with flash. The flash has produced a very contrasty image with a poorly exposed background. Natural light has produced an excellent image. I could have improved it further by a little 'gardening' to eliminate some of the background highlights.*

A tripod

Absolutely essential because magnifications in close-up photography mean that any movement of camera or subject will be greatly exaggerated. No motion is essential for good close-ups. Use a cable release, and trip the shutter as carefully as you can. Some early makes of single-lens reflex camera had such poor dampening on their mirror boxes and shutters, that a good, clear close-up was virtually impossible. They went off with a roar. Nowadays, most SLR shutters operate electromagnetically and they trip discreetly with a dull thud.

Some types of tripod have a lateral rail on which the camera sits. These allow you to set up the tripod and move the camera independently from the rest of the tripod mounting — without having to reposition the tripod each time.

Natural light and flash

Use natural lighting whenever you can. A sky with high overcast clouds is ideal because it helps reduce the contrast which is normally very strong in direct sunlight. Low contrast allows for good image detail.

Flash has essentially the same effect as bright sun, and should only be used as a last resort. With many small things, such as plants, it is their natural beauty that we want to capture on film. The emphasis is on *natural.* The subtle shades of colour in an orchid flower, the texture of a leaf or wing, these and many other delicate nuances of life can be blasted away by the bright light of one or more flashguns. Flash units also provide strong contrast between the lit surfaces and those that are unlit. This can create difficulties that test the ability of film to handle effectively.

Flash *brightness* is not the only problem. Because the *colour balance* of flash is calibrated to be close to direct natural sunlight, it may be completely different to the bright but diffuse shade of the forest floor, for example, where small plants are found. The resulting image looks bright, but is it true to nature?

I try to use available light whenever possible, and hope that any motion of my subject does not create blurring while my photograph is being taken. If air movement is a problem I would prefer to increase the shutter speed, and reduce my depth of field, rather than introduce a flash.

You can, of course, send a conveniently hefty spouse or other ally upwind to act as a windbreak. I once used a friendly Jersey cow that was peacefully sitting chewing her cud. By being there, she helped photograph one of the many clumps of buttercups scattered about her in the grass. She watched my every move with what appeared to be benign amusement.

BLUE CAP FUNGUS
Fujichrome 50, 90-mm macro lens, 32 secs at f 16.
A very long exposure was needed to capture this beautiful fungus hidden deep in the forest's shade. A flash would have destroyed both the delicate colour and texture of the subject. I used a long exposure in natural light instead.

Reflector boards

To help lessen shadows, you can use a small reflector board made up of a piece of 16 x 12 cm white cardboard, or one covered with aluminium foil with its bright side outermost. I use this to reflect sunlight back towards the subject, which increases the light by up to 2 stops, lessens the contrast, and gives more depth of field. You can also use thin sheets of white polystyrene. Move the reflector board about to model the light falling on the subject to create depth and substance. Having someone hold the reflector board can be a big help during the taking of the picture. Failing this, carry a 'bulldog' clip in your camera bag and use it to secure the reflector board to a handy object.

SPIDER'S WEB
Fujichrome 50, 90-mm macro lens, 1/2 sec. at f 11.
Early morning dew and a spider's web add up to magic for the photographer — a breeze is the greatest danger here: the quiet of daybreak is best. I used a reflector board to bring sunlight to the trapped water droplets.

Gardening

Before you take the picture, check to ensure that nothing extraneous is in the way of a good photograph. Check for any stray blades of grass, twigs and the like — anything that can be removed easily which might otherwise disrupt the integrity of your image. We call this 'gardening'. Gardening does *not* mean the wholesale removal of twigs or leaves that might conceal a bird's nest or small animal's lair. Do nothing to jeopardize an animal's or plant's survival.

Using flash

Sometimes a combination of dull light or a moving subject will require the use of a flash. Close-up photography is unusual in the context of picture-taking in that often there is too much light rather than not enough. The subject needs to be lit as naturally as possible, and flash light must be used carefully to prevent the image from appearing harshly and unnaturally lit. Most people expect to see photographs illuminated as if the sunlight was coming from over their left shoulder. You might consider this when positioning your flash.

There are several ways of using flash for close-ups. Some people use two small TTL flash units attached to the camera by means of a flash bracket, or alternatively the flashguns may be tied to a tent peg or something similar, enabling each to be independently positioned in the ground at any distance from the subject. All flash units off camera must be connected to it by a cable, unless you plan to use open flash, in which the gun is fired manually while the shutter is open. Twin-flash technique is good for small fungi and plants where modelling of the light can create effective pictures. If, however, your subject is watching you, two catchlights will appear in its eyes. You may find this unnatural and wish to avoid it.

NOCTURNAL FEAST
Kodachrome 25, 50-mm lens, hand held flash.
A small Pohutukawa tree rustles with life as many small geckos lick the nectar from the bright red flowers. Since geckos are disturbed by any torchlight, I positioned myself close to the tree in total darkness and waited until the rustling resumed, then I took this picture. It was a complete guess as I had no way of telling what would appear.

The simple alternative is to have a single TTL flash unit attached to the camera by a coiled extension cable. A small flash of lesser power will be perfectly adequate, or you can use a more powerful one, provided that you can reduce its power output. After setting up your camera on the tripod and focusing on the subject, you can hand-hold the flash unit and fire it from various positions relative to the subject. You may wish to use front lighting coming from close to the camera, in which case the best place to position the flashgun is about 10 cm above the front of the lens and slightly to one side. This will create a well-lit subject with a minimum of shadowing. John Shaw, *The nature photographer's complete guide to professional field techniques*, uses this approach, and in his excellent book he describes a very simple way of making a 'butterfly bracket' to hold the flash in this position.

Another approach is sidelighting, in which you 'skim' the light over the subject to highlight its texture. This is simple to do when handholding the flash. Try different ways of using your flash, and keep notes on what you have done. In this way you will soon find a technique you prefer. Experiment, and at the risk of becoming boring, always bracket your exposures.

Final Notes

John Ruskin (1819-1900) said, 'Fine art is that in which the hand, the head, and the heart go together'. His comment is one that photographers could, with benefit, bear in mind.

Whatever your photograph, take time out to create it properly. Most people fail to obtain good pictures because they failed to think carefully about what they wanted in the first place. A good image is created by the ability to look, plan, and consider before you shoot. It doesn't require a university degree, it requires creativity, and an irrepressible urge to seek a better way of photographing what you see. Thoughtful photographers are there at the right time, with the right equipment. This attention to detail often allows luck to intervene on their behalf.

Remember the three rules of good composition: simplify; simplify; simplify. A good picture that is easy to read is far more evocative than words.

Always be prepared to walk away from an indifferent shot. Occasionally I am led to think some people take photographs just to see whether their camera is working.

You will probably find, as I do, that photography is creative and exciting. Faced with a good shot, my adrenalin flows and I become acutely aware of the challenge before me. It's as though I'm being switched to a higher state of awareness. In spite of this 'high', try not to be carried away by the moment, so that you blaze away all your film in a brief photographic frenzy. Murphy's Law dictates that your subject will do something more interesting and thrilling when you have run out of film, or are in the middle of changing films, so pace yourself and check regularly to see how much film you have left.

Finally, try not to take your photography too seriously, don't be too protective of your work. Always be ready to do better next time. Enjoy nature photography for what it is: a marvellous journey to self discovery.

Suggested Reading

Adams, A. 1974. *Images 1923-1974.* New York Graphic Society, Boston, (Mass.).

Alinder, J., Szarkowski, J. 1985. *Ansel Adams: Classic Images.* New York Graphic Society, Boston (Mass.).

Angel, H. 1982. *The Book of Nature Photography.* Fountain Press, London.

Bond, C., Johnson, P., Siegfried, R. 1978. *Antarctica: no single country: no single sea.* Hamlyn, London, New York and Sydney.

Gibbons, B., Wilson, P. 1986. *The wildlife photographer. A complete guide.* Blandford Press, Poole, New York and Sydney.

Godwin, F. 1985. *Land.* William Heinemann, London and Auckland.

Landscape Photography. The art and techniques of eight modern masters. 1984. Amphoto, New York.

O'Connor, M. (Editor) 1983. *The image bank.* Amphoto, New York.

Moon, G.J.H. 1970. *Photographing nature.* A.H. & A.W. Reed, Wellington.

Poking, F. (Editor) 1986. *Nature photography 86-87 yearbook.* Fountain Press, Tolworth (United Kingdom).

Rowell, G. 1986. *Mountain light. In search of the dynamic landscape.* Century Hutchinson, Auckland and London.

Semenzato, C. 1982. *Photographing places.* Century Publishing, London.

Shaw, J. 1984. *The nature photographer's complete guide to professional field techniques.* Amphoto, New York.

Turner Ettlinger, D.M. 1974. *Natural history photography.* Academic Press, London and New York.

Warham, J. 1983. *The technique of Bird Photography* (4th edition). Focal Press, London and Boston.

Index

afternoon sun *49*
albatrosses *82, 99*
amber filter *22*
animal portraits *99*
animal's behaviour *99, 101*
animal's welfare *90, 127*
animals in motion *14, 99*
anti-cold battery pack *24*
aperture control priority *14*
auto focus cameras *17*

backgrounds *123*
backlighting *53, 72, 108*
bad weather *61, 64*
batteries *24, 105*
behaviour, animal *99, 101*
bird photography *7, 42, 94*
birds *92, 94*
blurred motion *14, 79*
bracketing exposures *15, 44, 66, 84*
buying second-hand lenses *21*
buying flash units *104*
buying cameras *14–17*
buying used equipment *20, 21*

cable releases *23, 124*
calcium fluorite lens elements *18*
camera accessory shoe *105*
camera bags *24*
camera equipment *13–26*
camera shake *20*
Canon cameras *14, 17, 119*
'catchlights' *106*
cats *84, 99*
changing weather *57, 61*
checking lenses *21*
checking shutters *20*
cleaning a lens *25*
close-up attachment lenses *118*
close-up photography *113–128*
'closure' *42*
coasts *73*
cold climates *24*
colour balance *18*
common faults in taking photographs *19, 20, 36, 90*
comparing viewpoints of different lenses *67*
complete lens kits *18, 19*
composition *30–45, 95*
creating your own guide number *106*
creatures of the landscape *81–101*

delayed action *15*
depth of field *20, 37, 95, 114, 120, 122*
depth-of-field preview button *15, 37, 120, 123*
designing a photograph *11, 29–45*
designing the frame *33*
dividing frame into thirds *36, 76*

Ektachrome 100ASA *24, 86*
electronic flashes *24*
exposure (value) or compensation dial (EVD) *15*
exposure control *15, 44, 90, 123*
exposure lock button *15*
extension bellows *119*
extension cables *105*
extension tubes *118*
eyepiece correction lenses *16*

fill-in flash *104, 108*
film *23, 24, 71*
film pressure plate *20, 44*
film speed dial *15*
filters *21, 22, 61, 124*
flare *19*
flash duration *111*
flash photography *103–111, 126, 127, 128*
flash units *104, 105, 106*
flashy filters *79*
focusing *18, 37, 99, 114, 122*
focusing screens *15, 124*
foregrounds *70, 76*
Fujichrome 50 and 100ASA *23, 67, 86*
fully automatic cameras *14*

'gardening' *127*
geckos *98*
grey card, for exposure calibration *84, 90*
guide numbers *106, 107*

habitat shots *95*
haze *76*
haze filter *79*
Heather Angel *37*
hides *94*
hiring lenses *20*
horizons *39*
horizontal and vertical formats *36*

improving photographic awareness *36*
instruction book, importance of *17*
insurance *25*

John Ruskin *129*
John Shaw *118, 128*
John Warham *82*

Kodachrome 200 *86*
Kodachrome 25 *23, 57, 66, 92, 106*

lakes *67, 70, 73, 76, 78*
landscape photography *63–79*
landscapes *52*
lens 'sharpness' *18*
lens brushes *25*
lens care *25*
lens hoods *21, 124*
lens tissues *25*
lenses *17, 18, 19, 20, 123*
light, the magic of *47–61*
lightmeter failure *25, 53*
lines, importance of in composition *42*
lithium batteries *24*
lizards *98*
low dispersion glass in lenses *18, 19*

macro lenses *17, 18, 21, 114–118*
manual cameras *14*
matt screens *15, 124*
metal shutters *20, 25*
Minolta cameras *14, 16, 17*
mirror lenses *19*
morning sun *48, 49*
motion in photographs *42*
motor drives *20, 99*

natural light and flash *126*
Nikon cameras *14, 17*
notebook *45, 67, 84*

Olympus cameras *14, 120*
out of date film *24*
overexposure *15, 53, 90*

pan and tilt heads *22*
panning *14, 42*
patterns *39*
penguins *38, 86–87, 89*
Pentax cameras *14*
people in landscapes *79*
perspective *42*
photographic competitions *45*
photographic equipment *13–26*
photographing animals *82*
photographing animals at night *108*
picture sequences *99*
planning photographs *45, 64, 67*
polarising filters *22, 67, 73*
pre-focusing *89, 98, 99*

'red-eye' *105*
reflections *76*
reflector boards *127*
reversing the lens *120*
rules for good composition *76*

seabirds *82, 98*
second-hand cameras
shadows *40*
shoulder 'stock' or rifle-grip *23*
shutter priority *14*
sidelighting *49, 128*
silhouettes *53, 108*
silver-oxide batteries *24*
skuas *90*
sky *71*
skylight filter *21*
soft-focus filters *22*
soft-touch release button *61*
split-image focusing screen *16, 124*
stability of camera *22*
storing lenses *25*
storing slides *26*
subjects *30, 31*
'sunny f/16 rule' *57*
sunrises *49*

Tamron lenses *19*
teleconverters/extenders *19*
telephoto lenses *18, 20, 37, 73, 84, 92, 94*
testing shutters *20*
the right lens *17*
the wrong lens *17, 83*
through-the-lens (TTL) flash *15, 105, 106*
through-the-lens (TTL) metering *15, 114*
tips for beginners *26, 33, 122*
tips on camera and lens care *25*
Tokina lenses *19, 117, 118*
travel photography *26*
tripod bush *15*
tripods *20, 22, 57, 79, 92, 124*
TTL flash *15, 105, 106*
twin flash techniques *128, 106*

underexposure *15, 73, 90*
using smallest aperture *123*
UV filters *21*

vignetting *21, 22*

'wait and see' photography *92*
water, photographing *42, 73*
white subject on dark backgrounds *90*
wide-angle lenses *18, 21, 38, 73*
wild animals *87, 88*

zoom lenses *19, 114*
zoos *86*